Studies in Animal and Human Behaviour

VOLUME I

KONRAD LORENZ

Studies in Animal and Human Behaviour

VOLUME I

Translated by Robert Martin

METHUEN & CO LTD
11 NEW FETTER LANE LONDON EC4

First published in collected form in Munich
by R. Piper Verlag under the title
Über Tierisches und Menschliches Verhalten
(Gesammelte Abhandlungen, Band I)
First published in Great Britain in 1970

Printed in Great Britain by
Butler & Tanner Ltd, Frome and London
SBN 416 41810 4

Contents

The papers included in this volume appeared originally in the following Journals and under the German titles shown (listed in the order of the Contents, above):

Beiträge zur Ethologie sozialer Corviden, *Journal für Ornithologie, 79, Heft 1, 1931*

Betrachtungen über das Erkennen der arteigenen Triebhandlungen der Vögel, *Journal für Ornithologie, 80, Heft 1, 1932*

Der Kumpan in der Umwelt des Vogels, *Journal für Ornithologie, 80, Heft 2, 1935*

Über die Bildung des Instinktbegriffes, *Die Naturwissenschaften, 25, Heft 19, 1937*

Taxis und Instinkthandlung in der Eirollbewegung der Graugans, *Zeitschrift für Tierpsychologie, Band 2, Heft 1, 1938*

Induktive und teleologische Psychologie, *Die Naturwissenschaften, 30, Heft 9/10, 1942*

Translator's Foreword

The translation of this first volume of selected papers has presented a demanding but very rewarding task. These early papers from the founder period of ethology are extremely interesting from many points of view, and perhaps the most fascinating aspect is the historical development of the terminology involved and the written style. These two aspects have also presented the most difficult problems in translation. Luckily, the major points of terminology and style have been covered in detailed discussions with Konrad Lorenz, and it is hoped that the English version here presented will provide a reliable reference source for students of ethology.

This first volume contains five papers covering the period 1931–1942, providing a rich source of information on the initial development of Lorenz's approach to the subject. These early papers are so full of novel ideas and comments of general interest that they should prove interesting even to many non-specialists. The study of animal behaviour has attracted widespread interest, and these documents present a stimulating supplement to general books on the subject.

Something should be said about the difficulties involved in the translation, and how they have been tackled. Some aspects, such as the representation or description of calls, have no real solution. The calls have been translated with the closest equivalent and the original should be consulted where necessary. Almost all of the other problems have revolved around the question of *terminology*. Where possible, modern terminology has been used, as long as this has not affected the sense of historical continuity. For example, the rather unwieldy phrase 'serving for survival of the species' has been replaced with the term 'adaptive' in many places. Generally accepted terminology has been used as far as appropriate, but there are some important departures. For example, Heinroth's term 'Imponiergehaben' is often assumed to imply a largely *dominant* quality. 'Imponiergebärde' is thus frequently translated as 'dominance display' or even 'aggressive display'. In other cases, translation has been too general and 'Imponiergehaben' is simply translated as 'display'. In some instances these terms are used in this way in the original German, but 'Imponiergehaben' as used by Heinroth and

Lorenz has a quite specific sense. Heinroth originally applied this term to a form of male display behaviour which *simultaneously* evokes a positive response from females and a negative response from other males. The motor activities concerned are frequently associated with the presence of a male 'Prachtkleid' (lit.: 'splendid attire'; translated here as 'display markings'). In order to convey the fact that the displaying male's behaviour serves to convey a strong visual impression to conspecifics, the term 'demonstrative behaviour' is preferable. This is more specific than 'display' alone and yet does not include assumptions about dominance.

There are more complex issues of terminology involved with use of the word 'Instinkt'. One major advance made by Heinroth and Lorenz was the refusal to use this term as a blanket to cover various different aspects of behaviour and their consequent emphasis on the description and evaluation of observed *elements* of behaviour – the *behaviour patterns*. A distinction was made between observed elements of behaviour and motivational factors affecting the probability and intensity of response. These factors were referred to as the *drive* (Trieb; Drang) by both Heinroth and Lorenz, and where Lorenz uses the term 'instinct' he uses it to refer to motivational factors and not to the actual response. Heinroth employed the term 'Triebhandlung' (lit.: drive-activity or drive-operated activity) for the observed elements of behaviour, and this term was later adopted by Lorenz. However, Lorenz later switched to using the term 'Instinkthandlung' (lit.: instinctive activity) and still later adopted the term 'angeborene Verhaltensweise' (innate behaviour pattern). In translation, the term *instinctive behaviour pattern* is used for the first two terms, since Lorenz uses 'Triebhandlung' and 'Instinkthandlung' synonymously and since it is the concept of the observable pattern of activity which is important. This variability in the terminology initially used highlights the gradual maturation of Lorenz's approach to the study of animal behaviour, but it should not be allowed to obscure the essential distinction between observed event and motivation.

From an historical point of view, it is essential to note that Lorenz was working in the middle of a controversy between the supposedly opposing vitalist and mechanist schools; his conclusions amount to a rational settlement between these two extremes. One of Lorenz's most influential teachers – Jakob von Uexküll – exhibited a combination of acute observational and analytical skill and adherence to the idea of the 'fundamental mystery' of biological happenings. Lorenz seems quite definitely to have inherited the former qualities and to have applied them to remove the last shrouds of mysticism surrounding the governing factors of behaviour. One of Lorenz's major contributions has been to underline the existence of a whole range of behavioural acts (innate/

instinctive; conditioned; insight-controlled) and to analyse the manner in which these may be interrelated in the total observable behaviour of an animal species. The difficulty in exactly separating these three classes of observable behaviour has led to a great deal of controversy, particularly with regard to definition of the 'innate behaviour pattern', but the initial aim was not to set up rigid distinctions – it was to examine the theoretical composition of an animal's behaviour. The question 'What is an innate behaviour pattern?' has still to be solved, in the same way that the actual processes involved in conditioning and insight have yet to be elucidated.

Whatever position one might adopt to the theoretical concepts developed in these papers, there is no doubt that they provide much food for thought. In presenting the present translation, it is hoped that these original papers will reach a much wider English-speaking public and provide a reference source for this still-growing science. Many of the ideas (e.g. the embryological analogy for imprinting) have not been as widely discussed as they might have been, and for historical reasons it is important to see the points at which other prominent ethological workers (e.g. Tinbergen, von Holst) began to influence Lorenz's writings. Overall, there is much to be gleaned from these papers, and they are offered here with the sincere hope that they have been accurately and constructively translated.

R. D. MARTIN

Introduction

There are a number of reasons why the reprinting of some of my oldest papers seems desirable. One of them is that some of the old discoveries of ethology, and indeed those which actually started its development as a new branch of biological science, have recently become the storm-centre of an anthropological controversy. This is not because new facts contradicting them have emerged, but because some anthropologists who had not hitherto grasped all the inferences of Charles Darwin's discoveries have at last realized that their own theories are being jeopardized. Indubitably, it was my book *On Aggression* that triggered off the new storm and so it is I who am accused of having committed the sacrilege of ascribing to man phylogenetically programmed instincts of the same sort as those which we can study in animals. In fact, following Darwin it was Oskar Heinroth who in his paper *On the biology and ethology of Anatidae* drew exactly this conclusion from the analogies found in the behaviour of birds and humans. Heinroth drew attention to the fact that these analogies are quite particularly striking in social behaviour. This was maintained in 1910 and everything which ethology has brought to light since, and particularly that which resulted from the progressive synthesis of ethology and psychoanalysis, has fully confirmed Heinroth's assertion.

The fact that the behaviour not only of animals, but of human beings as well, is to a large extent determined by nervous mechanisms evolved in the phylogeny of the species, in other words, by 'instinct', was certainly no surprise to any biologically-thinking scientist. It was treated as a matter of course, which, in fact, it is. On the other hand, by emphasizing it and by drawing the sociological and political inferences I seem to have incurred the fanatical hostility of all those doctrinaires whose ideology has tabooed the recognition of this fact. The idealistic and vitalistic philosophers to whom the belief in the absolute freedom of the human will makes the assumption of human instincts intolerable, as well as the behaviouristic psychologists who assert that all human behaviour is learned, all seem to be blaming me for holding opinions which in fact have been public property of biological science since *The Origin of Species* was written.

I make a habit of not claiming merit for ideas and opinions which are not my own but which I took over from great men. However, it is not only the creator of a new idea who acquires merit, but also the man who realizes its importance and who 'puts it on the map'. From this point of view it gives me justified satisfaction that it is against myself that the brunt of doctrinaire attack is directed. I seem to have stirred up a wasp's nest. The amusing paradox is that, as Philip Wylie has pointed out, the very core of this hostility is instinctual. He writes: 'No other force could account even for the fury with which men hold to their beliefs, against all evidence and all reason, including the still-prevailing belief that they are not instinctual.'

Indeed, the deep emotional disturbance so clearly apparent in many adversaries of ethological theory is far more intense than that ever elicited in one scientist by the errors of another, however stupid and however irritatingly presented they may be. Religious fervour like this is only aroused in men who sense the explosion of a long-cherished dogma. Ashley Montagu has formulated that dogma with all desirable clarity: man is devoid of instincts, all human behaviour is learned.

I may underestimate my adversaries as men of science, but I certainly do not underrate the persuasive power of the doctrine which they are defending. Considering its demonstrable untruth, its world-wide acceptance is rather surprising. It can only be explained, as Philip Wylie has pointed out, by the insidious and unperceived substitution of a fallacy for an indubitable truth: the rather obvious fallacy 'All men are equal' has been successfully camouflaged as the eternal and unquestionable truth 'All men ought to have equal opportunity'. This camouflage endows the doctrinaires with the possibility of employing a highly dangerous quasi-moral pressure to enforce their doctrine. It takes a lot of stamina not to succumb to mass suggestion of this sort. There are good men who have become very bitter in a life-long uphill fight against it. There are some ethologists who have become deeply unhappy and actually conscience-stricken on realizing that their own findings were incompatible with a doctrine which they honestly believed to be a moral and democratic truth.

This is a rather sad state of affairs and even so dauntless a fighter as Philip Wylie takes a pessimistic view of ever conquering the behaviouristic doctrine. He says: 'biology has proven that men are not equal, identical, similar or anything of the sort, from the instant of conception. Common sense ought to have made all that clear to Java Man. It didn't and still doesn't, since common sense is what men most passionately wish to evade.'

I profess to greater optimism, though I certainly do not underestimate the enormous dangers which our modern culture incurs by the belief that, given proper education, all men can be turned into angelically ideal citizens. If there is anything more dangerous than founding public policy on a lie, it is living in a fool's paradise. Still, I believe in the final victory of a very obvious truth and I feel confirmed in this belief by the fact that my views are considered to be 'appallingly dangerous'. I sincerely hope they are – to the dogma.

It is an old, old story that truth is helped along by its enemies and endangered by its supporters. If I have to confess to a sneaking liking for, and even to a feeling of gratitude to my adversaries, I think it only fair to confess that some of my allies make me squirm. I am not talking of authors of scientific works, particularly not of representatives of other branches of science who have cautiously made a legitimate use of my findings, as do psychiatrists like John Bowlby, sociologists like Ronald Fletcher, psychoanalysts like René Spitz, Anthony Storr and others, or pedagogues like Kurt Hahn or Hans-Joachim Gamm. It is the writers of popular books who, full of enthusiasm for ethological theories, are causing me serious alarm. Desmond Morris, who is an excellent ethologist and knows better, makes me wince by over-emphasizing, in his book *The Naked Ape*, the beastliness of man. I admit that he does so with the commendable intention of shocking haughty people who refuse to see that man has anything in common with animals at all, but in this attempt he minimizes the unique properties and faculties of man in an effectively misleading manner: the outstanding and biologically relevant property of the human species is neither its partial hairlessness nor its 'sexiness', but its faculty of conceptual thought – a fact of which Desmond Morris is, of course, perfectly aware. Another writer who makes me suffer with almost equal intensity, if for different reasons, is Robert Ardrey. I have a great and sincere admiration for him, as a man, a writer, and a popularizer of scientific facts. It is only when he popularizes my facts that I begin to perspire, much in the same way as one does when sitting in a fast car beside a driver who is taking a certain stretch of road very much faster than one would do oneself, rushing on where angels fear to tread. This simile may be graphically improved by adding that the dare-devil driver is blissfully ignorant of certain weaknesses of the car of which I myself am painfully conscious.

Notwithstanding the fact that my friend Robert Ardrey is fundamentally right and the doctrinaires of the instinctlessness of man are fundamentally wrong, they have one thing in common: none of them knows enough about the basis of observational facts on which ethological theory is built up. Hence none of them realizes where the strength and where the weakness of this theory lie. Animal and human behaviour is something rather more complicated than any of them seem to realize. Also, we know quite a bit more than our adversaries and even some of our supporters give us credit for. The original theoretical positions of ethology which were taken by Whitman and Heinroth more than half a century ago, and which have suffered no essential change since, owe their strength to the enormous breadth as well as to the solidity of the observational basis on which they are founded. The breadth as well as the solidity derive from the fact that *lovers* of animals observed for the joy of observing, *unbiased by the ideological disputes* which, even at that time, distorted the views both of vitalistic, 'purposivistic' psychologists, and of their mechanistic, 'behaviouristic' antagonists. Years ago, my highly respected friend Jan Verwey complimented me on the 'strength of character' which enabled me, even as a very young man, to write the first papers included in this collection quite independently of and uninfluenced by the antagonistic views of Watson and Yerkes on one side and of McDougall and Lloyd Morgan on the other. At that time, I had not even heard the names of these four gentlemen nor, for that matter, had my teacher Oskar Heinroth.

The only pioneers of ethology who were sufficiently well-read on psychological subjects to claim merit for the independence of their views, were Charles Otis Whitman and his pupil Wallace Craig, and the strength of their opinions lay in the simple fact that each of them knew more about animal behaviour than all the contemporary purposivists and behaviourists together – including William McDougall, who actually knew quite a lot.

If any scientist ever deserved the attribute of an *expert* on animal behaviour it is any of the three great pioneers of ethology: Whitman, Heinroth or Craig. An expert is, by definition, a man of large experience, which, in its turn, is defined in the Oxford dictionary as 'actual observation of facts or events'. In other words, an expert is a man thoroughly *familiar* with his subject, independently of whether or not he attempts to formulate any theory on the basis of the facts known to him. Being an expert therefore does not necessarily imply being a scientist, though it is, of course, one of the indispensable prerequisites of ever becoming one. In the study of behaviour, more than in any biological science, the modern contempt for purely descriptive observation and the tendency to premature experimentation does much damage by preventing many young research workers from ever becoming experts – and scientists. It is a fashionable fallacy to believe that operational and statistical methods are making knowledge and intelligence dispensable in the pursuit of science. As Paul Weiss has pointed out, this erroneous belief is engendered by confounding information with knowledge: random gathering of information procures, at best, the soil on which the tree of knowledge can – perhaps – grow. It takes quite a bit of intelligence and intuition (which is but another name for Gestalt perception) to winnow a few grains of knowledge out of a heap of information.

Ethology may be properly defined as that approach to the study of behaviour which, since the days of Charles Darwin, has become obligatory in all other branches of biological research. Ethology is the *comparative* study of behaviour. As in the terms comparative anatomy, physiology and others, the word comparative here denotes a *method* which, by investigating the similarities and dissimilarities of related species, strives to elucidate the evolutionary process by which living creatures and all their characters came to be as they are. All the important new facts brought to light by ethology, including those of a physiological nature (as for instance the endogenous production of stimuli underlying innate motor patterns) are due to the application of the comparative method *and could not have been discovered by any other technique*. More than any other method, that of phylogenetic comparison presupposes the knowledge of an expert. To become one in this field, it is necessary to become thoroughly familiar with a group of related animal species. Such familiarity is not easily achieved. In fact, it seems necessary to become emotionally involved to the point of 'falling in love' with such a group in the way many bird-lovers and aviculturists and other kinds of 'amateurs' do. Without this emotional motivation, no thorough knowledge of the comparable behavioural traits of any group of animals could ever be gained. Were it not for the unaccountable gloating pleasure some of us take in watching 'our' animals, not even a person endowed with the supernatural patience of a yogi could bring himself to stare at a fish, a bird or an ape with the unremitting perseverance which is necessary in order to perceive the governing principles prevailing in the behaviour of an animal.

We, the starers-at-animals, claim to be experts of a very peculiar kind. The Peckhams (who knew most of what there is to know about all the spiders there are); Jocelyn Crane (who knows practically all about all the fiddler crab species of our planet); the great community of waterfowl addicts – Heinroth, Delacour, Scott, Lebret, McKinney, myself and many others; Tinbergen's team of starers-at-gulls, and many other similarly occupied people are of course regarded as quite crazy by some scientists specializing in the now-fashionable operationalistic way of thinking. However, we can claim to possess quite a bit of knowledge which is relevant to the understanding of life's history and which is not accessible in any other way.

The first paper, the one on Corvidae, is included in this collection because I think it is quite a good illustration of how that kind of knowledge is gained, even though it may require a bit of reading between the lines to gather how much time I spent staring at those jackdaws. It does, however, give some idea of the pleasure it gave me to live with these birds, watching them day by day. The paper is just a plain story, pure description, with precious little theorizing. Heinroth's old concept of the 'arteigene Triebhandlung' – species-specific drive activity – is still retained unquestioned without any attempt at analysis or definition. Nowadays, most of the activities thus described would be regarded as the successive function of appetitive behaviour leading to the release of an instinctive motor pattern by means of an innate releasing mechanism. Neither this paper nor indeed the next one is any credit to my powers of analysis and abstract thought. Still, even in this first scientific paper that I ever wrote, there are indications that a number of phenomena and problems which are still occupying our research, were beginning to intrigue me. Imprinting is mentioned as early as page 2, whilst lowering of thresholds caused by effects of 'damming-up' of unused activities seems to have troubled me early in my career (page 37).

The perspicacious reader will soon realize that I was a firm adherent of classical Sherringtonian reflex theory. While it is some credit to my honesty that, in this and the following papers, I keep emphasizing the phenomena of spontaneity in innate motor patterns, it certainly is no tribute to my intelligence that I did not realize until about eight years later that my own observations were quite incompatible with Sherringtonian reflexology and particularly with Ziegler's theory that innate motor patterns were chain-reflexes. As an afterthought, it seems blatantly obvious that I should have reasoned somewhat like this: the motor patterns in question are rigid, unmodifiable characters of a species, quite as stable in phylogeny as any morphological character. Therefore, they are indubitably the functions of equally stable neural structures. Yet they cannot possibly be chain-reflexes because, unlike any reflex, they do not wait passively for releasing stimuli to impinge, but tend to break out actively whenever such stimuli fail to arrive for any appreciable period. In this case, first the threshold of releasing stimuli becomes lowered, then the whole animal is activated to search for adequate stimulation and, failing to find it, discharges the whole action pattern at a more or less inadequate substitute object, or even *in vacuo*. Do we know any other physiological mechanism which brings about muscular activity and which, though susceptible to stimulation, is able to cause rhythmically recurrent activity even when stimulation is withheld? The obvious answer to this question is: yes, indeed we do – this is exactly how the stimulus-generating 'centres' of the heart function. Similar performances were later demonstrated in the central nervous system itself by Adrian, Weiss, von Holst and even by Sherrington himself.

What prevented me from thus reasoning was that I suffered from the subconscious, and actually quite unfounded, idea that the explanation of innate motor patterns in any other way than by the chain-reflex theory implied a concession to vitalistic ways of thinking, of which – even at that time – I had a healthy abhorrence.

Otherwise, the second paper (dealing with the characteristic properties of instinctive activities) was still written without any knowledge of the theories held on the subject by purposivistic and by behaviouristic psychologists. The reader who knows more about these antagonistic theories than I did at the time, will realize that the properties of instinctive actions with which I was struggling are equally incompatible with both the purposivist's and the behaviourist's views. The fact that the innate motor patterns are rather constant and rigid, self-contained elements which can be built into various different behavioural mechanisms is clearly beginning to form in my mind without, however, being clearly formulated. With humiliating regularity, it is only in the next paper that I seem to arrive at a conscious abstraction of what was moving me while writing the preceding one.

The third paper included in the collection, the good old 'companion' paper, is an attempt, on my side, to bring some sort of preliminary order into a host of observational facts partly collected by myself and partly inherited from Oskar Heinroth. It was a typical case of beginner's luck that my first love, the jackdaws, are such highly social creatures. Since that time, my interest has been centred in the *social* behaviour of animals and men.

Before I started to write the 'companion', my psychological teacher Karl Bühler had made me realize that there were some other people besides Oskar Heinroth who had written on the subject of instinct. He made me read Spencer, Lloyd Morgan and McDougall as well as Yerkes, Watson and Jennings. I must confess that all these authors (with the sole exception of H. S. Jennings) caused me deep disappointment and disillusionment! They were no experts! They simply did not know enough about animals! They were ignorant of the phenomena and the problems with which I, young boy though I was, was trying to cope, even then. None of them was aware of what I had known even before Heinroth taught me: that innate motor patterns were just as reliable comparable characters of species, genera and even larger taxonomic units as were the structures of bones, teeth or feathers. The purposivists thought that all innate behaviour was variable, which I knew it wasn't; the behaviourists believed that innate behaviour didn't exist, which I knew it did. Youth is intolerant and I simply refused to concede any scientific value to the writings of the authors just mentioned. I had not then grasped Hegel's great discovery that thesis and antithesis, however fallacious each of them may appear when taken separately, may contain some valuable truth when considered jointly. The only author who impressed me by his knowledge, and whom I immediately acclaimed as an expert of Heinrothian merit, was H. S. Jennings. His concept of the 'system of actions' characteristic of each species is one which ethology could not do without.

For my part, I had discovered Wallace Craig. He became one of my most important teachers and I owe him an immense debt of gratitude for the patient correspondence with which he honoured me and which had a very obvious beneficial influence on the theoretical discussions at the end of the companion paper.

However, the real benefit of Wallace Craig's influence becomes apparent only in the next paper dealing with the establishment of the instinct concept. It is hardly an exaggeration to say that this paper is about half written by Wallace Craig himself because it is really the distillate of the extensive correspondence we had in the years 1935–37. This paper contains a critique of Spencer's and Lloyd Morgan's theory that instinct, by becoming more and more strongly influenced by learning and intelligence, merges into the latter ontogenetically as well as phylogenetically. The same facts on which this critique is based are also used to disprove William McDougall's theory that instinctive activities are essentially purposive: the consummatory act represents the purpose at which appetitive behaviour is directed but is definitely non-purposive in itself. Thirdly, the phenomena of spontaneity which I have already mentioned are used in a critique of the chain-reflex theory. I did not, then, think it worthwhile to offer an extensive critique of the 'instinct theory' of behaviourists. It is mentioned only on the first pages and in a rather cavalier manner: 'The "instinct theory" of behaviourists (in the narrower sense of the word), whose main proponent may be taken to be Watson, need only be fleetingly touched upon in this introduction. Complete ignorance of animal behaviour, which ails so many American laboratory research workers, is a necessary requirement to justify the attempt to explain all animal behaviour simply as a combination of conditioned responses. The presence of more highly-specialized patterns of motor co-ordination is denied flatly and with a certain passion by the behaviourists. Since this denial is based on a simple lack of knowledge, a detailed refutation can be taken from the outset as unnecessary.' By so much did I underrate the power of a doctrine and the perseverance of superstition!

However, the dominance of a preconceived idea is sadly demonstrated at the end of this paper by myself, in the weak and unconvincing attempt I make to defend – against the evidence of my own findings – the chain-reflex theory of instinct!

Again, the conclusions which really ought to have been drawn in the summary of one paper are to be found in the introduction of the next, which deals with the behaviour pattern involved in egg-rolling by the Greylag goose. Here, for the first time, it is clearly formulated that orientation responses or taxes are truly reflex-like in nature, while innate motor patterns are something physiologically quite different, being effected by endogenous generation of stimuli and central co-ordination. The fruitful synthesis of Wallace Craig's and Erich von Holst's findings was thus reached for the first time. It was definitely with the help of these two men that I at last arrived at the clear conceptualization of appetitive behaviour, innate releasing mechanism and innate motor pattern. I have to admit that even at the time I wrote this paper I was not yet quite clear concerning the fact that these very same physiological mechanisms of behaviour could be integrated into functional units different from the sequence just mentioned. For example, it remained for Monika Meyer-Holzapfel to point out the very frequent occurrence of appetitive behaviour directed at a stimulus situation which, so far from releasing a consummatory act, results in nothing but the cessation of appetence.

The last paper included in this volume (on teleological and inductive psychology) contains my defence against the critique which Bierens de Haan, a definitely vitalistic student of behaviour, launched at ethological theories. I had been inclined to regret the inclusion of this paper in this collection because it seemed that any argument against the vitalistic use of anticipating explanatory principles was redundant nowadays. But the appearance of Mortimer S. Adler's book *The Difference of Man and the Difference it Makes* convinced me that this was not so. The author comes to the clearly formulated conclusion that the uniqueness of man is due to a præternatural factor which neither stands in need of, nor is accessible to, a natural explanation. As this book has evidently found a wide distribution, the refutation of vitalism given in this last paper may as well stand as it is.

If I am violently attacked by the behaviourists who contend that all human behaviour can be explained as a bundle of learned responses and if, on the other hand, anthropologists who are idealists and vitalists, attack me with equal violence because, in their opinion human behaviour cannot be explained at all, it would seem that my theories hold the narrow but golden mean of plain commonsense. The papers here included and written a very long time ago contain, even in their original form, all the essential facts which arouse the doctrinaires of two still older ideologies and, simultaneously, contain all the answers to their critiques. I am very glad to have these papers reprinted, and I am quite particularly grateful to my friend Bob Martin for the meticulous care he has given to their translation.

Contributions to the study of the ethology of social Corvidae (1931)

Inspired by the experiments which I carried out in 1926 with a tame free-flying jackdaw (*Coleus monedula spermologus*) – see *Journal für Ornithologie*, October 1927 – I decided to accustom a number of these birds to free flight conditions in the following year. My first jackdaw, referred to here as 'Tschock', showed a number of instinctive behaviour patterns indicating that the ethology and sociology of the species is quite complex. As is so often the case, the isolated captive bird exhibited a number of apparently purposeless instinctive social activities which can naturally only be understood through the reactions of conspecifics.

In order to save time, I at first attempted to use jackdaws from a dealer, but in vain. Such birds are always shy wild captures or individuals hand-reared from an early age in isolation. Of course, the former are completely unsuitable, and the latter, quite apart from the fact that they are usually physically handicapped, almost always show re-direction of their instinctive behaviour to the human being. The human subject, and not a true conspecific, is regarded as a species companion. (This feature was discussed in detail in my earlier article.) It is therefore particularly surprising that such birds, which show no trace of the normal species-recognition behaviour, do develop a strong gregarious drive in one situation – in flying. Jackdaws and even other Corvids are followed with obvious zest. As in other cases, it appears that the behaviour is strongly dependent upon a primary impression. The first Corvids with which Tschock became acquainted were hooded crows, and he continued to fly with the wild hooded crows even after having a great deal of contact with jackdaws at home. This behaviour only changed when he adopted a young jackdaw, which he fed and guided. After the feeding drive had subsided, Tschock once again kept company with the crows. In anthropomorphic terms: He considered himself as a human being during the courtship phase, as a jackdaw during the rearing phase and as a hooded crow for the rest of the year.

It is nevertheless interesting that Tschock adopted and fed a jackdaw and not a young hooded crow available at the same time. But this can easily be explained. It is not vitally necessary for the survival of the species that species-recognition should be innately determined, since determination of such recognition by primary impressions (usually deriving from parents and siblings) is quite sufficient. But the responses to young birds of the same species must be innate, since the first young that a jackdaw sees are its own offspring.

It is obvious that no insight into the sociology of the species can be obtained from birds showing such abnormalities in the recognition of conspecifics. For this reason, I hand-reared fourteen young jackdaws in 1927, believing that such a large number of birds together would exhibit undisturbed retention of normal species-recognition. This proved to be the case. However, a young magpie reared together with the jackdaws also retained normal species-recognition completely unimpaired, although it first encountered conspecifics as an adult. This just goes to show how differently closely-related genera may react in this context.[1]

The attic of our house was suitably converted to serve as a home base for the jackdaws. I was aware that birds usually encounter difficulty in finding their way back through a window which they have previously used as an exit (presumably because the window looks different when seen from outside), whilst they regularly recognize a wire cage from the outside. I accordingly constructed an aviary in front of the attic window. A broad brickwork gutter (5 feet wide) served as a base for the cage, while the hind wall was provided by the inclined roof. The cage extended almost the length of the side wall of the house, covering a length of over 30 feet, and measured over 6 feet in height. The jackdaws were able to leave the attic and enter the aviary by means of a small window and I was able to enter the cage from outside through a vertical door leading in from the gutter. The outer aviary was divided into two unequal sections by wirenetting such that the attic window and the door led into the larger section. Using this arrangement, I could always hold back some of the birds as a lure for their free-flying companions, by enticing them into the smaller section and closing the connecting door before opening the main door. I regarded this measure as essential, since I could not be sure that so many simultaneously-reared birds would be sufficiently attached to me for me to represent the leading adult as I had for Tschock. Young jackdaws are completely helpless without a leader. In a flock of young birds of this kind, each bird seeks leadership from another with the result that no one individual will decide to land or find its way back home. The outcome is that the jackdaws drift around in a tightly-packed flock, giving distress-calls, until they eventually become completely lost.

The aviary described above was at first occupied solely by Tschock, who was completely 'flown-in'. The young jackdaws and the magpie were introduced later, after they had fledged. During the transfer, one of the jackdaws escaped. When this jackdaw heard the others calling in the flight chamber, however, it spontaneously headed towards them and allowed itself to be enticed inside without difficulty. The young birds soon became accustomed to their new living quarters and adapted readily to the method of feeding. In order to keep them tame, I fed them exclusively by hand. Since this resulted in them all trying to settle upon me at once, one was often forced by lack of space to land on the back of another. Feeding was thus accompanied by much fluttering and crash-landing, and my hands and face were permanently covered with scratches. Incidentally, I have noticed with other passerine fledglings that one sibling will often land upon the back of another even *without* space restrictions. The perching bird is always dislodged as a result and the landing bird takes its place. This occurs *much* too frequently to be dismissed as a chance phenomenon. Perhaps this type of landing is really specified for flight onto the parents, who might employ this method (voluntarily or involuntarily) to indicate suitable perches to their offspring. As soon as the young have become just a little more accurate in landing, this behaviour disappears altogether.

When the young jackdaws were able to fly properly, I began the free flight experiments. First of all, I marked the individual birds on both legs with coloured celluloid pigeon-rings. (These birds, in contrast to those ringed as adults, paid absolutely no attention to the rings.) I then chose the two tamest birds, Blue-Blue and Blue-Red, for the first experiments. After testing them both sufficiently in flying after me from room to room inside the house, I took them out into the garden (at first singly and later together). On one occasion, Blue-Red drifted high into the air on an up-current, did not venture to land again and finally flew off. After two days, Blue-Red returned of his own accord. He had evidently flown by chance into hearing-distance of the captive jackdaws and thus found his way back to the house. At a time when I could rely upon the aerial following abilities of Blue-Blue and Blue-Red, I allowed them to fly from the roof together with two other young jackdaws and Tschock on their first trial together. I drove the other birds into the sealed section of the aviary and kept them there. As I opened the main door, Tschock naturally flew out straight away whilst the four young birds stared in fright and alarm at the unusual sight of the open door. But as Tschock flew past outside the open door all four stormed out in a tightly-packed mass. They flew after Tschock, but (since he took no notice of them at all) they lost him as soon as he performed his first dive. The young birds at once began to utter distress-calls, but not one of them dared to dive after him. A strong up-current formed over the house by the customary west wind carried them higher and higher into the air. This is a phenomenon which I have often noticed with captive birds on their first flight. Such birds, though fully fledged, have no experience of manœuvring in the wind and easily become frightened when involuntarily carried upwards. In this frightened state, the birds are at least prevented from developing the determination necessary for a 'downhill' flight and may even be driven to climb still higher. At the altitude ultimately reached, the birds are stimulated to fly so far that they cannot find their way back home. Almost all of the birds which I lost on their very first free flight were victims of this effect. I have never observed a bird which would not have wanted to return to its cage had it had the ability to find its way back.

My four young jackdaws climbed higher and higher, their calls increasing in frequency. When the members of a flock of flying jackdaws utter the social contact call in ever-increasing frequency, it can only mean that the flock is preparing to fly some distance. Social contact is increased so that no individual will go astray.

As can be expected, the still captive jackdaws reacted by calling as well, indicating that they were motivated to accompany the others. Eventually, the four liberated jackdaws slowly descended with great caution and, after a number of indecisive attempts, landed on the cage. If the number of liberated birds had been only slightly larger, the flock would certainly have flown off. A flock of old birds which have been companions for some time exhibits much tighter coherence. It can be observed in the field that a large flock of old birds can be prevented from taking off by a minority which chooses to remain on the ground. I have often seen this with migrant flocks of jackdaws resting and seeking food on the Tullnerfeld. When several satiated birds take off and utter the gathering call (social contact call) described above, the birds still on the ground join in the vocalization without actually taking off. Even when the latter are in the minority, the flock regularly settles again. A direct contrast to this is seen in the response to an alarm. One bird which has sensed danger takes all the others up with him, showing how exactly one bird recognizes *why* another takes off. Consideration towards all members of the flock functions to preserve the species by holding the flock together and ensuring that individual members are not deprived of food or rest. But, as is the case with many instinctive social activities, the phylogenetic origin of such behaviour is difficult to explain, since the responding bird obtains only extremely indirect benefits.

The four liberated young jackdaws subsequently spent the rest of the day on the roof, part of the time staying on the wire-netting above the flight chamber. After their lucky return from the air, they showed a definite disinclination to venture their way back into the open sky. They first acquired the inclination to fly and play in the air a few weeks later.

This does not typify the natural behaviour, since this effect was based upon the fact that my young jackdaws of 1927 were enclosed in a very small space at the time when they would normally have begun to fly. Under natural conditions, little is seen of the 'learning' phases of flight, since this goes hand-in-hand with the development of the implements of flight, i.e. the keratinization of the flight feathers. The observer is inclined to attribute the imperfections of the flight of a young bird to the immaturity of its definitive plumage. In any case, a bird learns to fly very rapidly at the appropriate physiological age, whereas the process is very slow once this stage has passed unexploited. It is as if the co-ordination of the complex motor components of flight, which are doubtless extensively pre-formed in the central nervous system of the young bird (possibly in the form of an inherited kinaesthetic mental[2] image) were to break down at a later stage. A bird which has never flown appears to *forget* how to fly. The first flight pattern to be developed, and the last to be forgotten, is that of horizontal and climbing flight, i.e. the most primitive and probably the oldest phylogenetic form of flapping flight. This clumsy ancestral form of flapping flight is one of the factors which leads a bird with highly-specialized flight apparatus into the involuntary upwards drift described above.

Subsequently, I allowed the other young jackdaws to fly two-at-a-time with Blue-Blue and Blue-Red, the two which were accustomed to aerial following. Finally, all of the jackdaws were more or less 'flown-in' as a result of this procedure and they regarded the attic as their centre of activity. It was only then that I attempted to entice them down to the garden, something which I had avoided previously.

Like Tschock, the young jackdaws responded from the outset to my imitation of the species' social contact call. As I called to them from the garden, it immediately became obvious that they were too attached to their home base to venture so far away. Every time I called, Blue-Blue and Blue-Red flew down towards me, but they always exhibited uncertainty half-way down and eventually returned to the roof after circling a number of times. Finally, I carried them both on my hand through the house and into the garden. Since the stairway was unknown to them, they did not venture to fly away from the only familiar object (myself). On the few occasions when they attempted to do so, they immediately returned to my arm. They still remained closely attached to me down in the garden, since this area was also unknown territory. After this first excursion, they returned to home-base independently when I called to them from the roof. Following a few excursions of this kind, they eventually responded to my call from the garden and flew down from the roof unaided. I had to call them from the highest points on the steep upward slope of our garden, since for some time they continued to prefer to fly a long horizontal stretch rather than fly down a steep slope, however short this may have been. My elevated position reduced the necessary descent to a minimum.

The other juvenile jackdaws were released in small batches and they soon became accustomed to following Blue-Blue and Blue-Red down to meet me in the garden. Since I was by this time able to risk releasing several birds simultaneously, an imposing flock of jackdaws would follow me around whenever I went into the garden. I thus had the opportunity to observe that Blue-Blue and Blue-Red remained attached to one another even in the presence of the other twelve jackdaws. This is certainly in correspondence with the behaviour of siblings towards one another.

At this time, I first observed the young jackdaws performing an instinctive pattern which I had previously observed with Tschock and had described (although somewhat incompletely) in my first publication. One of the birds was unwilling to be enticed into the cage, so I simply grasped it in my hand. The other jackdaws, at that instant all on perches, were immediately alerted and flattened their feathers. One began to utter a metallic *churr* ('rattling call'), immediately followed by the others, and all leant forward and began to beat their wings. Tschock was initially in the inner section of the attic cage, but he immediately stormed out when the rattling began and attacked my hand, pecking and clawing. For some time after I had released my hold on the young jackdaw, all of the birds remained agitated and shy towards me. Even the magpie had approached and uttered the alarm-call of its species. The next day, I deliberately elicited a *rattling chorus* of this kind whilst some of the birds were flying free, since I wanted to see whether they would react with approach or retreat to the capture of a companion. As soon as I had one of the jackdaws in my hand, another jackdaw began to rattle, as before. At once, all of the free-flying birds stormed up from all sides. Just as on the previous day, the magpie reacted too. This behaviour of the magpie does not seem so remarkable as at first sight when one considers that small birds of extremely different types respond to (i.e. 'understand') each other's warning calls as to their own, despite the large differences which exist. Once, on seeing a Blackcap (which had discovered a perching kestrel) utter its 'ticking' call, I was convinced that a pair of greenfinches nesting nearby approached in direct response. The latter birds joined in with the noisy vocalization of the Blackcap by uttering a drawn-out whistle with a querulous ring (the alarm-call of the species). On the basis of this reciprocal understanding, small birds form a kind of organization against the predators which threaten them most, above all against owls. The only features peculiar to the jackdaw are the mode of elicitation of the response (which is not evoked by the sight of the predator itself but by the sight of a fellow Corvid in distress), and the violence of the ensuing attack. The jackdaws circled around my head as a tightly-packed flock, manœuvring abruptly and carrying out occasional mock dives directed at my head. Only Tschock worked up to actual attack on this occasion, though the attacks were directed at the hand holding the jackdaw and not at my head. The magpie did not take off but tried to attack me from a perch behind me. The attack from the rear is extremely typical of both magpies and ravens.

Since my jackdaws were still agitated long after I had released their captive companion, and since they exhibited distrust, which made it very difficult to entice them into the cage that same evening, I did not dare to deliberately elicit this rattling-reflex so soon again. The birds would otherwise have become shy in a very short space of time. In any case, I had plenty of opportunity to observe this response in the course of the next two years. Evidence that a purely innate instinctive pattern is concerned is predominantly provided by the circumstances in which it is elicited. The occurrence of the reaction appears to be dependent upon a situation where a Corvid (alive or dead) is *carried* by some other animal. Strangely enough, it does not seem to matter which species is carried. The reaction is *not* elicited if a jackdaw is in distress in some other way. For example, on one occasion when a jackdaw caught its hind toe in the wire-netting of the cage, tore off its claw and then began to shriek in pain and fright (an expression of extreme distress) the response was not elicited. None of the jackdaws paid the slightest attention to the plight of their companion, but a rattling chorus immediately commenced as I rushed up and grasped the bird in my hand in order to free it! This one observation alone would seem to establish that the defence of a conspecific is not a voluntary and rational response among jackdaws but a purely instinctive response. This assumption is amply supported by additional observations. In the 1929–30 winter, one of the jackdaws died and the corpse remained unnoticed by the jackdaws. However, when a hooded crow which lived together with the jackdaws as a familiar and unfeared member of the colony simply turned over the corpse with its beak, a mob of rattling jackdaws descended on him. It is by no means necessary that a jackdaw or its corpse be carried around for the rattling response to be elicited. The same reaction was evoked when I held out a dead magpie towards the jackdaws and even the presentation of a single, large black feather sufficed to provoke a rattling attack. In fact, the black feathers appear to be the decisive factor, since my birds did not react to presentation of a jackdaw plucked of all its feathers. Later on, a jackdaw pair failed to react when I held out their unfledged young on the palm of my hand and yet promptly began to rattle and attacked angrily when I tried the same experiment a few days later after the feather sheaths had ruptured to expose the fledglings' feathers. In Spring 1929, I witnessed an unusual miscarriage of this instinctive pattern. A rattling chorus was elicited, although not followed by active attack, by a jackdaw carrying a crow's pinion to its nest! The sight of dark feathers is not absolutely essential, however, as one incident demonstrated. I provoked my jackdaws to launch a rattling attack on me completely unintentionally by walking through the garden carrying my wet, black bathing trunks in my hand. The colour and the limp, dangling nature of this object seemed to be sufficiently similar to a Corvid corpse to elicit the same response from the jackdaws. It is therefore striking that the jackdaws did not respond to a coal-black pigeon which I held out to them as they would have to a Corvid. One would imagine that the pigeon has more in common with a Corvid than with black bathing trunks. The jackdaws responded more rationally by flying up and pecking the pigeon without uttering the rattling-calls. All Corvids of my acquaintance have attempted to 'help at the kill' when I hunted, captured or held other birds (or, for that matter, other animals) in my hand. Thus it would appear that the rattling response of jackdaws is suppressed by characters of the pigeon distinct from Corvid characters, despite the large number of common features they share. The characters of the pigeon obviously elicit a distinct, characteristic response operating to the exclusion of the rattling response. Apparently, the few characters which are common to the bathing trunks and a jackdaw corpse are sufficient to elicit the response adapted to the latter.[3]

This rattling response, which can be summarized as being elicited by 'black objects carried by *any* animal', is of course fairly specifically 'intended' for predators which have seized a jackdaw or some other Corvid and are in the act of carrying off their prey.

The purpose of the instinctive pattern is obviously less the rescue of the victim from the predator than prevention of the predator from enjoying his spoils and thus deterring him from future predation of Corvids. The question consequently arises as to why the jackdaw's attention has become fixed upon the sight of the killed, or at least carried, object regardless of the nature of the predator. Take, for example, the attack on the female jackdaw carrying a black feather. A jackdaw is most unlikely to direct a serious attack at a predator which does not elicit the rattling-reflex by carrying off a dead Corvid. At the most, jackdaws may direct a largely playful attack at a flying bird of prey. Only parents of young birds appear to depart from this rule.

On the other hand, ravens, magpies and probably crows as well will attack any hairy or feathered predator which comes into sight, possibly in order to deter predators from their living areas. But the jackdaw is certainly less suited for such attacks than these other Corvids. The ravens and the crows are larger in size, while the magpie, with its incredible adroitness and extremely agile take-off, can torment large predators still better than the physically bigger raven. Since the jackdaw possesses neither the power of the raven nor the speed of the magpie, it is limited to attacking defenceless predators carrying away their prey, at which times it attacks with almost unparalleled ferocity and tenacity.

I do not know whether crows possess a reaction analogous to the rattling response of the jackdaw or whether they are at least comparable to the magpie in responding to this reaction. I regard both of these possibilities as highly probable, since crows are probably more closely related to the jackdaws than are the magpies and are certainly more similar in general biology. Unfortunately, all of the crows which I have kept to date were so far from being typical representatives of their species that I am unwilling to draw any conclusions from their behaviour. It is apparently unbelievably difficult to rear completely healthy crows, and especially rooks. I have never myself removed young crows from the nest and the young birds which I did obtain were already ruined for life. Anyone who would doubt that crows are more difficult to rear than jackdaws, ravens or magpies should go to the nearest zoo and compare the species kept there. An unkempt raven is a rare sight and the sight of a crow with sleek plumage is still more uncommon. In order to analyse such a complex response, observations in the field are sorely needed; particularly on crows. I would be *extremely grateful* for any information on this score.

I do have quite definite information about ravens, however. They do not respond to the capture or carriage of Corvids of other species with any defensive mechanism. Instead, if a human being acquainted with them should hunt or capture other Corvids they will actually join the hunt at once and try to 'help at the kill'. This probably ties up with their behavioural response to conspecifics engaged in hunting. Thus, the raven has no part in the offensive and defensive alliance of these other Corvids. This must also apply to the jays, which differ from the other Corvids morphologically and show even greater differences in ecology and behaviour. Consideration of the relationships of the socially co-operative Corvids to one another and to the raven gives rise to the impression that the non-social behaviour of the latter species, which certainly exhibits the most specialized family structure, is secondary in nature.

Ravens vigorously defend conspecifics when they are captured. The anger-call which is uttered at such times in fact has a certain similarity to the rattling-call of the jackdaw, but the defensive act seems to be less compulsive in the raven since the defended conspecific must apparently be a known companion of the defender. I was able to do what I liked with a young raven which I reared this year without producing the slightest arousal in those reared the previous year. In fact, the latter were more prone to pounce upon the young bird! A further indication that personal friendship is more important than species relationship in the raven is the fact that my oldest raven defended me in the same fashion against an attacking cockatoo. He also took action against a housemaid who was trying to extricate his consort from a tangle of wool, in which she had become trapped through playing with my mother's knitting. The attack-call of a raven defending a friend is the same as that uttered on seeing a dog or a cat or on seeing an armed man (or even a man dressed like a hunter). This call differs from the normal fighting call of the raven in being shorter and in having a harsher, nasal quality. A similar call, but much more nasal and far higher in pitch, was uttered by nesting hooded crows when they attacked my ravens in an attempt to drive them away from the vicinity of their nest. The rattling-call of the jackdaw resembles the attack-calls of the raven and the hooded crow in its harsh, nasal quality but differs from both calls in that it is uttered persistently rather than in short bursts.

In my first paper, I expressed the opinion that the attack-call of the jackdaw is identical both to the 'grumbling' tone which is heard when they attack a bird of prey or playfully buffet one another and to the fright call of a bird which is (or believes itself to be) in extreme danger. However, closer acquaintance with these calls shows that all three can be distinctly separated. The fright call is characterized by a rather more shrieking quality, while attack rattling and 'grumbling' when buffeting (seriously or playfully) differ from one another in that the former is uttered persistently – as stated above – whereas the latter is uttered only as a burst of a few short calls in rapid sequence. The 'grumbling' call also sounds somewhat mellower, deeper and less nasal than the attack rattle. It is common to the jackdaw, the crow species and the raven, has the same significance in all and is very much the same in all species except for differences in pitch.

Since the above account indicates that the defence of comrades by the raven cannot confidently be ascribed to a species-specific instinctive-mechanism, whereas the rattling response of the jackdaws most definitely represents an instinctive behaviour pattern, it would be very interesting to know how crows behave in this context.

An absolutely characteristic example of the almost reflex nature of the rattling response was provided by Tschock's behaviour at the time when I first came to comprehend this instinctive act and to understand its significance more exactly. Tschock was, as a general rule, absolutely devoted and affectionate towards me and passionately hated the young jackdaws. From his behaviour, it appeared that he did not regard the other jackdaws as conspecifics. Nevertheless, whenever I grasped a young jackdaw in my hand he would at once attack me unconditionally! In all other situations, Tschock's behaviour towards the young birds was none other than hostile. He always chased them from my shoulder and persecuted them in every conceivable fashion. The young jackdaws did not learn from these experiences, however. They always ran towards Tschock and gaped whenever they saw him – especially when he flew in from some distance away. The young jackdaws must therefore have recognized instinctively the slight differences between the juvenile and adult plumage of their species, since they never tried to beg from one another. It is worth noting that they may have retained a certain impression of their parents, since I had acquired them at a fairly advanced age.

After a while, Tschock quite suddenly altered his behaviour towards one particular young jackdaw – Left-Yellow. He adopted, guided and led this particular bird. I was then able to hear the feeding call of the jackdaw, which is also uttered by a jackdaw male when feeding his consort. This call is one of the few animal communication signals whose origin is immediately apparent. It is, in fact, the normal jackdaw guiding call 'kia' considerably altered by the fact that the bird's crop[4] is so full that it is unable to open its beak without dropping some of its load. When calling its young, the jackdaw utters the same call even if there is so little in the crop that it could easily produce the normal 'kia' with its beak open. In fact, the jackdaw will even call its young with the feeding call when the crop is demonstrably empty, thus more or less guiding the young under false pretences. They are taken in every time. But if Tschock really had a full crop, Left-Yellow would notice this immediately and begin to beg even if the feeding call was not uttered. Left-Yellow followed Tschock like a shadow; he was never farther than six feet away from Tschock on the ground or in the air. If Left-Yellow did on occasion lose sight of Tschock, he became distressed and helpless; he would wander around with head raised, looking about him and calling continually just like a lost gosling. This dependence upon the guiding parents can only be compared with that of many young nidifugous birds. Most remarkable is the fact that this dependence first appears when the young begin to search for food together with their parents. Until this stage is reached (i.e. long after fledging), the young jackdaws show no motivation to fly after the parents. Instead, they sit close to the nest and wait until the parents return to feed them, as do most other passerines. This nidifugous type of guidance of the young consequently first makes its appearance at a time when other passerines are becoming self-sufficient. This explains the drastic difference between the responses of young and old jackdaws, particularly with regard to the low reactivity and limited intelligence of the young. The only other endemic Corvid whose fledged young exhibit such awkwardness is the rook. This has led me to assume that the latter species is very similar to the jackdaw in reproductive biology, something which is quite probable, simply because the rook also breeds in colonies. All other Corvids reach an approximately adult level of intelligence as soon as they are able to fly properly. For this reason, a comparison between a two- to three-month-old jackdaw and a magpie of similar age greatly favours the latter. A year later, the jackdaw excels the magpie in memory, cage detour experiments, radius of activity – in fact in almost every respect.

The inability of young jackdaws to solve problems of any kind particularly impressed me because Tschock, and later the young magpies, were always present as an object for comparison. The young jackdaws were, for instance, unable to grasp the significance of the cage-door, through which they flew out to the open air every morning. If they became thirsty or sleepy outside of the aviary and happened to land on the cage roof, they would try to fly straight to the attic window. It never occurred to them to make a short detour through the door. The harder they tried to reach the inside of the attic, the more difficult it became for them to retreat from the window in order to find the door in the wire-netting. Only when they had given up their attempts as worthless did they occasionally chance to approach the door, sight the attic window without any intervening cage-wire and then fly inside. But, despite the frequent occurrence of such chance successes they never did learn to find their way systematically. As can be expected, they were also unable to follow Tschock's example. He, of course, flew in and out of the door with ease every day. It would have been very simple to construct a trap-door over the attic window so that even the young jackdaws would have inevitably found their way in, but I deliberately avoided doing this since I wanted to determine the age at which the young birds would be able to solve this problem just as well as adults. Tschock, on introduction to the freshly-constructed aviary, had immediately mastered the problem of the door without any apparent intervening learning process. The answer to my 'when' question arrived in a most impressive fashion. As I shut the jackdaws in on the 7th of August to leave them shut up while I was away for several weeks, the behaviour of the young jackdaws towards the door had not changed at all. When I released them after my return on the 2nd of September, they mastered the problem of the door just as rapidly and thoroughly as Tschock had previously. This transition took place just at a time when the birds had no opportunity to 'learn' about the door. This is an especially clear indication that the abilities of the bird are not determined by external circumstances. The intelligence of a healthy jackdaw experiences an enormous increment at this time of the year, a fact which can be observed in situations other than that of the cage detour. Since the *following response* disappears at the same time and since a noticeable tendency towards alienation from the human foster-parent then appears, it is reasonable to assume that this is the time when jackdaw families disintegrate. (The intimate relationship between Tschock and Left-Yellow also disintegrated at this time.) Even the members of old-established pairs become cooler in their mutual relationships.

In order to remain in temporal sequence, I must record the following event: In the early Summer I received two more young magpies, which obviously came from a late clutch since their tails were still short. They were initially very tame, but at once became estranged when I placed them with the other birds. This happens very easily with magpies if such a change in environment occurs, unless drastic counter-measures are taken. Their shyness was very unwelcome because they of course refused to eat out of my hand and I was forced to leave a continual supply of food in the cage. This was a bad educational influence for my jackdaws. Soon afterwards, one of the two shy magpies escaped. I did not expect that an arboreal bird such as the magpie, which never enters crevices in the wild, could be persuaded to return to the attic, since any tree would provide equally suitable sleeping quarters. The escaped magpie not only returned, it promptly found its way back through the door – in contrast to my young jackdaws. I first noticed that the magpie was perched quietly on its customary sleeping-perch when manœuvring the jackdaws into the cage that same evening. I was most surprised that this shy young magpie, still in possession of its short tail, was able to achieve so much. I was thus provoked to release my tame, fully-adult magpie; something which I had not dared to do previously. As soon as I opened the door, the magpie flew out. At first, he merely ran around and flew to-and-fro with the jackdaws on the roof. But he soon became excited; just like a jackdaw which intends to fly some distance, he flattened his feathers. The next moment, he shot off and disappeared on a straight course into the distance. The jackdaws flew with him as far as their still limited radius of activity allowed and then returned to me. In spite of my experience with the young magpie the previous day, I could not free myself of the prejudice that magpies must respond similarly to jackdaws. I regarded a spontaneous return as impossible, so I packed a small bag of mealworms in my pocket to use as a lure for the bird. I was just about to set off in pursuit when I spotted the magpie through a window. He flew up high in the air, flying just as straight as on the outward journey, and dived elegantly as soon as he reached the house – to land less than three feet away from me. I had the impression from his behaviour that he must have done this many times before. In a cage, magpies can develop their flying abilities more fully than jackdaws, which have a characteristically less rapid wing-beat and inferior manœuvrability. The magpie on its first free flight was consequently no less able than Tschock, despite the latter's year of flying experience. The radius of activity shown on this first excursion was not much shorter than the longest flight made by Tschock. I had already learned from Tschock the previous year how little the orienting ability of a young jackdaw is improved by experience. I also observed that a few days after their very first release, my jackdaws were able to excel Tschock both in radius of activity and orienting ability, despite his longer experience. But I would never have expected such a performance from an eight-week-old magpie.

I had conjectured earlier on that young magpies show a lesser and briefer dependence upon their parents than jackdaws, and this was confirmed by the behaviour of this magpie. Admittedly, he did make social contact with me whilst I was in his home range, but he would just as often fly off abruptly. When I was lost from sight, he showed no dependence upon me. At least, I was unable to perceive any discomfort on his part. He was much tamer in the open than he had been in the cage. We christened him Elsa and he soon learned to respond to this name, responding to this better than to the jackdaw guiding call with which I summoned the jackdaws. As soon as he was completely 'flown-in', his 'range' proved to be smaller than I had supposed on the basis of his first excursions. Once the range had been selected, he observed its boundaries very precisely. Although he readily accompanied me within the confines of these boundaries, he at once left me as soon as I exceeded them and flew almost hurriedly back to the house, the centre of his range. There was good reason for this behaviour; if he overstepped the boundary, he was at once fiercely attacked by an old magpie pair whose range obviously included part of our garden. Elsa was only able to maintain his ground because these resident magpies did not venture closer to the house than indicated by these boundaries, which *they* and not Elsa had determined. This enmity disappeared in early Autumn and Elsa thereafter began to spend a good deal of time in the company of his former persecutors and their offspring. The next Spring, however, Elsa was even worse off. Our house now lay on disputed territory between the ranges of two magpie pairs. The pair on the east side literally waited for Elsa to leave the house; the two birds often sat for hours in the neighbouring tree-tops continuously watching Elsa as he roamed about on the roof of the house. The two magpie pairs often forgot their dispute with one another, so that all four went after Elsa. In Spring 1929, Elsa was two years old. He had by then conquered his own territory and he (obviously a male) had come to share his territory with a magpie reared in 1928 (evidently a female). Although the pair built a nest, the female was too young to breed in 1929.

In passing, it is worth pointing out that the courtship of the male magpie bears no resemblance to that of the male jackdaw. The male magpie hops around with his head raised and tail inclined. His neck feathers are markedly ruffled and he utters a low-pitched song in accompaniment. This is reminiscent of some small passerines rather than of other Corvids. Elsa and the female differed externally in that the secondaries of the female bore a green gloss along their entire length, whereas those of the male bore a broad, reddish-violet iridescent transverse band. I do not know whether this is a constant feature differing between the sexes; I have found both types in young birds discovered in the same nest.

After my experiences with Elsa, I am inclined to believe that young magpies do not follow their parents in flight in the manner typical of jackdaws. The necessary drive is completely lacking. To counteract this, the magpies are virtually equivalent to their parents in orienting ability as soon as they are able to fly properly. At least, this is true for the fairly narrow range to which a magpie pair restricts activities during the mating season. A young magpie therefore experiences little discomfort when its human foster-parent is out of sight, thus contrasting with jackdaws or other young animals equipped with a *following drive*.

Since the gaping response typically disappears earlier in the magpie than in the jackdaw, parent magpies would also appear to feed their young for a shorter period. It must also be remembered that artificially-reared birds usually gape for a longer period than free-living individuals.

Elsa soon came to dominate all of the jackdaws, including Tschock. When Elsa was engaged in a fight with a jackdaw, it could clearly be seen that he was more adroit in flying within the confines of the cage. He easily excelled his opponent in climbing ability and would climb above him in a flash and force him to the ground. Elsa especially took to pursuing Tschock, who was dominant over the remaining jackdaws, whereas the magpie's relationships with the young jackdaws were quite amicable once he had convinced them of his superiority. These young jackdaws discovered how to exploit the feud between Elsa and Tschock in an extremely interesting fashion. Whenever I held out the food-bowl in my hand, Tschock would chase away all of the other jackdaws except his adopted 'son'. After Elsa, in his turn, had flown up and chased Tschock away, the others would immediately exploit the opportunity to satisfy their hunger, always free from any interference from Elsa.

As Yellow-Green, the strongest male among the young jackdaws, seized dominance over the others in Autumn, Elsa abruptly switched his enmity to him. In fact, the dominant member of a flock of jackdaws at any given time is only aggressive towards 'pretenders to the throne' and amicable to those birds low down in the rank-order. If one forces birds with no strong social characteristics to live together in a small enclosure (e.g. an aviary), a type of 'peck-order' will also tend to appear, but in this case the stronger members will most readily attack the weakest. I would have expected the magpie to exhibit this latter type of behaviour, not supposing that a species which is only social at specific times would exhibit an inhibition of this kind. The jackdaws possess an instinctive behaviour pattern leading to active suppression of tyrants in addition to the inhibition which serves to protect weaker individuals. This will be described more fully in the appropriate part of this chronological account.

This relative compatibility of the magpie with the jackdaws disappeared as the birds entered an active reproductive condition in Spring, 1929. The magpie then killed a number of weak young jackdaws straight off; in fact, the drive to clear his 'territory' of fellow inhabitants had emerged.

Elsa had always been very prone to attack anything which was not a fellow Corvid. Quite early on, he had exhibited an obvious tendency to attack any animal he encountered, both mammals and birds (whatever their size), and would carry out his first trial attack from the rear. He shared this drive with my ravens, but not with the jackdaws. He would always approach his victim from behind, hopping up sideways with a characteristic crouched posture indicating preparation for take-off. If he managed to approach the victim in this way unheeded, he would peck with all his might and almost simultaneously give a rapid wing-beat and retire two feet away. If the victim then fled, he would renew the attack immediately with increased boldness, i.e. he would usually fly on to the victim's back from behind. If the victim offered resistance, however, he would not fly far away, as ravens would do in a similar situation. Instead, he would utter a loud cackling call, hop a little to one side and persist in trying to reach his opponent's back. Elsa brought my free-flying Greater Yellow-crested cockatoo to the verge of distraction in this way. After attempting a number of ferocious attacks accompanied by terrible shrieks, the big cockatoo regularly gave up and flew so far away that his shrieks of anger would fade into the distance.

Our dog naturally attempted to catch the cheeky magpie at the beginning. The magpie led his pursuer round and round in tiny circles until he was quite worn out. As soon as the dog came to a halt, Elsa would once again sit less than a metre from his nose and make a pronounced attempt to provoke him to further pursuit. This he did by repeatedly flying off, uttering the cackling call, and immediately returning to sit in front of the dog's nose. The magpie's amazingly short response time meant that he could permit the carnivore to approach incredibly close without actually being caught.

A film which I took of the described scene involving the dog showed very nicely how exactly the bird always maintained the same short gap between his tail and the dog's nose. This is demonstrated particularly well in successive frames, where one shows both animals sitting on the floor with scarcely sixteen inches between them and the next shows the two at full impetus, still separated by the same gap. Another very tame magpie had the unpleasant characteristic of giving the same response to human beings as well. For some time, this magpie had the habit of coming into my bedroom at the crack of dawn and unbearably molesting me. When I attempted to chase the magpie out, he took up the same posture towards me as Elsa had to the dog and nothing I could do would drive him off. I was unable to scare this magpie away with missiles and even an upturned stool – which was the embodiment of all that was dreadful for Tschock – failed to frighten him.

In the same Summer of 1927, a lucky chance observation on wild birds confirmed my assumption that the described pattern of provocation and hounding of predators represents a species-specific instinctive behaviour pattern in magpies. On a meadow, just at the edge of a field of tall corn, I saw a flock of excitedly chattering magpies alternately flying up into the air and landing again. Under cover of a deep ditch, I stealthily crept up and managed to observe the following: Fourteen magpies were squatting on the meadow with their heads turned towards the corn. They slowly hopped towards the corn with the same crouched posture observed in Elsa. Suddenly, a weasel took a tremendous leap out of the corn at the nearest magpie. But he was just as unable to reach his target as my dog had been and he was just as effectively harried. As soon as this small carnivore gave up the hunt and returned to the corn, the entire flock of magpies instantly followed him. The weasel must have been very hungry, since he sprang out of the corn a dozen times, leaping almost six feet into the air. But he was just as unable to reduce the constant distance between him and a magpie as the dog had been. Alongside the magpies, this accepted symbol of agility seemed almost clumsy.

I can quite well imagine that the magpies in this way render their territory so uncomfortable for any predator that it will prefer to hunt somewhere else. This would be sufficiently advantageous to act as the goal of this instructive behaviour pattern in the magpie. In contrast to jackdaws, magpies are easily able to catch grasshoppers by virtue of their short response time. Both birds are equally incapable of discerning these cryptically-coloured insects as long as they sit still, but magpies are able to spurt after a leaping grasshopper and catch it the moment it lands.

Magpies generally seem to be much more adapted for a carnivorous diet. Mine, at least, seemed to miss animal food more than the jackdaws.

It is worth noting here that ravens can even spot immobile grasshoppers immediately. Since it can scarcely be assumed that these birds possess better powers of colour-discrimination than other Corvids, I am inclined to the opinion that the central exploitation of the image is much better developed in ravens. This agrees very well with the prevalence of ravens in other behavioural features.

Everything that has been said leads to the inevitable conclusion that the jackdaws were almost dismally inferior in any comparison with magpies of the same age. Unfortunately, dependence upon parental guidance had the effect that an extremely large proportion of my orphan, leaderless young jackdaws met with mishaps. Three flew away in June alone, presumably because they had happened to stray further than usual from the house. I chanced to see them circling in a tight-packed group and calling in the characteristic fashion some distance away. They subsequently flew off down the Danube. One of the three returned two days later, but the others were lost for good. The returned bird had evidently drifted randomly into hearing distance of the house and then responded to the social contact calls of its captive companions.

The next jackdaw to succumb was Blue-Red. We found him the next year mummified in one shaft of our domestic heating system. Unfortunately Blue-Blue (the second of the specially tamed young jackdaws) disappeared shortly after the first. Before my departure on the 7th August 1927, Left-Green and Right-Blue also disappeared in rapid succession. Since these last three birds disappeared one by one, I am convinced that they were the victims of mishaps. I find it almost impossible to believe that a young jackdaw would fly away on its own. These unfortunates were probably devoured by our cats.

A further indication that the lack of leadership was the cause of this high mortality was the fact that there were no further mishaps following the transition in the reactions of the jackdaws which took place in August. At least, this proved to be true of the period from the 2nd September until the 4th November, on which date I locked up the birds for the whole winter since they showed an alarming tendency to join up with itinerant flocks of migrating jackdaws.

During his second period of free flight, Tschock again flew continuously in the company of a small flock of hooded crows which roosted nearby. Following his release each morning, Tschock would climb high into the air and steer purposively in the direction which he knew would lead to the crows. He always returned at lunchtime in order to take his accustomed meal with us at our meal-table. After a short pause for digestion, he would fly off once again to join the hooded crows. Tschock took no further notice of the young jackdaws, not even Left-Yellow, whereas the young birds were influenced by him to the extent that they made their excursions in the same direction.

In exact correspondence to Tschock's behaviour when he was out with me, the young jackdaws showed greater cohesion as a group with increasing distance from their centre of activity. In more exact terms, cohesion increased with increasing unfamiliarity of the environment. They nevertheless failed to show the intimate, unconditional cohesion that is to be observed in Autumn migratory flocks. Such travelling flocks, in contrast to groups of resident birds, cannot have specific feeding, drinking and resting sites in the unfamiliar surroundings and their feeding and resting periods are determined by what they encounter in the way of suitable localities at any given time. Thus, any jackdaw which loses contact with the migratory flock has little hope of finding its companions again. Seen in this light, the almost fearful cohesion of migratory flocks is entirely understandable.

After some experience with the motor patterns of the jackdaws, it is possible to recognize at once the migrants' lack of experience with their surroundings. Apart from their tendency to follow a specific directional course, as long as they are flying high up, such birds behave in almost exactly the same way as leaderless young jackdaws. As soon as they alight to seek food or to rest for the night, the same symptoms of insecurity and indecisiveness are noticeable. Typical circling in a tightly-packed group and continual interchange of calls can be observed. After they have all managed to land successfully, the birds regularly take off again a number of times before finally calming down. If, whilst still on the wing, they are lucky enough to spot another Corvid squatting on the ground, they are spared the difficult decision and rush down almost blindly to join the squatting bird. Whenever I take my tame raven with me down to one of the sand-banks lining the Danube on any October evening, I am given a convincing demonstration of the incautious manner in which otherwise timid crows and jackdaws will alight alongside me if they happen to fly along the river. It is often as if my raven were serving as a sort of magic hat, rendering me invisible. In fact, a method used for catching crows on the Kurische Nehrung is based upon this characteristic of migrating Corvids. It is not immediately obvious why jackdaws, and crows in particular, should exhibit this conspicuous uncertainty. Many other birds which migrate in flocks fly with utter decisiveness, as if inspired by a communal will. Take, for example, the co-ordinated manœuvres of starlings or sandpipers. To the human observer, the manœuvres of the individual birds appear to be absolutely synchronized. Since I am not inclined to assume that all members of the flock react with such certainty to the same external stimuli at each instant with equal and simultaneous effect (which would be the basis of the alternative explanation), I prefer to believe that one bird initiates the manœuvre and that the others follow in a time-interval too short to be perceived by the human eye. This reminds me of the film of the magpie and the dog, where it is also impossible to perceive which of the two animals starts to move first.

But even large birds with a relatively long response time frequently show a far better 'organization' within their migratory flocks. Flocks of wild geese, for example, move with complete purposiveness, resembling Corvids within familiar territory. That this oriented appearance is not an illusory effect, as it is in the case of flocks of small passerines, would seem to be proven by the fact that the orientation can be *lost*. In foggy weather (a frequent feature of the Danube valley in Autumn), flocks of geese can often be sighted. When the weather has cleared, single individuals or small groups can be seen, their behaviour in every way resembling that of disoriented Corvids. They have obviously become detached from the flock in the fog.

Thus, the behaviour of a flock of wild geese differs from that of a flock of resident jackdaws only in the unqualified cohesion within the goose flock, which is necessitated by the fact that many of the birds in the flock are unoriented, just like the detached groups of three or four birds found after the fog.

I am well aware that the assumed existence of leaders possessing knowledge of a given area conveys an air of anthropomorphism. But since experiencing the promptness and exactitude with which my ravens found their way back along a long pathway which they had only followed once under my leadership, I am quite prepared to credit an old greylag goose with the immense feat of memory necessary to recognize every sleeping and grazing site between Lapland and the Danube Delta. It is widely recognized that animals are most prone to exhibit amazing feats of memory in connection with localization.

In spite of the intimate relationships binding the members of a jackdaw flock together, no such cohesion is seen in the group during the period of occupation of breeding territory. Outside of the migratory season, Corvids in fact have such fixed habits and such accurate daily schedules of activity that cohesion within a flock is rendered unnecessary. Each bird can at any time seek out the limited number of sites at which his companions are sure to be found. Of course, jackdaws alter their preferred locations according to the seasonally-determined abundance of prey, but such changes are never accomplished so quickly that it is impossible for all of the birds to keep pace.

This seeking out of specific sites was particularly obvious when Tschock set off in search of the hooded crows each morning after release. I was able to follow his course without difficulty with a pair of binoculars from the roof of our house. His rapid and decisively straight flight to a specific spot in the wood was very impressive, as was his response to finding the site unoccupied. With undiminished certainty of success, Tschock would veer off almost at right angles and fly in an equally straight line to visit a certain spot on the meadow where he almost always located the crows. Every normal jackdaw behaves in exactly the same way towards companions in the colony.

After their release every morning, a large contingent from the resident flock of young jackdaws, whose behaviour by now no longer differed from that of adult individuals, flew off to the fields after performing the departure formalities described above. I never saw a single bird fly off in advance. Whenever a jackdaw did set out to do this, it would immediately turn back to the house uttering contact calls. On such occasions, I saw a behaviour pattern which I had observed with Tschock the previous year and which I had erroneously interpreted as a female courtship pattern: The originally departing jackdaw glided quite low over the heads of his companions on the roof, freed its tail from its aerodynamic functions by retracting its wings and then waggled the slightly elevated rudder to-and-fro in a horizontal plane. This acts as a summons for the others to join the flying bird in the air, a signal which is infallibly followed by every jackdaw concerned. I noticed this display pattern to be particularly frequent between members of an established pair, but the pattern is exhibited by both sexes and certainly has nothing to do with the female invitation to mating, although the latter display is actually similar in form.

I was utterly flabbergasted at a later date by a very tame, free-flying cockatoo, which exhibited the same pattern in exactly the same way towards me. Since it seems extremely improbable that two groups should have developed this instinctive behaviour pattern in an absolutely similar manner, as it is an effect which cannot be explained as a product of parallel adaptation, the only logical assumption is that the pattern is an extremely ancient phylogenetic feature. It would appear probable that this pattern might be a feature characteristic of many other birds if it is performed by two such distinctly separated groups as the passerines and the parrots. I succeeded in taking a film of Tschock in which it is easy to see how he flew up towards me, dived as he approached (wagging his tail at me) and then turned sharply to fly off. In spite of my efforts, I did not succeed in filming this behaviour in the cockatoo. This bird is unfortunately now estranged from me and it no longer performs the response.

The extremely distinct turn away from a companion summoned to join the flying bird in the air (as seen with Tschock in the film) is quite often exhibited by the jackdaws *without* the tail-wagging component, particularly when the companion has once taken to the air. Whenever a jackdaw simply flies away from the colony without making such an attempt to induce his companions to accompany him in flight, as Tschock did every morning, then it is quite safe to assume that the jackdaw concerned is flying off to meet other companions and knows exactly where these are to be sought. This search after an initially invisible and distant goal is always extremely impressive.

Throughout the entire day, individual birds or small groups would fly back and forth between the colony and the site preferred for food-hunting at any given time. It is in fact a general rule that the birds would return home more or less on their own. But it never occurred that a single bird flew away when all the others were at home. Since all of the jackdaws at first spent the midday period at home, just like Tschock, a formal mass departure just like that in the morning occurred early every afternoon.

This all goes to show that members of the colony are aware whether or not birds are still absent. Since counting of the number of companions at home at any given time can be excluded, I believe that each jackdaw virtually keeps an account in its head of each colony companion that flies off. This task is greatly facilitated by an extremely pronounced group-formation process, which separates the large number of birds into an easily assessable small number of discrete units. In human terms: The bird does not say, 'There are twelve birds at home, so three are absent'. It says instead, 'Group A, Group B and jackdaws X, Y and Z of my group are outside'. The cohesion within the separate groups is particularly obvious when the groups prefer different destinations on excursions, as is sometimes the case. It is noteworthy that a jackdaw in such a colony does not even glance after birds of its own group when they fly away (i.e. apparently ignores them) and instead continues its own particular activity unperturbed. However firmly the jackdaw's attention, and thus its visual focus, may be fixed in another direction, it nevertheless notes exactly the passage and course of its companions when they fly past. After a while, often several minutes later, the squatting jackdaw will abruptly take off and follow the others (by then well out of sight) so exactly that his course does not differ by more than a metre from theirs. I still cannot decide whether a bird is able to follow the passage of his companions so exactly by indirect observation or whether their flight-paths are so fixed by habit that the following bird is consequently able to follow the course with such accuracy. This latter assumption is not so improbable as it may seem;

many facts indicate that birds do not initially take in the spatial structure of their surroundings on a rational basis, as we ourselves do, but that their mastery of their environment – strictly speaking – amounts to the sum of intersecting learned pathways. For instance, my ravens slavishly and exactly retraced a path which they had taken only once, even when the first excursion was made by following me as I rode on my bicycle along a winding road and the second flight to the given destination had to be made without me. This behaviour had a particularly nonsensical appearance because the destination was visible from our house and the birds could have reached it by a direct flight through the air. However, when the birds had visited the same destination along several different routes, they subsequently chose the most expedient (i.e. the shortest) quite rationally.

Jackdaws nevertheless remain predominantly attached to habits once acquired, even when this dictates a considerable detour to reach a given destination and when the direct route is apparently clear to see. For example, the pathway around our house was at first partially hidden from Tschock's view. When he took off from the window of the room which he then occupied and flew around the outside of the house, he always returned along the route which he had taken initially. This even occurred if he had already traversed three sides of the house, when a return along the fourth side would have been one third of the actual return distance. It was characteristic that Tschock did this in both directions, thus covering two sides of the house from either direction. The spatial structure of the closed circular pathway around an opaque obstacle was something which the jackdaw could not grasp. However, when Tschock was transferred to the roof the following year he was more easily able to survey the spatial arrangement of the house and he did in fact abruptly master the problem.

Although path-conditioning in this way eventually led on to rational treatment of simple spatial structures, this was by no means always the case with somewhat more complex structures. A very good example of this was provided by the behaviour of an old-established jackdaw pair, Yellow-Green and Red-Yellow. These two birds bred in 1929 whilst in my care, constructing their nest in the hindmost section of the roof aviary. They had quite admirably solved the resulting detour problems, which were in many instances quite difficult. When they flew up to the aviary from behind, they had to solve a task which presents most birds with insurmountable difficulties – that of reaching the door when the nest was a direct distance of only ten metres away. Yet these two solved the problem every time without the slightest hesitation. Since they arrived from every conceivable direction and yet immediately found their way purposefully to the cage-door and on to the nest, they really gave the impression of insight into the spatial structure of the cage. However, when the offspring first left the nest and perched just a few metres away, the slight alteration in the nature of the problem sufficed to cause an initial breakdown in the parents' behaviour. This would not have been possible had insight actually been operative. Nevertheless, the adults took less time to adapt to the change than would have been the case with an entirely new problem of comparable complexity. These birds in fact substitute a capacity for self-conditioning for insight. As Pilcz quite rightly states in speaking of the mentally disabled: 'They learn the facts of life by heart.' Of course, problems with the particular degree of complexity presented by such a cage scarcely occur in the wild, so the purposive back-and-forth flights of my birds seemed very 'intelligent'. This effect was particularly striking when the flock flew home from some distance away and all the birds dived down rapidly from high in the air to land close to (or even upon) me.

But, in contrast to Tschock, these jackdaws did not seek contact with me whenever I encountered them in the fields. They were by no means as shy towards me as they were towards strangers, showing that they had actually recognized me, but they seldom allowed me to approach closely. In the presence of strangers, they would fly off at roughly the same distance as wild crows. Towards the middle of October, they showed a continually increasing inclination to join company with migrating jackdaws and even with other Corvids (particularly rooks), an inclination which they had never shown previously, Tschock excepted. In the course of time, their radius of activity increased markedly and they returned home much less frequently, even abandoning the previously regular midday rest-period. I was anxious that they might stay away completely one fine day, so on the night of the 3rd November I shut the cage-doors which had been open for all the birds since the 2nd September.

In this latter period, I had had so little opportunity to observe the jackdaws that I only noticed that pronounced reproductive motivation had already emerged as I locked them in. Two birds in particular, Yellow-Green and Red-Red, behaved like genuine partners; they were always close together, they fed one another with their crop contents and scratched the backs of one another's heads. Usually, the male would perform mainly the first activity whilst the female typically performed the latter, exactly as in pairs of pigeons and parrots. There was scarcely room for doubt that Yellow-Green was the male of the pair. He exhibited the tense, ostentatious behaviour typical of so many male birds (and of male animals in general), the main characteristic feature being that he performed the smallest movement with an excessive expenditure of energy. He ran continuously around the female with his crown feathers ruffled and responded extremely sensitively to a close approach made by any other jackdaw. This increased aggressivity was probably one of the reasons for the reshuffling of the rank-order which took place among my jackdaws in the initial period following confinement. Yellow-Green deposed Tschock from his ruling status – a rare occurrence, since a subordinate hardly ever ventures to revolt against a bird which had previously dominated him. In this case, I incline to the explanation that Tschock had had so little contact with the other jackdaws in the preceding period that the birds virtually established new acquaintanceship after enclosure, such that Yellow-Green prevailed in a renewed phase of combat. I have already mentioned that the magpie Elsa left Tschock alone from this point onwards and persecuted Yellow-Green. I experienced another interesting case of re-shuffling of the rank-order in Autumn 1929. At this time, a male jackdaw which had been absent for some time returned and defeated Yellow-Green after an embittered struggle. The returned jackdaw thereupon became betrothed to a very small, somewhat wretched female. This all took place within the course of the day following his return. The next day, Yellow-Green at once made way for the bride at the food-bowl, although she had been penultimate in the rank-order only two days previously. This is an apt demonstration of the rapidity with which the rank of a dominant male is extended to include his consort.

Yellow-Green, just like the magpie, was most aggressive towards the bird most likely to act as crown pretender throughout his reign. He was actually very good-natured in his relationships with the birds far beneath him in the rank-order. In general, there is always a very tense relationship between jackdaws close together in the rank-order, whilst any bird will avoid individuals of much higher rank without friction.

It would seem appropriate here to give a somewhat more detailed account of the jackdaw's display patterns, especially of those related to agonistic interactions. If a bird intends to chase away another (usually a subordinate), it will usually adopt an erect posture and approach the other bird with its beak extended obliquely and its feathers markedly flattened. The assumption of this posture definitely represents an intention movement. It is nothing other than a rudiment of upwards take-off towards another bird, a pattern which we know so well from many other fighting male birds, such as the domestic cock. If the attacked individual does not obligingly give way, but itself adopts this posture, there is regularly a smooth transition to a specific type of fighting. The opposing birds become more and more elongated and eventually fly at one another. Each bird attempts to attain a greater height than his opponent and to throw him onto his back. In jackdaws, this is the usual form of an encounter concerned entirely with personal rivalry. A quite different threat posture is adopted in the nesting phase when an attacked jackdaw is unwilling to give way to an attacker. This is particularly prevalent when the former is in the vicinity of his nest (or simply close to a potential nesting site) and consequently places particular value on retaining his perch. The attacked bird will then markedly ruffle his feathers (especially those on his head and back), lower both his head and his partially-spread tail and usually accompany this with a twist of the tail to the side from which the attack is coming. This defensive posture is accompanied by a special call – a high piercing 'tsick-tsick', which is accompanied by a twitch of the tail and an apparent tremor of the entire body. This all conveys the approximate meaning: 'I am roosting here. This site is my nest; I shall under no circumstances take off and I intend to defend this roost to the last.' If the attack was of purely personal nature, the attacker will immediately leave peacefully. If he himself has designs on the particular nest-site, however, he may either fly onto the other bird's back (which generally achieves nothing) or he may – as in the majority of cases – abruptly pass from the extended attack posture into the described defensive posture. The two birds then squat opposite one another for some time, both uttering the 'tsick'-call and performing threat patterns. Sometimes, when the two birds are squatting very close to one another, they will also exchange pecks, though almost always without actually striking one another, since under such circumstances neither bird will move even a centimetre from the spot. This display is the nest-defence pattern, and each bird sits just as fast as if it were already brooding eggs in the nest. During the phase where my young jackdaws were occupied with the choice of their nesting sites, the 'tsick'-calls were almost unceasing. But this would seem to be an exclusive feature of jackdaws which have not yet formed stable pairs. Members of old-established pairs will, in the same situation, utter another call which (as we shall see later) is followed by a specific response. The 'tsick-tsick'-call is largely a feature of young bachelors. When such birds have appropriated a nest-hollow, they will sit there and utter the 'tsick'-call, at which stage the call probably has the function of attracting a female to the nest. Red-Red often slipped into the nest-box to join Yellow-Green when he gave the 'tsick'-call.

These two birds remained together until shortly before Christmas, at which time they abruptly paired off with other mates. I was unfortunately unable to observe this event, since I was living in Vienna at the time. The jackdaws were looked after by an acquaintance and I only returned on Sundays to ensure that all was in order. On one

Sunday, the two were close companions; when I returned to see the jackdaws one week later, Yellow-Green was accompanying Red-Yellow instead of Red-Red and he behaved with distinct hostility towards his former consort. The latter, in her turn, was betrothed to Yellow-Blue – the strongest male. The reason for this interchange remained a complete mystery to me, but I did have the impression that the separated partners were more earnest about their new relationships than about the initial betrothal. Yellow-Green and Red-Yellow were virtually inseparable and they in fact remained faithful to one another until the male died. It struck me as particularly remarkable that birds which first engage in reproduction at the age of two years should enter a definite betrothal at scarcely eight months of age.

In January 1928, my flock of jackdaws sustained further lamentable losses. A storm tore the snow-laden wire-netting from one of the supports. Left-Yellow, Right-Yellow, and unfortunately Tschock as well, escaped through the hole. The jackdaws were evidently unable to find their way back to the inconspicuous narrow gap from the outside, and they probably drifted off with migrating jackdaws or crows. In any case, I have never heard or seen anything of them since. The result was that only six of my original stock of fifteen jackdaws were left. Of these, one was excluded from my observations because of its inability to fly. This bird had obviously suffered some illness whilst still a very young nestling and consequently retained deep fissures in its adult feathers. In the course of a nocturnal frenzy of fluttering in the aviary, this jackdaw had subsequently broken off almost all of its primaries at the sites of these fissures.

The reproductive motivation of the five unscathed jackdaws abated somewhat following the onset of colder weather at Christmas-time, but returned with renewed vigour as soon as milder weather arrived. I always get the impression that many birds would proceed to breed in late autumn if the weather were to remain warm, i.e. that the reproductive motivation induced by internal secretory processes arises just prior to the moult. The fact that only external effects based on climate subsequently restrain the bird from breeding until the spring is indicated by the observation that many birds begin to sing in a mild autumn and that some actually do proceed to reproduce.

As the cold spell abated at the end of February and the jackdaws once again exhibited courtship, they demonstrated a novel instinctive behaviour pattern, at first incompletely but later with increased clarity. In my opinion, this pattern is the most interesting in the repertoire of the species. It has certain parallels to the 'rattling' response discussed previously.

I first saw the new pattern on the 4th March, when Yellow-Green was attacked by Elsa (the magpie). On this occasion, the attacked jackdaw emitted a new call, one which is difficult to represent in writing. I think that the best way to give an indication of the nature of this call is to mention that I attempted on that occasion to represent it in my diary as 'Jöng', whereas I later recorded it as 'Jüp'. Yellow-Green flew to the nesting-box which he and his consort favoured at that time, repeating this call in a rapid staccato. He alighted on the landing platform, turned abruptly to face the pursuing magpie and adopted the defensive posture described above, whilst still uttering the novel call. At the same instant, Red-Yellow approached uttering this same call and alighted quite close to her consort. She adopted the same defensive posture and joined the male in facing the magpie. As the magpie squatted threateningly opposite the two jackdaws, which continued unremittingly to utter their calling concert and yet did not quite dare to take action, the other jackdaws became strongly aroused. One after the other, they also began to utter this new call. They all flew up to the nest-box, assembled around the squatting pair and adopted agitated threat postures whilst joining in with the calls of the pair. It was clear to see that the birds were enraged in a quite reflex manner by the calls of their companions. They had no idea what the object of the threat was. At least, they did not direct their threat postures towards the magpie, nor did the response lead on to actual attack. The manner in which the birds streamed together and began to call sufficed to make the situation so uncomfortable for the magpie that the latter rapidly withdrew.

This instinctive behaviour pattern first attained full development when the birds had achieved full sexual maturity. By this time, the jackdaws which streamed up to assist actively attacked the original attacker. This response was by no means confined to reactions to the magpie; they also attacked any jackdaw which so intensively pursued another that the pursued bird was driven to utter the 'Jüp'-call. At this time, however (i.e. in the late winter of 1929), the mischief-maker was almost always the magpie. It was very impressive to observe how efficient this reflex response proved to be against the magpie, even though the latter was far superior to any individual jackdaw in a fight and was exclusively concerned with the elimination of all jackdaws from his territory. Since fights *between* jackdaws can similarly result in a calling concert, accompanied by assembly of the entire flock, and frequently involves actual attack on the unruly member, I believe that the instinctive response is not directed against enemies of other species as is the 'rattling'-reflex, but rather performs the function of preventing the emergence of any individual bird as a tyrant preventing other members of the colony from breeding successfully. If one were to force a similar number of non-social birds (such as magpies) to live together in a space as restricted as that presented within a jackdaw colony, one pair would certainly emerge as a despotic alliance. Even if it were possible to partially reduce the drive towards territorial demarcation by accustoming the birds to one another over a long period of time, so that the 'top dogs' would make no serious attempts to kill off their companions in the aviary, they would still destroy the nests of the other birds, or at least never permit those on the nest to enjoy the peace and quiet necessary for successful breeding. Birds are naturally unable to restrain unruly individuals by consciously-conspired mutual action, in contrast to human beings. Thus, every colonially-breeding species or genus must achieve this by means of inherited involuntary drives. It is quite obvious that the main factor concerned is the protection of the individual nests.

Many birds are in fact far more courageous than is otherwise usual when in the vicinity of the nest or actually squatting on it. Above all, they have an almost insurmountable inhibition against rising from the nest as long as they feel themselves to be threatened. In many colonially-breeding birds, this behaviour appears even more conspicuous and is further modified such that the partners of a pair never leave the nest together, so that one always remains behind on watch. This seems to be the case with herons and possibly with rooks as well. Exact observations on the drives and inhibitions which ensure the protection of individual nests in other colonial birds are sorely needed.

However, the jackdaws do not show even an indication of a tendency towards alternating watches on the nest. Perhaps they are not suited to such long periods of immobile squatting, or perhaps half a day is not sufficient for a jackdaw (bearing in mind its high metabolic turnover) to supply itself with food. It is also possible that a supplementary rôle is played by the fact that the limited number of possible nest-sites available for this hole-nesting species dictates competition. This, in itself, would necessitate a special mechanism (which cannot be provided by the individual bird alone) to protect broods already in progress. It is a definite fact that *only* jackdaws which are reproductively motivated will respond to an attack with the 'Jüp'-call, and this only when they are in occupation of a hollow into which they can retreat whilst still uttering this call. Further, the ease of elicitation of the 'Jüp'-reflex increases with the proximity of the bird to its own nest. In the immediate proximity of the nest-box, my old-established breeding pair Yellow-Green/Red-Yellow would react with this response to a mere harmless approach made by one of the other jackdaws when young were actually present in the nest in spring. But this only occurred when a quite specific, rigidly-demarcated boundary was violated.

Of course, the others rapidly learned to avoid this extremely dangerous area, to the extent of shunning it even when both adult birds were absent. In fact, it can be demonstrated with other birds (by regularly chasing them away from particular sites in a room) that they generally associate just the site and the unpleasant experience, without paying any attention to the presence or absence of human beings. Ravens and large parrots, on the other hand, do record this latter feature in a logical manner – as soon as the keeper turns his back, these birds are more than likely to visit the forbidden area. But nothing would have induced the jackdaws to approach the dangerous area of the nest, however far away the owners may have been. The owners seemed to be entirely confident of this fact, since they frequently left the nest untended for long periods as soon as the offspring had developed to some extent. Since the only other male jackdaw from the 1927 batch (Blue-Yellow) flew off in 1928, I was only able to carry out observations of the behaviour of nesting males towards one another in the short period prior to his disappearance. In my opinion, however, the behaviour of these immature males does not differ from that of adults.

In general, the two males tended to avoid one another under conditions of 'armed neutrality' whenever they met. In encounters, both would assume a specific intimidation posture in which the neck was held erect and the beak was pointed downwards, consequently causing the nape to bulge and display the long, lightly-pigmented bristly neck feathers. Incidentally, the members of a pair may sit alongside one another and move along together displaying this posture. The inclination of the beak against the thorax is an expression of non-aggressive intentions, whereas the other features of the display have an intimidatory appearance.

In subordinate jackdaws, particularly juveniles, and also in pairs where the partners are engaged in mild squabbling, a kind of appeasement or submissive posture may be seen. The beak is likewise lowered and simultaneously turned away from the attacker in a very expressive manner. The bird invokes, so to speak, the social inhibitions of the attacker by rendering itself defenceless. This is always successful; I have never seen a jackdaw peck at the nape of a companion when presented in this fashion. But, apart from the lowering of the beak, this submissive posture is very different from the 'threat-appeasement' posture displayed in encounters between males in reproductive condition. Whereas a jackdaw begging for mercy retracts its intertarsal joints to squat with its body almost vertical (giving the same wretched impression as during a heavy downpour), a bird exhibiting the latter posture holds its body decidedly horizontal and stands with its legs fully extended. This factor, in conjunction with the upwards extension of the neck and the contrasting effect of the lowered beak, lends such an uneven contour to the body that an impression of a voluntarily-contrived tense posture is given. Even the uninitiated observer would immediately interpret this posture as ostentatious.

On the infrequent occasions when the two males did actually come into physical contact, the result of the combat depended predominantly on the site of the struggle. If one of the nest-sites was closer than the other, the owner always emerged victorious. If the combat took place on completely neutral territory, however, it was evident from the start that Yellow-Green would be the victor. I believe that the predominance of the bird close to its nest would be still more pronounced in adult, successfully breeding jackdaws than was the case with these immature animals. Yellow-Green, when breeding with Red-Yellow in 1929, was far more prone to exhibit the 'Jüp'-reaction than in the prior, more individual contests with Blue-Yellow. One factor which dictated that defence of the nests started in 1928 was unable to play such an important part was the fact that my jackdaws at first exhibited a remarkable aberration in the nest-building drive as they began to carry nest-material in the middle of March. Each pair assembled nest-material *at several different sites*, without deciding upon one specific site. Only when one of the nest-sites had reached a certain degree of completion were the other sites completely abandoned. Subsequently, only the selected nest was developed further. However, the pair Yellow-Green/Red-Yellow possessed two nest-sites in two adjacent and quite similar hollows close to two consecutive badger sets. The jackdaws were apparently unable to distinguish between the two sites, since both birds carried their nest-material indiscriminately into either of the two holes. I observed a similar case a few years back, with a pair of Black Redstarts which nested under the long roof of an indoor bowling alley. The cavity under the roof was divided into a large number of similar sections by rafters and three adjacent rafters each bore a nest. The pair had bred in the middle nest. The degree of completion of the empty nests showed that they had been abandoned only just before the eggs were laid. It may even have been the first egg which rendered one of the nests individually recognizable to the members of the pair. Unfortunately, the jackdaw pair employed another, less ambiguous hollow for breeding the following year, thus shirking solution of the problem.

Whereas normal mature jackdaws do not progress to nest-construction whilst unpaired, the superfluous female Left-Green built a reasonably complete nest entirely for her own uses. This female's behaviour was also interesting in other respects. In the depths of winter, she fell in love with Blue-Yellow, who at that time was already betrothed with Red-Red, and virtually persecuted him with her affections despite the fact that he was not interested in her. Left-Green did not select the strongest male, thus contrasting with the assumed normal behaviour of females. Green-Yellow was the undisputed leader in the colony and Blue-Yellow's relationship to him was undoubtedly inferior. In spite of this, Left-Green never tried to estrange Green-Yellow from his mate as she did with Blue-Yellow. Left-Green repeatedly squatted next to Blue-Yellow with her crown feathers ruffled so that he would scratch her head. Instead, Blue-Yellow would always respond by pecking at her angrily and chasing her off. She repeatedly tried to caress him, but with no greater success. Left-Green eventually found a way around this: Whenever Blue-Yellow allowed himself to be caressed by his legitimate mate Red-Red, she would move up silently and rapidly and caress him from the other side. At such times Blue-Yellow was so absorbed with submitting to caresses that he did not notice the deception. If he did chance to look up, he would always chase the impostor away. Red-Red was herself quite obviously aware of the situation and she persecuted Left-Green with unmitigated and steadily growing hatred. Since Red-Red was stronger (or, to be exact, higher ranking), she often chased Left-Green around so long that the latter would eventually do what her jackdaw companions in the colony would always do when persecuted: she crept into her nest-hollow and uttered the 'Jüp'-call. As described above, this reliably brought an end to the quarrel. However, Blue-Yellow slowly altered his behaviour towards Left-Green. He allowed her to caress him with increasing frequency without attempting to peck at her and would even permit this when his mate was not caressing him. One day, I eventually saw him feed Left-Green; but his love for Red-Red nevertheless prevailed. If he were together with his mate when Left-Green approached, he would 'fail to recognize her' and chase her off, whereas he would feed her when on his own. In the presence of his legitimate mate, particularly in the vicinity of his nest, he responded towards Left-Green as towards any 'strange bird'. When alone and on neutral territory, on the other hand, he would respond to her female characteristics. I was surprised beyond measure when, on the 22nd April, I finally saw the two females sitting together quite amicably in the nest belonging to Blue-Yellow and Red-Red, despite the fact that I had been unable to detect any gradual decrease in their mutual feud prior to this. If the two females had been sexually mature, it would have been interesting to see whether they would have laid eggs in the same nest.[5] But it is quite possible that inconsistencies of this kind are in fact typical of *immature* birds.

At the time when the reproductive motivation of my jackdaws had reached its zenith and the birds were flying to-and-fro the whole day long carrying nest-material, a strange jackdaw seeking contact with the others arrived. The ferocity with which my five jackdaws persecuted and eventually drove off this conspecific was amazing. This prompts me to ask whether the members of a more populous colony recognize a stranger and unite against him. The sixteen jackdaws in my possession in 1930, which were strongly reproductively motivated during the prevailing warm weather, certainly united in this way.

As the nests were abandoned in May, Blue-Yellow continually remained attached to Left-Green and in fact flew more in her company than with Red-Red. I had to undertake a journey towards the end of May and when I returned on the 3rd June both Blue-Yellow and Left-Green had disappeared. I do not believe that they met with an accident; in my opinion, they flew off together, for they had conducted noticeably long excursions together in the latter stages.

The jackdaw with the broken primaries was subsequently devoured by a cat, leaving me with only three of the original fourteen birds that I had reared in 1927. At this point, I introduced three new jackdaws, which I had brought from various sources, to join the three survivors and Elsa. After initial contests, the new arrivals rapidly settled down; but they never established intimate relationships with the resident jackdaws and they were never treated as full members of the colony. Of these three individually-reared jackdaws, one (presumably a female) immediately began to flirt with Elsa; one made approaches to me, and the third exhibited most remarkable behaviour. This latter bird had grown up on a farm and at first did not want to stay and live in the elevated home provided by the attic. This jackdaw disappeared as soon as it was liberated for the first time and the same afternoon I was informed that the bird had turned up at a farmhouse about a mile away and invited himself to lunch. He was unconditionally tame towards all members of the house, whereas his behaviour towards me had always been somewhat reticent. In the evening, the jackdaw was caught and returned to me. This sequence of events occurred a number of times until the jackdaw tired of being captured every evening. He flew off one day and returned spontaneously in the evening! From this time on, he would sleep along with the other jackdaws and spend the entire day among strangers. (He had become acquainted with even more people in the meantime and was known and respected in the surrounding dwellings. The people involved believed this to be Tschock, who had already established a certain fame in the area.) This jackdaw was particularly prone to visit the bathing areas along the Danube. On one occasion when I went to find him there, he recognized me immediately; that is, he was almost insultingly timid towards me. Unfortunately, this jackdaw fell victim to a disease, together with seven other jackdaws, in January 1929.

In summer 1928, I reared sixteen young jackdaws and another magpie. Instead of introducing all of these birds to the free-flying jackdaws in the attic straight after fledging, I first accommodated them in an indoor bowling alley, where they had enough room to fly. I fitted a trap-door in the roof of the aviary at the front of the attic, attaching it directly over the attic window so that even the most stupid young jackdaw would be able to find it. After the old birds had overcome the shock of this innovation, I transferred the first young jackdaw equipped with a numbered ring to the attic on the evening of the 19th June. The last adult birds returned home before the young bird found the trap-door, so I latched the door behind the returning bird. I noticed no signs of developing enmity towards the young jackdaw in the behaviour of the adult birds. The next day, I first unlatched the trap-door late in the afternoon. As the adult birds rushed out, full of the courage born of confinement, the juvenile followed the last of them adeptly through the trap-door and at once soared high into the air. He ventured a dive and landed again on the roof of the cage. In fact, like all of the jackdaws from this year's batch, he flew much better than those of the previous year, for the simple reason that the new jackdaws had begun their flying in a much more spacious chamber. In this way, I introduced the young jackdaws individually at two- or three-day intervals to the adult birds in the attic. On one occasion, I took three at once and I was forced to regret this immediately. As in the case described above, they gathered in the air seeking leadership and drifted off.

With the exception of three weakly individuals, which succumbed, the 1928 batch of jackdaws contained individuals which were physically better than those from 1927, probably because the second batch had been reared on a diet containing more heart-meat and ant pupae and lacking horse-meat. The juvenile birds were not only able to fly better, they also showed somewhat more intelligence in comprehending cage-wire, in finding doors and in their radius of activity. In addition, the abrupt increase in intelligence occurring in August (see above) was far more distinct in these birds.

At this point I should like to say something about variations in intelligence among members of the same species. Nowadays, one frequently reads and hears that animals vary just as much among themselves as do human beings. This is very probably true of the highest evolved mammals, but it is absolutely erroneous when applied to birds, at least in the sense that different individuals of the same species exhibit various degrees of 'talent' from the very beginning. The majority of differences observable in captive specimens (especially in the highest evolved animals) are not attributable to predisposition but to variations in previous history and health and to variations in constitution of the animals concerned. An incredibly small difference in the physical condition of two birds can often give rise to an amazingly large difference between their responses. One of the main symptoms of a constitutional deficiency, however small, is delayed extinction of all drives directed towards the parents and correlated appearance of supra-normal tameness. A bird in really perfect condition cannot be tamer and more trusting towards me than he would be towards conspecifics under natural conditions, provided that there are no sexual or other factors determining a special relationship with me. Small and unintelligent birds naturally provide an exception to this rule, since they do not regard their human foster-parent as an individual but as some innocuous and possibly food-providing object in the environment, such as a tree-trunk. Such birds might, for example, flirt with a human face and yet angrily peck at the same person's hands.

Surprisingly, sexual impulses develop in such weakly individuals early in the juvenile stages. For example, Tschock greeted me with wing- and tail-trembling scarcely three weeks after fledging, whereas normal jackdaws first exhibit this behaviour subsequent to mating after the first moult (although this itself is long before the attainment of sexual maturity). As mentioned above, the healthier 1928 batch of jackdaws correspondingly began to exhibit courtship later than those of 1927. But even purely psychological responses are extremely dependent upon physical constitution, always with the effect that constitutionally inferior animals *lack* certain responses which are vestigially present and typical (in their fully-developed form) of wild birds. It does not, for instance, occur that an individual innately responds to a given stimulus in a fundamentally different way. Naturally, responses which are complex or those which represent relatively recent acquisitions of the species are most prone to exhibit organic deficiencies. Since behaviour patterns involved in the care of the young exhibit a number of such refinements, it is easy to understand why so few birds entering full reproductive condition in captivity actually proceed to breed successfully.

Perhaps it would be best at this point to give a few examples of instinctive behaviour patterns which may fail to appear in weakly individuals: Of the jackdaws which I had purchased as adults after they had been individually reared by laymen, none performed the attack pattern described above as the 'rattling'-reflex on seeing a dead jackdaw carried or after seeing a living jackdaw captured. After they had spent the summer under free-flying conditions and had passed through the first adult moult, however, they abruptly exhibited faultless performance of the described reflex. (Incidentally, I have repeatedly observed that the first adult moult or the correlated appearance of sexual maturity results in an improvement in the general health of weakly Corvids.) In addition, only the biggest and most immaculate of my adult jackdaws would exhibit softening of food in water, whereas all other jackdaws vainly attempted to devour a hard bread-crust. Of course, those birds lacking this instinctive behaviour pattern are completely incapable of utilizing the example set by a companion performing the pattern correctly. This would require mental capabilities of a quite different nature. Softening of the food in fact appears to be a fairly recent acquisition of the jackdaw, since Corvids exhibiting specializations in other typical Corvid characters are far less prone to lose this particular instinctive pattern than are the jackdaws, which are comparatively primitive. I have observed hooded crows and ravens which were obviously in bad condition perform this behaviour pattern flawlessly. The situation is almost identical in the case of the pattern of hiding food by covering it with inconspicuous objects.

A very good example of the loss of instinctive behaviour patterns is provided in the following case: An old, beautifully-developed jackdaw, which I had obtained from the Schönbrunn Zoo and which proved to be a perfect jackdaw in every respect, exhibited a specific instinctive pattern when presented with a medium-sized bird's egg (the best object being a pigeon's egg). The jackdaw would carefully peck a small hole in the shell, grip the egg by pushing the tip of the lower bill-tip into the hole and then carry the egg (with the hole directed upwards) to a safe place. Once there, he would set the egg down, still with the hole directed upwards, and carefully devour the contents. Two factors indicate that this instinctive behaviour pattern is probably innate. In the first place, the solution of the physical problem of keeping a fluid in a vessel by holding the latter with the aperture upwards is too complex for a jackdaw brain. As yet, I have only ever seen a chimpanzee solve this problem by means of insight. It would thus seem to be ruled out that an individual of the genus *Coleus* would be able to invent the procedure. Secondly, my other jackdaws did in fact exhibit the same pattern, but with another object. Admittedly, they would bite pigeon's eggs in the clumsiest fashion and thus lose three-quarters of the contents; but they did employ the entire behaviour pattern described above when dealing with *plums*, which they would only carry home in the manner described. I never observed one of my jackdaws simply carrying a plum in the angle of its beak, even though this would have been more efficient, since the skin which they grasped to carry the plum often tore so that the fruit was lost. Several more examples of the lack of appearance of specific instinctive patterns in weakly individuals could be given, but at this point I am more concerned to mention an example of the converse type of behaviour, in spite of (or rather because of) the fact that this instance does not readily fit into the view just presented. Tschock, as mentioned previously, was purchased from an animal dealer and he was by no means a 'perfect' jackdaw; yet he habitually carried appropriate objects in his claws whilst still in the juvenile phase (i.e. before the first moult of the auxiliary feathers) and would always transfer the object from his beak to his claws when in flight, just as ravens do.[6] I can think of no other explanation for this strange emergence of an instinctive pattern which is otherwise typically absent from the species than that an originally typical Corvid pattern has been lost in *Coleus* and that this pattern reappeared in a more or less abnormal fashion in this particular individual.

In 1928, I let all of the jackdaws free simultaneously, adopting a procedure different from that of the previous year since I could naturally rely upon the adult birds, which were unconditional leaders in the colony. At this point, I observed that the young birds were evidently less nervous and fickle in their general responses to external stimuli. On the other hand, they would respond extremely readily to the slightest symptom of unease in their leaders. This factor had produced a quite unique form of behaviour in the leaderless birds of the previous year. Although their flock was extremely bold and remained unfrightened by things which would normally alarm adult jackdaws, they often dived away in extreme fright without any apparent reason. On closer observation, I determined that a panic of this kind was always elicited when one of the birds suffered a fright because of some minor mishap such as a stone toppling. The symptoms of fright in one bird would induce a somewhat stronger response in the adjacent bird and this effect would mount into an avalanche response as the response passed from one bird to the other at lightning speed. The flock would then take off, taking with it the original bird, which was itself completely unaware that the original frightened response to the toppled stone (which would never have induced him to fly off had he been alone) was the cause of the general escape response.[7]

What is remarkable is that young jackdaws with adults as leaders are well aware of the signals to which they must respond, and they only respond to fright in the adults and not to that in their equals. In this particular case, courage and not just fear (as is otherwise typical of birds) can also be transferred by example, as was demonstrated by juvenile birds born in my colony. These birds, following the example of their parents, ate out of my hand despite the fact that they had never seen me before leaving the nest. This greatly surprised me, since I had not expected them to be the slightest bit tamer than free-living young birds of the same age.

Whenever the mixed flock of jackdaws reared from 1927 and 1928 flew some distance away, the adult jackdaws were usually at the fore and flying below the level of the young birds. The older birds, with their developed flying skills, almost always dived sharply at the take-off in order to build up speed smoothly and rapidly. The youngsters on the other hand, would at first flap off from the most elevated take-off points in just the same way as when taking off from a flat substrate at ground-level. In general, the young jackdaws spread the wings and the tail more, with the result that the consequent reduction in surface load produced a lesser flight tempo. I intend to leave open the question as to whether this wide spread of the lifting surfaces is the *cause* or the *effect* of the general 'upwards tendency' of young birds with little flying experience. The greater speed of the adults is certainly facilitated by the fact that they are actually well into the adult moult at the time when they are leading the young and that their minimal speed is thus increased. This all results in the adults exerting a (definitely entirely involuntary) directional influence on the flock.

Since I had attributed the great losses of 1927 predominantly to cats, I attempted to avoid this hazard in 1928 by leaving the birds locked up during the early morning hours when cats are still on the prowl. I first liberated them at around 10 a.m.; a measure which proved to be entirely successful. Every evening, when the birds were already asleep, I had to climb quietly up to the roof and carefully latch the trap-door. Every sound evoked a distressing frenzy of fluttering, since the jackdaws (in contrast to my ravens) were very nervous in the dark – even towards me.

The 1927 jackdaws did not pay much attention to their younger companions. At least, none adopted a 1928 youngster in the way that Tschock had adopted Left-Yellow the year before. The young birds, for their part, flew after the adult jackdaws with great constancy and copied their daily habits and customs from the very beginning. For example, they immediately flew down to the ground with the adult birds, whereas the birds in the previous year had not ventured to do this for some time. They also rapidly ceased to exhibit fright in response to my large free-flying cockatoo. The 1927 batch of jackdaws had taken a long time to become accustomed to this big bird, which often flew after them uttering spine-chilling screeches in an overflow of high spirits.

If I had not dissolved the company of the young birds and attached them one at a time to the adult birds, such complete adaptation to the habits of the latter would surely not have developed.

The formation of groups among the young birds after the transfer to the attic was most interesting. The groups corresponded exactly to the temporal order of transfer of the individual birds and thus to the numbers printed on their rings (since each had been marked with one of a consecutive series of numbered rings on transfer). No. 1, for example, was well able to remember No. 2 when the latter was transferred two days later, whereas he did not recognize No. 4, whose transfer first took place three weeks later. The birds with adjacent numbers thus kept close company, while the widely-separated individuals always remained aloof from one another. Indeed, this effect persisted for years. At this point, however, I must emphasize that adult, mutually-acquainted jackdaws will recognize one another at once even after months of separation. Incidentally, a further factor may have contributed to the described process of group-formation. The birds confined in the bowling alley became more and more timid as a result of the continual capture of their companions, since they naturally 'believed' the latter to have been subsequently devoured. Birds which had already been transferred, on the other hand, retained their former degree of tameness. Birds with low numbers were always tamer than birds with high numbers. This may have been an additional influence in unifying the separate groups.

Throughout the Summer of 1928, my jackdaws enjoyed complete freedom, apart from their confinement during the period from sunrise to 10 a.m. Despite this, I did not lose a single bird. I cannot confidently say whether this was due to the morning confinement period or to the leadership of the experienced adult jackdaws.

As the birds once again began to exhibit a tendency to join company with strange Corvids towards the end of October, I once more locked them up to prevent them from migrating

At the beginning of January 1929, the three remaining 1927 jackdaws already began to show signs of pronounced reproductive motivation, but this abated during the excessively cold weather which followed soon afterwards.

During the most extreme phase of the cold spell, my jackdaws unfortunately fell victim to a disease accompanied by severe diarrhoea. This cost me no less than three birds, although this number luckily included none of the two-year-olds. I had the impression that the jackdaws suffered greatly because the water provided froze so rapidly. In particular, those animals occupying low-ranking positions (who were the last to be allowed to drink) had insufficient time to quench their thirst. These birds were probably dependent upon eating snow. This might explain why the old, high-ranking birds remained unafflicted by diarrhoea.

In the spring of 1929, I liberated the jackdaws quite early on (25th March) in the hope that they would recover more rapidly from the stress of the Winter. At this time, the Tulnerfeld was swarming with non-resident jackdaws and rooks. One fine day – the 9th April – a calamity occurred. My jackdaws became so closely intermingled with a migratory flock containing several hundred birds that they were unable to break away in full strength. They had in fact lost their mutual cohesion in the swarm of foreign birds. Single jackdaws repeatedly returned to the house, calling despairingly after their companions, only to return to the migratory flock when they failed to make contact. This was the first occasion on which I noticed that guiding call of a squatting jackdaw which is attempting to summon flying companions has a different quality to that of the normal social contact call, particularly when the guiding call is uttered from the home base. Whereas the contact call sounds like a shortened 'kia', the signal 'come home' is a somewhat drawn-out 'kioo'. The latter call is also uttered by flying jackdaws which are attempting to summon back companions which have flown further away from the colony than they themselves have. Of course, I had often heard the guiding call, but this was the first opportunity to hear it repeated in such rapid sequence. It frequently occurred simultaneously with the social contact calls of the strange flock, so that the difference and the consequent significance of the 'kioo'-call eventually made itself obvious.

Yellow-Green, the 'chieftain' of the resident jackdaws, was particularly active in calling back the members of the colony. I do not believe that this activity is generally typical of the alpha-animal, however. I am much more inclined to the opinion that all adult males in a jackdaw colony behave in this manner in such a situation. The behaviour of Yellow-Green during this incident was extremely interesting – he acted in very much the same way as a well-trained sheepdog does when attempting to reunite a fragmented flock of sheep. He repeatedly flew low over one of my jackdaws as it was sitting or flying in the midst of the strange birds, continually uttering the 'kioo'-call. He would then induce the straying bird to fly with him, employing one of the methods described above, and more or less 'tow' him back home. The other two adult birds exhibited the same response, but so ineffectually that they failed to achieve any definite success. The unfortunate part was that the one-year-olds which had been safely brought home were by no means motivated to stay there. After a while, they would once again become restless and fly back down to the migratory flock on the meadows. Nevertheless, Yellow-Green brought the deserters home at a somewhat higher tempo than that at which they were departing, so that the migratory flock only took two 1928 jackdaws with it as it moved on that same evening. I am convinced that without the presence of the sexually mature male and his concerted activity not a single one of the one-year-olds would have returned home.

I found it remarkable that the birds should have apparently experienced the entire episode as a catastrophe, just as I had (for obvious reasons), since they were evidently extremely agitated the whole time. They also seemed to miss their two lost companions afterwards; the 'kioo'-calls continued practically unceasingly during the next few days. In addition, the jackdaws were nervous for some time afterwards in just the same way as they were when companions were captured and taken off. This demonstrates that this reaction is a purely instinctive response to the absence of companions, since in this case the survivors should have known that the two absentees had flown off unharmed under their own steam.

It is equally interesting that all of the one-year-old jackdaws from the 1928 batch gradually left the colony later on, during the breeding season, without evoking the 'kioo'-reaction in any of the sexually mature birds. The fact that they had not been driven away by the old jackdaws was proved by the fact that a somewhat poorly juvenile remained with the colony throughout the summer. I regard this abandonment of the colony as normal behaviour, believing the reproductive motivation exhibited by the 1927 birds during their first spring to be an effect of captivity. My main reason for this assumption is the fact that all of the departed jackdaws returned in October. I regard it as highly probable that the one-year-old jackdaws (i.e. those which have not yet attained sexual maturity) avoid the colony during the breeding season so that they do not deprive the large number of breeding pairs (which are in any case crowded into a small area) of food for themselves and their young. In view of the generally high degree of differentiation of the sociology of *Coleus*, I think it is very likely that an instinctive response of this kind would exist. This isolated observation is of course insufficiently convincing on its own, especially since a proportion of the birds returning in the Autumn had lost the marking rings from their legs.[8] Incidentally, this loss of rings also occurred as a common feature with the birds which remained at the house. Since the attachment of the rings was the same as that used at the ornithological station at Rossiten, I am inclined to believe that a good proportion of the Corvids ringed there are similarly prone to lose their rings.

I should like now to conduct exact observations on a colony of tame jackdaws so large that I could permit the birds to migrate in the autumn without fearing the extinction of the colony. By using reliable marking rings, I could investigate the truth of my supposition that the one-year-old jackdaws abandon the adult colony during the breeding season and then – so to speak – come back to pick up the mature birds for the winter migration. It could also be determined whether I am correct in assuming that newcomers are taken into the company of the colony companions during migration. In the vicinity of the home base, my jackdaws always ferociously persecuted any new arrivals seeking contact. The uneventful acceptance of the 1928 jackdaws on their return in the autumn of 1929 was just as good a proof to me that they were the same birds as those which had flown away in the spring as was the correspondence in number and in the marking rings still present on the legs of some of them.

The behaviour of the three sexually-mature jackdaws in the reproductive phase of 1929 was quite different from that exhibited the previous year. The unmated Red-Red showed absolutely no sexual activity and reproductive motivation was far less evident in the young jackdaws than it had been in the young birds of the preceding year. Even the pair Red-Yellow/Yellow-Green exhibited little obvious sexual activity until the end of March. These two birds first began definite nest-building at the beginning of April. Following a number of bitter struggles, they had succeeded in upsetting the formerly prevailing rank-order to dominate over the magpie. This provided a further rewarding opportunity for me to observe how this asocial bird was forced by the effective co-operation of all the jackdaws to adapt to their social system. But since the magpie persisted in repeatedly approaching the jackdaw nest, in accordance with the defiant psychological make-up of this species, I eventually removed him so that the breeding pair would be free of the disturbance. Since Red-Yellow and Yellow-Green never encountered serious resistance from the young jackdaws, the 'Jüp' and 'tsick' anger-calls were barely to be heard by this stage.

The pair chose a very small nest-box for the construction of the nest. As in the previous year, the male interested himself mainly in the construction of the foundation, thus selecting fairly thick twigs. But the nest-box was so small that there was scarcely room for the nest-cup itself. The female (the architect of the nest-cup) appeared to grasp this fact. In any case, she repeatedly ejected the twigs which the male had collected and herself carried in soft nest-material – mainly straw and newspaper. This continued as long as the nest-building drive was in operation. I subsequently examined the nest and found that the activity of the male had been almost completely superfluous, in fact almost exclusively disruptive. In the period from the 30th April to the 2nd May, Red-Yellow laid four eggs. On the first day of laying, she actually left repeatedly to eat, but she had already begun to restrict the absences to periods of a few minutes. After laying the last egg, she was seen markedly less outside the nest. Yellow-Green did not immediately accept the absence of his consort without resisting; instead, he began to call to her continually. However, since she failed to respond to this, his continual utterance of the social call-note gradually changed to song. His longing for his mate often induced him to sit close to the nest. Since he was alone and 'bored', he sang a great deal, just as a bird confined in a cage will sing. The male did not, of course, think in terms of pleasing his mate with the song.

At this point, I should like to say a few words about singing in jackdaws. The song consists partly of imitated sounds, but it also contains (surprisingly) notes taken from the 'colloquial language' of the species. The sessile social call-note 'kia', the aerial guiding call 'kioo', the 'Jüp'-call and the rattle-call can all be heard in the muddled monologue which the bird utters. Remarkably, the bird adopts the appropriate characteristic posture when uttering sounds with a 'communicative' function. For example, the jackdaw will bend forwards and flap its extended wings when rattling, or cower when giving the 'Jüp'-call as if it were sitting in a confined space. Like an orator, it accompanies its utterances with the appropriate gestures. To my ears, the display calls uttered by the bird in song appear to be the same as those emitted in a 'serious' context. I have repeatedly leapt to the window in order to see what was wrong on occasions when a bird uttering a gently undulating song abruptly emitted the loud rattle-call. But I never saw another jackdaw misled in this way, even when the song *began* with a rattle-call, as was often the case. When one considers how promptly and comprehensively the response to a jackdaw rattle-call is given in a serious situation, this fact appears particularly striking.

The brooding female was generally provided with food by the male. He visited her in the nest-box at short, irregular intervals, always arriving with a full crop and re-emerging with his crop completely empty. Sometimes, she would emerge as he flew up calling to her and she would then take the food from him outside the nest. On these occasions, I was able to observe that she quite definitely recognized him by his voice. In such cases, she would always fly off afterwards and the male would follow. If she herself had not returned within a few minutes, the male would return alone and quietly clamber into the nest-box. Of course, I do not know whether he brooded whilst in the nest, but this would be a reasonable assumption to make. The longest period that the male spent alone in the nest while under observation was eight minutes. The female always returned from such excursions with additional soft nest-material for the nest-cup. The material was always bundled so that the entire beak, from the base to the tip, was full. Even at this stage, she sometimes carried a small lump of dried clay in the tip of the beak, to round off the bundle as it were. Later on, these transported lumps of clay became bigger and bigger, while the bundles of moss and grass became smaller and smaller. Eventually (after the young had hatched), no more nest-material was carried into the nest-box. Only lumps of clay were carried back. These lumps were apparently pulverized by the birds, since the nest (and the young, at a later stage) was continually covered with dry, powdered clay. As long as the female was brooding, the male took absolutely no part in carrying nest-material and clay into the nest. He contented himself with feeding the female and relieving her from brooding now and then. When brooding, the male was less rigidly attached to the nest than the female. At least, he would immediately emerge when I tempted him with mealworms, whereas the female had to be coaxed for some time before emerging. Strangely enough, both birds failed to show the slightest symptoms of alarm when I examined the nest and they consequently failed to defend it. This provided a great contrast to the behaviour of the adult jackdaws from the Schönbrunn Zoological Garden which nested in a cage in 1927. The male of the pair viciously defended even the first rudiments of his nest, using his claws to attack anyone who approached. This occurred even though the male did not feed out of the hand like Yellow-Green and must therefore have had a greater inhibition against physical contact than the latter bird.

At 10 a.m. on the 17th May, I discovered Yellow-Green squatting on the eggs. As explained above, he usually left the clutch straight away when I enticed him, but this time he remained sitting on the eggs for some time. I was beginning to think that neither parent was in the nest-box, since I could see the female sitting on the weather-vane; but as I went through the door of the aviary, Yellow-Green emerged from the nest-box. I subsequently examined the nest and found one egg to be chipped. At 3 p.m., I checked again and found a freshly-hatched nestling. No fragments of the shell were to be seen, however.

I am sure that the parents began to feed the nestling the same afternoon. As I tested the male by offering him mealworms, he did not simply take them into the crop as he did when he intended to take them to the female. Instead, he tore them into little pieces before taking them into the crop and followed this by wiping his beak. I had already noticed a few days previously that both parents had begun to wipe their beaks before entering the nest-box. The drive to clean the bill would thus appear to emerge somewhat earlier than necessary. On this occasion, the male went into an actual paroxysm of beak-wiping before clambering into the nest-box to join the female and the nestling.

Judging from the sounds which emerged, I assume that the male transferred his crop-contents to the female and that she then fed the young. If the female did not herself devour any of this carefully prepared food (which I regard as likely since she had a number of times eaten large quantities of food from my hand during the afternoon), then the nestling received ten mealworms during the first afternoon after emergence.

The next day, the second young jackdaw had already emerged by the time I examined the nest at 10 a.m. When I did not provide them with any food, the parents fed the young predominantly with green caterpillars which they had collected from nearby lime-trees. They fed the young just as irregularly as the male had fed the female. They could nevertheless be stimulated to feed the offspring at any given time by provision of mealworms or ant pupae, though other types of food did not have this effect. They tore the mealworms into tiny pieces as described, but the ant pupae were simply split open before the parents took them into the crop. I was amazed at the quantities of food which they crammed into the young.

In order to keep to the actual chronological order of events, I must mention at this point that the two jackdaws reared in 1928 which had flown off on the 1st April returned this same day. They disappeared again a few days later, taking with them two companions of the same age. From this time onwards, there was a gradual departure of the one-year-old jackdaws, as described above. As mentioned previously, all of these juveniles returned one after the other by the 1929–30 winter.

The two remaining eggs hatched on the 19th and the 20th May – both hatching in the early hours of the morning just like the second egg. By the time the last nestling had emerged, the first was almost twice as big. The parents continued to feed their offspring with green caterpillars; I was able to observe this exactly, since the adults always emptied the crop in front of the nest-box and fragmented the caterpillars before performing the described beak-wiping and taking the food to the young. Even the ant pupae were at this stage taken into the crop from my hand and then fragmented on the landing platform of the nest-box. On the 2nd June, I saw the male carry a small bundle of moss soiled with a minute faecal smear away from the nest. The faeces of the young were always removed by the parents, even later on, by removal of pieces of nest-material with the faeces adhering. The faeces hardly ever came into contact with the beak while this was done. If this did somehow happen, the parent carrying the faeces would frenziedly shake its head and wipe its beak – a clear indication of aversion. This is particularly striking since crows apparently carry off the nestlings' faeces in the crop. The jackdaw parents seldom carried the polluted nest-material far away; they usually laid it with surprising care on the frame of the trap-door to the aviary, which eventually became virtually decorated with the soiled 'nappies'. This naturally led to the gradual dismantling of the nest, a process which went so far that the nest-hole was completely empty by the time the offspring left the nest. But the interior remained completely unsoiled.

On returning home on the 23rd May, after an absence of two days, I saw one of the one-year-old jackdaws which had remained in the colony flying around with a dead nestling in its beak. I chased the jackdaw off from the corpse and discovered that the latter was barely injured, though well on the way to putrefaction. In my opinion, the jackdaw had not robbed the nest; the nestling had most probably died in the nest and had first been carried out by the parents. When I examined the nest, I found that the two oldest young had grown amazingly and were almost equal in size. The third, judging from its size and that of the dead nestling, was definitely No. 4. It was extremely retarded and barely larger than it had been prior to my departure.

Some time before, I had noticed with virtually all my observations of successfully-reared clutches of passerines that the youngest of the clutch always remain smaller, weaker and more susceptible than the older individuals. I can remember one case where a younger individual did actually overtake an older sibling, but the human keeper of course involuntary feeds the weakest when feeding nestlings, because of his 'sense of fairness'. The parent birds certainly do not do this. Their behaviour is quite the opposite, since they will cram food into the first beak to be thrust forward far enough, and this is usually the beak of the older nestling, I had always imagined that this retardation effect was an artefact of captive conditions, produced by the fact that older nestlings have spent a greater proportion of their developmental phase under relatively natural conditions than had the younger ones. It is of course a well-known fact that passerines, up to the point where the escape response appears and they are no longer willing to gape for a human being, are easier to rear with increasing age at removal. In fact, the nestlings receive certain storage substances (which the human foster-parent finds difficult to provide) from parental feeding and these substances last longer the later the nestlings are taken into human care.

But on observing rearing of the young by my jackdaws I began to doubt the validity of this interpretation. As early as the 23rd May, I had the impression that the mother was remaining away from the nest longer and more often than was good for the youngest nestling. After noticing this, I felt the young birds while the female was away and found that the smallest was in fact appreciably cooler than the other two. On the same day, I saw the old female fragment ant pupae for the last time. Thereafter, both the pupae and the mealworms were fed to the offspring whole and the correlated beak-wiping pattern also disappeared by this time. On the 23rd, I also noticed remains of hemp-seeds clinging to the gape of the oldest nestling, so the parents had begun feeding plant material at this early stage.

On the 24th May, the difference between the two biggest and the smallest nestling was even more marked. One eye of the biggest was just opening. On this day, the mother only warmed the offspring for brief periods of a few minutes, which was definitely too little for the smallest nestling. It would therefore seem that the parents adjust their feeding pattern more to the older nestlings.[9] In this particular case, the effect had possibly been accentuated by the chance loss of the third nestling. By the 25th, the youngest nestling had disappeared without trace. Each of the two older nestlings had both eyes open and the dorsal skin was beginning to darken owing to the emergence of the flight feathers beneath the skin. This gave me the idea that the sight of the young in my hand might now elicit the 'rattling'-reflex in the parents; but this did not prove to be the case. The parents were not at all aroused as I held out the young towards them on my hand. On this date, I once again saw the parents carrying in lumps of clay, something which I had not seen since the hatching of the young. The male, in particular, carried in lumps so big that he could just manage them in his beak. The skin at the corners of his mouth was stretched so tightly that it became translucent. I had never seen a jackdaw open its beak so widely before, and I did not know that this was actually possible. By the 30th May, the tips of the feathers had to some extent penetrated the feather-sheaths on the shoulders and the wings, so that tiny parts of the feathers were visible. When I took the nestlings in my hand and held them out to the parents at this stage, this immediately elicited a joint 'rattling' attack. This was in fact just what I had expected, but it is nevertheless striking that an intelligent bird such as the jackdaw should protect its young in such a reflex manner. The same day, my freshly-fledged raven flew up on to the roof of the house for the first time. The parent jackdaws immediately attacked him, this being the only case that I observed where the 'rattling' reaction was elicited by something other than a dead Corvid or a substitute for the latter. The parents eventually forced the raven to flee and they would not tolerate him on the roof even later on. Their attacks first abated long after the young had left the nest; but even then they exhibited not the slightest fear of the raven.

On the 18th June, as the nestlings were thirty-two days old, I saw the elder of the two young jackdaws sitting on a perch in front of the nest. In the evening, the jackdaw was not to be seen, having probably crept back into the nest-box. The next day, I did not see either of the two young jackdaws, and I first saw them both on the morning of the 20th, when they were both perching in the aviary. That afternoon, one of them found the way through the trap-door into the open, evidently by chance, and I spotted him perching in a big elm on the other side of the house. The parents attempted to entice the fledgling back home, but their efforts remained unsuccessful for some time. The young jackdaw followed the parents just as imperfectly as a hand-reared jackdaw of the same age will follow its human foster-parent. In addition to the previously mentioned methods which the jackdaw can employ to induce another bird to take off and join it in the air, I saw a new, more drastic technique which is employed only by parents with freshly-fledged young. This instinctive behaviour pattern took the following form: The young bird must be perching high up, as is almost always the case since freshly-fledged jackdaws are shy of the ground. The adult bird flies up from behind just low enough to graze the fledgling across the back and dislodge it from its perch and yet still maintain enough speed after the impact to arrive in front of the young bird in the air. The fledgling will then follow. This extremely interesting instinctive behaviour pattern is rarely elicited, appearing only when the adults themselves are greatly alarmed and motivated to flee without leaving their young behind, or when a fledgling (with its typical over-confident behaviour) has settled on a perch which the parents regard as particularly dangerous. The response may also be elicited, as on this occasion, as the ultimate consequence of extended stimulus-summation. Both parents exerted themselves the whole day in this manner until they had succeeded in guiding the fledgling away from the elm, around the house and onto the lime-tree standing close to the house on the side bearing the jackdaw aviary. The fledgling subsequently spent three nights in this tree. The second fledgling first found its way through the trap-door into the open on the 25th. By this time, the two young birds were already following the parents somewhat more reliably, but only if the adults were already on the wing. At that stage, they still did not follow the parents on the ground, even when they were extremely hungry and begging for food. They made no more attempt to gape and approach the parents gaping than do the majority of smaller passerines. As soon as one of the young jackdaws took off, however, one of the parents would follow immediately, fly low over the younger bird to overtake it and then reduce speed in order to stay just in front. This latter feature was apparently essential because the following response of the juveniles at that time would immediately evaporate if the parent were more than a few metres ahead. The parents appeared to keep constant watch on their young in order to prevent them from straying off. They never allowed one of the youngsters to fly unaccompanied, even for a matter of seconds. The young themselves seemed to pay little attention to this accompaniment in the first stages. At all events, they repeatedly flew off blindly without the slightest utterance of guiding calls or social contact calls.

Until the end of the month, the juveniles spent the nights in the company of the raven (to whom they had become completely accustomed) in a group of pine-trees some distance away from the house. It was only after this phase, when their general liveliness and inclination to fly was increasing, that they developed the characteristic nidifugous type of attachment to the parents. This change was coincident with the keratinization of the primary flight feathers and other signs of physical development.

Since the young now began to follow their parents to the nocturnal roost in the inside of the attic and everywhere else as well, they consequently approached me much more often and they surprisingly turned out to be far from shy. I had naturally expected that these young birds, which had been reared by their parents and had only seen me when I examined the nest, would have been almost as shy as young jackdaws of similar age which had grown up in the wild, without regard to the tameness of their parents. I had in fact experienced this with all the small bird species that I had bred. The young jackdaws approached me quite closely without hesitation and they soon ate ant pupae out of my hand in the company of their parents. When I laid my hand flat on the ground, they would even climb onto my hand to eat. It was nevertheless impossible to induce them to fly onto my hand as the adults did. I knew just how strongly young birds can be influenced by indications of fear in their parents. I had also had the opportunity to observe how the young jackdaws in the previous year, spurred on by the example of a companion one year older than themselves, had become accustomed to the cockatoo much more quickly than had the older bird himself. All the same, I would never have believed that such blind trust in the example set by the parents could conceivably be present.

In the same year (1929), I also reared a number of young jackdaws which I intended to incorporate into the flock one by one just as I had the year before. But this time, in order to bypass the above-mentioned intimidatory effects of continually trapping individual birds, I accommodated the young birds in a large chamber in the attic. From here, I could allow them to fly through a connecting door to join the free-flying birds without capturing any of the young birds. In addition, those birds which were still confined at first were able to see and hear their free-flying companions afterwards. This arrangement proved to be excellent, both in that year and in the following year (1930). Unfortunately, one of the young jackdaws which had been born in captivity met with an accident as a result of this set-up. The jackdaw managed to pass through the door to the hand-reared jackdaws and naturally refused to gape towards me once there. Amidst the general confusion, I failed to notice the bird until it was already extremely weak from hunger. I immediately released the young bird, but it had scarcely settled on the roof when the raven seized and killed it before I could do anything to stop him. Up to that time, no raven had ever made a serious attempt to catch a jackdaw and this was something which would in any case have proved impossible normally. This sensitive response to symptoms of illness in animals which would never serve as prey when healthy is thus an innate feature of the raven. Thienemann has described a similar phenomenon in the goshawk and the response is very probably a characteristic of many large, predatory birds.

In the winter 1929–30, following the return of the 1928 jackdaws which had been absent during the 1929 summer, my colony contained twenty birds. I intended to rear a suitable number of additional young jackdaws in the summer of 1930 and then to suspend the confinement of the entire flock in the following autumn so that I could study their departure and return. However, while I was lying in hospital in March 1930 after a car accident, the colony was almost completely wiped out by an inexplicable catastrophe. I do not even know whether the birds flew away because of some alarming event or whether they succumbed. This disaster was aggravated by the death of one of the two surviving birds soon afterwards, and the only jackdaw remaining from my magnificent colony was the adult female Red-Yellow. I do not think that I am inclined to anthropomorphic, sentimental commiseration with animals, but the way in which this solitary jackdaw searched the entire area for her lost companions, continually uttering the 'kioo'-call, was in itself sufficient to induce me to acquire more jackdaws. Like all isolated birds, Red-Yellow sang almost continuously after she had overcome her initial despair. I have already mentioned that the jackdaws incorporate vocal displays in their song. But it was for me completely new that the jackdaw's 'mood' can be expressed in the song. The song of Red-Yellow was almost completely composed of 'kioo'-calls. She would repeatedly interrupt her song and fly off searching the countryside whilst uttering genuine 'kioo'-calls which were no longer components of a song. She gradually gave up searching for her lost companions, but her song has remained predominantly composed of 'kioo'-calls right up to the present day.

When I introduced four young jackdaws to her this year, she paid hardly any attention to them and it is still rare that all five are seen flying together. I only once saw an exception to this general indifference, at the time when the new batch of young birds had just been released. Two of them strayed and Red-Yellow brought them back home with an extremely impressive, quite typical and perfectly developed 'kioo'-response.

Although my jackdaw colony is therefore not quite extinct, the prospects for observing the process of migration, particularly the departure of one-year-olds from the resident area, has been set back by two years.

Summary

If the facts which have emerged from the above observations of the social behaviour and ethology of the Corvids, in particular of the jackdaws, are briefly summarized, the following overall picture emerges:

Within a flock, whether a migratory flock or the entire assembly of the members of a resident colony of one of the colonial species, a perfectly balanced rank-order prevails within each individual species. This presupposes that the animals are capable of individual recognition, a fact which is further demonstrated by the observation that the members of a jackdaw colony immediately recognize a strange intruder as such and drive him off. The vehemence with which this is done indicates that new members, in so far as they are not born within the colony itself, are probably only absorbed into the flock during the winter migratory phase.

The disappearance of a member of the colony is at once noticed and this evokes extreme alarm and general instability. This is most likely to cause a wandering flock to leave the vicinity. The evident formation of sub-groups, especially within fairly large colonies, apparently facilitates the registration of the presence of each individual member within the colony. The alarm exhibited by a group which has experienced a loss will infect the whole flock. The response does not appear when an individual becomes ill and dies, however.

The adult males will make particularly active attempts to prevent a member of the jackdaw colony from straying. The straying bird is followed and the pursuer attempts to induce the former to follow him back to the colony whilst uttering a specific guiding call, which contrasts to the normal jackdaw social contact call 'kia' and has more the character of a drawn-out 'kioo'. This response seems to be especially important when strange migratory flocks threaten to carry off individual members of the colony. The response does not appear when the one-year-old sexually-immature jackdaws depart from the colony area in the breeding season. This apparent rule has obvious advantages for the species. The one-year-olds subsequently return of their own accord in the autumn and they are then accepted without friction (a sure indication that they have been recognized; see pp. 39–41).

If any Corvid, with the probable exception of the distantly-related jays, is seized and carried off by any predator, the jackdaws (and probably all of the Corvine species) will react with a vicious attack on the predator. When doing this, jackdaws emit a quite characteristic call – a metallic 'rattle'. According to the description given by Löns for the corresponding response in ravens and hooded crows, these two species would appear to possess an essentially similar call. This would seem to be extremely probable since the jackdaws, the crow species and the raven have another attack-call in common – the deep *'grumbling call'* which accompanies strikes upon predators in attempts to drive them off and which can also be heard during playful buffeting. Crows and jackdaws, and to some extent magpies as well, form an offensive and defensive alliance to drive off predators. I believe that the 'rattling' attack itself is aimed less at saving the seized bird than at preventing the predator from enjoying his spoils. This would render the area, and probably future predation on Corvids, unattractive. The response is purely instinctive, at least in the case of the jackdaw, and it in fact has an almost reflex quality such that it can be elicited fairly easily by 'erroneous' stimuli (pp. 6–11).

The raven seems to have departed secondarily from the Corvid system. This bird does actually still possess the relevant instinctive patterns, but it apparently employs them only for the defence of personally identifiable conspecifics. In this case, the response would thus appear to be modifiable by insight.

Within a breeding colony of jackdaws, the heavily-marked differentiation in rank appears to necessitate a special form of protection for the nests of low-ranking pairs. This is achieved by means of a very unusual instinctive behaviour pattern. As long as the jackdaws (particularly the males) are still in the process of pair-formation, they utter a call, which sounds like a short, high-pitched 'tsick,' in or close to suitable nest-holes. This call apparently serves the double function of enticing the female to the nest and acting as a defiant display to all other males. This does not prevent seizure of the hole from a weaker (or, more exactly, lower-ranking) male, but this does become virtually impossible as soon as stable pairs have formed. As soon as a jackdaw in this situation feels itself to be seriously threatened, it will utter a special 'S.O.S.-call', which can best be represented as 'Jüp'. This call is initially copied by the consort, and the other colony members soon join in. All of the jackdaws, particularly the former, will then attack the troublemaker. Even if the majority of birds employ this response, the swift assembly and the vicious attack made by the consort will nevertheless suffice to still the quarrel. The original attacker usually provides a demonstration of the instinctive basis of the entire response, and his ignorance of his own part in eliciting the response, by joining in with the 'Jüp'-calls as if some other bird had been the cause of the response. The response first appears in young jackdaws at the time when pairing first takes place (pp. 27–29, 42).

Young jackdaws first become paired during their first Autumn, when they are just five months old, although they do not attain reproductive maturity until two years old. Some individuals wait until their second

Autumn to become 'betrothed'. Since a low-ranking bird automatically moves up to assume the rank of the partner with which it becomes paired, the pairings often give rise to fairly pronounced changes in the rank-order and these are immediately recognized by virtually all members of the colony. The first sign that I observed indicating that a pairing had taken place was in many cases the sight of a previously high-ranking bird giving way to a former subordinate (pp. 25–26).

Both partners take part in nest-building, which can occur incipiently soon after 'betrothal' if the Autumn is very mild. The male conspicuously occupies himself with the foundation of the nest while the female is largely concerned with moulding the nest-cup. The work to be carried out by the male varies greatly with the size of the chosen hollow (pp. 42–43).

My jackdaw female 'Red-Yellow' laid four eggs in the early morning hours of the last two days in April and the first two days in May. Although she interrupted the brooding periods frequently during the first day, it can nevertheless be stated that she began brooding as soon as the first egg was laid. The oldest nestling hatched around midday on the 17th May and the other three followed in the early morning hours during the three succeeding days. The parents fed the nestlings during the first days of their life predominantly with mealworms, ant pupae and green caterpillars. The insect food was always torn into fine pieces before being transferred to the crop. Before feeding, the adults would carefully clean the beak. After the first few days, the parents had already ceased to prepare the food in this way and the young were soon given coarser food. The faeces of the young were unfailingly removed with the nest-material to which they adhered (p. 45).

After leaving the nest, young jackdaws remain close to the nest and the colony as a whole for some time. Only later, at a time when other passerines are achieving independence from their parents, does the drive to fly after the parents appear in the young jackdaws. The intensity of this instinctive following tendency probably has no equal except in some young nidifugous birds. Since the young jackdaw virtually follows in the shadow of its parents, it needs no strongly-developed escape response and it only becomes alarmed when it sees its parents to be alarmed. For this reason, young, hand-reared jackdaws seem extremely over-confident at this age. In actual fact, this *following drive* of the young jackdaw at this age supplants so many other responses, that the juvenile will appear extremely artless and helpless (as compared with crows or magpies of the same age) when deprived of its leader. Only the rook is likely to be similar to the jackdaw in this respect. The all-powerful following drive of the young jackdaw is not extinguished until the approach of Autumn, at which time the young bird suddenly starts to behave just like its parents. In my opinion, this late behavioural maturation of the jackdaw (and probably of the rook, as indicated above) is a fairly unique phenomenon among our altricid bird species (pp. 11–13, 98ff).

Since few people are likely to have been able to rear so many juvenile birds of one species in such a relatively short time as I have done, for the purpose of observing later behaviour, I should like to emphasize once more at this point that one cannot be too careful about overestimating the variability in behavioural abilities of birds of a given species. This may apply to the most highly-specialized mammals, such that they differ from one another just as much as 'we human beings' do, but it certainly does not apply to birds. It is not that one jackdaw is 'shy and distrustful' by nature, while another is 'tame and self-confident'. What is possible is that a difference in age of four days (as may occur between siblings in a nest) may determine that one bird will *still* learn to gape to a human foster-parent whereas the other *no longer* can. The first will become tame, while the other will literally remain hopelessly shy throughout its life. A similar lasting effect on the behaviour of a bird can be determined by whether it was reared in isolation or together with conspecifics. Species-recognition is *not* an innate feature of most passerines and this probably applies to many other birds as well. This means that when one rears them in isolation from a very early age they will not recognize conspecifics as such. This effect is also expressed in their sexual behaviour. The 'restriction to conspecifics' takes place at different phases in different birds at a time characteristic for each species. In jackdaws, this occurs at a relatively very late stage, at about the time of fledging. After this time, no change can be effected in the 'species-restriction' of any given bird. However long the individually-reared bird is kept together with conspecifics, it will never regard them as such. Conversely, a bird reared in the company of conspecifics will not become attached to man. It should be borne in mind just how dependent a basic and lasting difference in the behaviour of an animal may be on a difference of three or four days in the time of arrival in human care. Many species which are regarded as untameable only appear so because 'species-restriction' occurs at a very early stage.

I should like once more to emphasize the great effect which the physical condition of a bird may have on its general behaviour. In particular, animals retarded by defective rearing always appear more artless than normal specimens. In one respect, the difference in intelligence is only apparent, because of the fact that species-specific responses employed by normal healthy individuals fail to develop in the weakly individual, as in the case of softening of hard food morsels or the hooking pattern for carrying eggs. On the other hand, there is a retardation in the actual powers of intelligence of the constitutionally weak individual. I can maintain that despite the fact that I have now reared over forty jackdaws and observed their later behaviour I have never seen a difference between the birds of the same sex, except where this could be traced back to differences in past history of the individuals concerned. I must mention, however, that such differences do in fact occur in the raven (which has far greater intelligent capacities), though these are very slight in comparison with the differences found among higher mammals[10] (pp. 34–37).

Table showing the period of captivity and life-history of jackdaws frequently referred to by name

NAME	YEAR OF BIRTH AND LIFE-HISTORY	FATE
Tschock (?)	1926: Adopted Left-Yellow in Summer 1927	Escaped through a gap in the cage-wire in January 1929
Yellow-Green (♂)	1927: Betrothed with Red-Red in Autumn 1927. Mated anew in January 1928 with Red-Yellow and remained with her thereafter. Successful clutch in 1929. Dominant member of group from November 1927 until dethroned on 15.12.1929 by a male from the 1928 batch	Lost in March 1930
Red-Yellow (♀)	1927: See Yellow-Green. Widowed in March 1930. At present (Autumn 1930) about to become betrothed with a male from this year's batch	The only jackdaw remaining from those described in the text
Red-Red (♀)	1927: Betrothed with Yellow-Green in Autumn 1927. Mated anew in January 1928 with Blue-Yellow; abandoned by him in May 1928. Betrothed with a male from the 1928 batch in Autumn 1929	Lost in March 1930
Blue-Yellow (♂)	1927: Betrothed in January 1928 with Red-Red. Initiated relationship with Left-Yellow in March 1928	Flew away with Left-Yellow in May 1928
Left-Yellow (♀)	1927: See Blue-Yellow	See Blue-Yellow

In conclusion, I should like to express my conviction that all of the above instinctive behaviour patterns, as observed in tame, free-flying Corvids, are characteristic of *all* healthy, free-living birds of the species concerned. It is naturally quite possible that some responses of the latter may have failed to appear in my birds (which had of course grown up in captivity), and that such responses therefore escaped my notice.

A consideration of methods of identification of species-specific instinctive behaviour patterns in birds (1932)

I Definition of the concept of the instinctive behaviour pattern

Before turning to the methods of identifying and analysing species-specific instinctive behaviour patterns in birds, it is probably advisable to give a comprehensive definition of my concept of the innate behaviour pattern.

In the following discussion, the word *instinct* itself is avoided and replaced with the concept of the *instinctive behaviour pattern* (*Triebhandlung*) because the former term has previously been employed to cover so many different concepts that to use it might well give rise to misunderstandings. Apart from the ambiguous nature of the term *instinct* and its origin from a foreign language, there is the fact that the new German term *Triebhandlung* (lit. 'drive activity') in fact conveys more about the character of the process involved.

The sense in which I intend to apply the term instinctive behaviour pattern is expressed in the definition given by Ziegler in the following passage defining the instinctive pattern: 'I have defined the difference between instinctive and intelligent activities in the following way: the former are based upon *inherited pathways* and the latter on *individually-acquired pathways*.[11] The psychological definition is thus replaced by a *histological* definition.' In his attempt to exclude all subjective factors as far as possible, Ziegler defines all behaviour patterns which are at all affected by individual variation as intelligent activities, without further differentiating between activities involving insight and conditioned responses and ultimately restricting the concept of intelligent behaviour to activities based on *insight*. I should most emphatically like to make this latter distinction, following the example set by Köhler.

One is far less inclined to include conditioned responses in the same category as intelligent responses after having personal experience of the gigantic part played by self-conditioning in the behaviour of animals with poorly-developed nervous systems, and particularly when one has had some experience of the overwhelmingly reflex performance of these 'acquired automatisms' (Alverdes). Alverdes quite rightly points out that even in man conditioned responses of this kind actually tend to be *less* open to voluntary control than the instinctive behaviour patterns themselves.

However much I am inclined to follow Ziegler's example in *not* employing purely subjective estimations of the voluntary or involuntary character of an element of behaviour and the presence or absence of purposive factors as criteria distinguishing intelligent behaviour from the performance of instinctive behaviour patterns, I must nevertheless stress my conviction that conscious insight is the *definitive* feature of intelligent behaviour even if we are seldom able to use this as a practical identifying feature. In this respect, I am once again in agreement with Köhler's views.

I propose to demonstrate with two examples the extent to which the solution of a detour problem by self-conditioning may differ from solution of the same problem by means of insight. Both examples were provided, initially without any intervention on my part, by my free-flying jackdaws. The first problem was the following: A jackdaw pair kept in a very narrow, long aviary had built a nest inside a nest-box close to one of the narrow side-walls of the chamber. The trap-door leading to the open was situated in the roof of the chamber, just above the other side-wall. If the parents flew up from the side where the nest was located, they were consequently forced to make an acute-angled detour since they of course had to fly almost directly away from their goal in order to fly through the trap-door. The birds solved this problem – an extremely difficult one for a bird – without the slightest hesitation every time they were thus confronted. An impartial observer would doubtless have been convinced that the birds' actions were based on a genuine insight into the spatial relationships within the cage. However, I knew that no such insight was involved and that this behaviour was quite simply based on a perfect path-conditioning process, since I had witnessed the development of the solution of this problem by the jackdaw pair. During their initial attempts to reach the nest-site, the birds had only found the door by chance after repeatedly running up and down along the roof of the cage above the nest-site and eventually giving up their attempts to reach the nest. The duration of the to-and-fro phase gradually decreased with increasing experience until this activity disappeared entirely. Thereafter, the birds found the door straight away on every attempt.

The fact that the birds had not actually grasped the entire spatial arrangement within the chamber but had only learnt the various ways leading to the nest 'by heart' was evident from their behaviour at the time when the young left the nest. When the offspring sat near the nest-box in the chamber, the parents were suddenly unable to find the door again and they fluttered against the cage-wire from outside. There was no indication of insight. The parents had to learn this (only slightly modified) detour problem from scratch. The similarity to the previous detour problem was evident only from the fact that the new problem was mastered in only a fraction of the time taken to learn the first detour. Unfortunately, the young birds soon left the aviary and they only returned when already able to fly properly, so they no longer represented a fixed goal for detour problems. Had this not been the case, it would have been possible to observe very neatly how far the adult jackdaws would have learnt *all* of the possible detour problems set by the various positions of the perched offspring 'by heart' and thus compensated for the absence of insight into the spatial structure of the aviary.

At a time when the young jackdaws had left the aviary, I carried out detour experiments in the same chamber with a Greater Yellow-crested Cockatoo which had been free-flying for some time and which was extremely active (both mentally and physically). I then discovered that this bird ceased to run searching to-and-fro along the cage-wire after it had found the door *once*. Wherever I affixed the goal within the cage, the cockatoo would immediately head for the door, achieving the same degree of success each time. He had thus exhibited a 'perceptive leap' similar to that demonstrated by Köhler in chimpanzees. I should never have believed that my cockatoo would be capable of displaying a process so unusual in birds, had I not seen how decisively a *single* perceptive feat determined his future behaviour in a number of different tests. The cockatoo admittedly failed to master detour problems in some cases, when they were too steep or too long, but where such failures occurred *no attempt* was made to solve the problem. This was probably due to the fact that it is difficult to provide a really strong incentive as a goal for a perpetually satiated herbivore.

In the case of my jackdaws, self-conditioning to a particular detour broke down not only when the detour was slightly altered, but also when *the same detour* was provided leading to an entirely different (but none the less strongly stimulating) goal. In Summer 1931, it was my habit to feed the jackdaws I had reared that year in the open. These jackdaws, which had already been flying free for some time, would all seek out the interior of the attic to pass the night in the company of my older jackdaws. They had already completely mastered the simple detour involved in slipping through a trap-door provided at an appropriate point on the external aviary and subsequently passing through a window into the attic. It had already been some time since I had last seen a young jackdaw still outside in the evening. However, when I tested them by allowing them all to become reasonably hungry while they were outside and then attempting to entice them to come into the attic from outside for food, they utterly failed to solve the same detour problem that they had solved every night. All of the jackdaws fluttered against the cage-wire near to the door and they even pushed their heads through the mesh, just as they had done when attempting to reach the interior of the attic in the evenings. Since my aim was to feed the jackdaws without leaving the house during the daytime as well, this provided me with the opportunity to observe that the young jackdaws learned to associate the new goal with the old detour in somewhat less time than it took to learn the original detour.

However minor the part played by insight in such self-conditioned responses might appear on closer examination, these responses may nevertheless replace genuine insight to a great degree and even give the illusion of insight until their mode of development is known. Since such a sudden change is seldom forced upon the bird in solving the problems encountered in its natural surroundings, the bird can actually get along quite well with such (to our eyes) inferior mechanisms.

A very important feature of self-conditioning is the fact that it is quite likely to be based upon an original manifestation of insight. Frequent repetition, accompanied by increasing reinforcement of the relevant pathways, may lead to gradual loss of the original insight component – as Köhler has demonstrated with convincing examples. I nevertheless believe that it is quite consistent to dismiss any explanation of the origin of instinctive behaviour patterns from activities involving insight (the assumption made by Lamarckists).

In fact, when a species possesses the ability to solve a given problem by insight, or simply by means of self-conditioning, an instinctive behaviour pattern is *unnecessary* for the purpose, since the resultant behaviour pattern should normally be the most favourable for the survival of the species. Quite often, we can observe behavioural chains which are not inherited as a unitary whole but always possess 'gaps' which are appropriately filled up by self-conditioning or insight behaviour (usually the former) during the ontogeny of the individual. Of course, appropriate filling of the gaps will only occur when the individual lives under 'normal' conditions, i.e. under those conditions for which the species concerned has developed the particular behaviour patterns observed. For this reason, an animal reared in captive isolation often demonstrates such 'gaps' in instinctive behaviour chains very distinctly. The gaps may be left unfilled or they may be filled by acquired activities which are quite inappropriate to the instinctive behaviour patterns concerned and therefore appear quite irrational.

For instance, the recognition of the 'appropriate' object for particular behaviour is frequently *not* inherited as an integral part of instinctive systems adapted to respond to a given object. Instead, there is an instinctive tendency to try out various objects, the range of which is gradually restricted to appropriate objects. Heinroth has demonstrated a particularly fine example of this effect in the development of the prey-impaling pattern of the Red-backed Shrike.[12]

This raises the question as to which factors eventually condition the individual to respond to the 'appropriate' object for which the instinctive response of the species has been 'designed'. It is easy to answer this question when, for example, a raven eventually restricts its drive to conceal all possible objects to the rational activity of hiding food, or when a Red-backed Shrike learns to impale insects in the correct manner on thorns so that they are not lost. It is more difficult, however, when fairly long chains of innate responses are concerned, where the attainment of the consummatory act is not so direct. When such chains incorporate links with a variable conditioning component, self-conditioning usually appears to occur in such a way that the conditioned response (or the conditioning of a response to a given object) ensures that the chain of behaviour patterns concerned is performed in a manner typical of the species. It is not my intention to discuss whether conditioning occurs as a result of the rewarding effects of a 'functional appetite' or because of punishment effects originating from aversive responses produced by rupture of the chain.

Gaps of this type, filled by individual experience, can occur unpredictably at extremely varied sites in different behavioural chains. I intend to use the nest-building chain of the Corvids as an example, since I am very familiar with this particular behaviour. The first instinctive component to appear in the complex instinctive nest-building chain was the drive to *carry* all manner of objects. This was true both of my pair of ravens and of the jackdaws, all of which carried the objects concerned much farther than was actually necessary for nest-building. The drive to *fly* with nest-material can even be observed in canaries, which otherwise exhibit little inclination to fly. When canaries have collected a bundle of the preferred nest-material in their beaks, they carry out upwardly-directed flying movements just like those of a freshly-captured bird attempting to fly through the roof of the cage. Even such a long-domesticated bird as the canary has been unable to adapt to the fact that the distance to be flown with the nest-material from the collection site to the nest-site is only a matter of inches.

At the very beginning, there was no recognizable preference for suitable materials to be seen in the choice of objects carried by the ravens and the jackdaws. Both species initially carried fragments of roof-tiles more than anything else, these being the objects which they most frequently encountered during their spells on the roof close to the cages. This occurred despite the fact that twigs suitable for nest-building were readily available. The birds first began to direct their attentions exclusively to twigs and to ignore the tile fragments, pebbles and so forth when the drive to *secure* the nest-material in the nest developed. In this latter drive, which is probably characteristic of all birds which carry material to the nest, the material is fitted into the nest with the familiar quivering, sideways thrusting motions. The behaviour pattern concerned is not fitted for the securing of inappropriate objects. However, this did not prevent my raven from carrying into the virtually completed nest flat *pieces of ice*, after breaking them off from the edges of the hole which I had hacked for the ducks in the thin layer of ice covering my duckpond. It would therefore appear that it is the opportunity for satisfying the 'securing drive' in the raven which restricts the 'carrying drive' to those objects for which the instinctive behaviour pattern concerned was adapted during the evolution of the species.

In birds with lesser developed mental capabilities, in contrast to the Corvids, recognition of nest-material to be used for the nest is innately determined down to the finest detail. In actual fact, such birds can often be stimulated to start nest-building by presentation of appropriate nest-material. In some birds, the attainment of full reproductive condition may be dependent upon the presence of material in the cage, as is the case with some small foreign Estriblinae. Similarly, I once saw the nest-building drive elicited in a newly-fledged Night Heron. The bird happened to land on a thick horizontal branch in a plane tree in such a way that a thin, branching twig protruded between its legs. The Night Heron diligently tried to secure the twig to the branch beneath him, using the gentle lateral thrusting movements described above. A stork, one week after fledging, was once observed to seize a piece of dried, twisted grass from a raven which was playing with it. The stork obviously believed the plaything to be edible, since it pounced upon the raven uttering the typical squeaking begging-call of its species. After the stork had played around with the knot of grass for a while, however, its behaviour suddenly switched. The root was taken firmly into the beak and the stork began to stride off with its head held high. It was immediately obvious that the bird was motivated to take off, and the bird did eventually take wing to fly in a wide arc to its artificial nest on a low-lying roof nearby. The grass was then incorporated into the nest in the typical manner. In this case, too, the behaviour patterns appropriate to nest-building activity had been elicited by chance acquisition of suitable material. The fact that young birds of the stork/heron group carry out nest-building activities whilst still in the nest is already a well-known phenomenon. It presumably has biological significance in that the nest is better maintained. I was once able to observe young Night Herons which showed great interest in twigs which were about to fall out of their artificial nest. They pulled the twigs back and secured them. The observation that the carrying drive itself can be elicited in a freshly-fledged stork was very important to me, since a stork will not take off without considerable stimulation of some kind (internal or external). If no specific external stimulation occurs and the stork is merely motivated in a general way to take off and fly home (e.g. as evening draws on), then the bird will wander about with its neck extended and its wings distinctly raised, maintaining the take-off posture for a period of minutes before actually taking off. A long prior period of stimulus-summation is obviously necessary for such a take-off. This fact lends great significance to the observation outlined above and definitely excludes the possibility that this was a chance effect.

It is quite easy to imagine that the Corvids can 'afford' such gaps in their instinctive behaviour chains (mechanistically speaking), since they will ultimately always arrive at the 'correct' type of activity as a result of their undoubtedly great learning abilities and their extremely characteristic tendency to try out novel patterns of activity. In other birds, the restricted capacity for conditioning or the highly-specialized form of the nest necessitates more exact 'programming' of the animal's activities within the framework of innate behaviour patterns.

In spite of this consideration, the non-innate quality of the recognition of nest-material for the foundation of the nest in ravens and jackdaws is extremely unusual, since recognition of the soft materials suitable for lining the nest-cup after the rough foundation has been prepared (largely as a result of the male's activities) is an *immediate, instinctive response*. The female raven was not only stimulated to line the nest by the presentation of suitable material (as would be seen in any female canary) – she also showed a special preference for *the one substance* which (according to Brehm) represents the normal lining material of raven nests in the field: strips of raffia. In fact, the female was able to obtain the raffia with obviously instinctive skill by peeling off the bark of suitable branches.

Apparently, many birds exhibit a relatively extensive independence of the instinctive behaviour patterns of nest-construction from those of nest-lining. This makes it much more understandable that the raven is able to incorporate a 'conditioning gap' in the former while leaving the latter entirely instinctive. For instance, I once observed a male Blackcap which was building alone in an aviary and found that the completed nest consisted entirely of relatively thick grass roots. It was completely transparent and possessed absolutely no internal lining of the finer material which is to be found (admittedly very scantily) in normal Blackcap nests. It is possible, however, that the female of this species is also responsible for constructing the soft lining. A solitary male jackdaw in breeding condition similarly constructs an unlined nest-cup, whereas a solitary female jackdaw can construct a complete nest (i.e. she possesses both the instinctive pattern for lining and the pattern for construction of the rough foundation). On the other hand, I observed a pair of Zebra-Finches (*Taeniopygia castanctis*) which exhibited a complete absence of the behaviour pattern for forming the rough foundation, whilst retaining the instinctive lining pattern. I had provided the birds merely with a cup-shaped nest-support consisting of coarse wire-netting affixed to the cage and had then offered every conceivable type of nest-material. The birds persistently attempted to line the wire-netting entirely with the softest materials available, and they did not even construct a proper nest-cup. This type of behaviour is generally typical of domesticated canaries, but if they are allowed to 'run wild' in a large aviary it can sometimes be observed that they will abruptly begin to build in the manner typical of the species. I should emphasize, in saying this, that the recognition of the foundation material is just as innately determined as is recognition of the lining material in canaries.

The fact that the use of twigs for construction of the rough foundation of the nest is not innately determined in the raven is highlighted all the more by the apparent presence of another, quite specialized, innate pattern for the *acquisition* of the twigs. I had already observed at Christmas that my pair of ravens repeatedly clambered about in a semi-withered oak-tree. In complete contrast to their normal behaviour, they would perch on dead twigs and fall down with them as they broke off. Ravens can generally gauge quite accurately the carrying capacity of branches and twigs – an ability which must be individually acquired. My first reaction was therefore to assume that this was a new sport, forming part of a frequently changing series of 'fashions' (adopting Köhler's term for a similar phenomenon in chimpanzees). But I soon became puzzled at the persistence with which the birds continued this activity. Soon afterwards, I distinctly saw how each raven would hurl itself down upon a chosen twig, snap it off and then tumble down out of the branches without losing its hold on the twig. Usually, the raven would then dive steeply away, in order to gain momentum, before transferring the twig (which was generally quite heavy) from the claws to the beak. Only if the twig was very big did the raven rapidly abandon such attempts and fly on with the twig in its claws.

At about the same time, the drive to carry objects around made its first appearance. This later developed into the carrying-to-the-nest drive. In free-living jackdaws, which make far less contact with inappropriate objects than was the case with my captive birds, it may well be that the twig-snapping drive plays a part in guiding the carrying drive towards the appropriate objects.

I regard nest-building in the raven as a very good example of the intercalation of fixed action patterns with conditioned elements of behaviour. I propose to call this type of behavioural control *Erbtrieb-Dressurverschränkung* or *Instinkt-Dressurverschränkung* (variable component intercalation). Nevertheless, I feel it is necessary to emphasize once more just how *little* plasticity and variability is present in the nest-building behaviour of this bird, despite the fact that the raven is so superior to other endemic birds in intelligence that it really belongs in a class apart. I should also like to emphasize that the inherent basis of such plasticity can only be regarded as self-conditioning. Insight is not involved.

However, I do believe that I have seen one slight *indication* of *insight behaviour* in my male raven during nest-building. At a time when some form of foundation for the nest had already been prepared, he began to fly up to perch on the edge of the nest with the twig in his beak and *survey* the nest for a few moments each time, rather than thrusting the twig into the nearest available position. As long as this behaviour persisted, he actually did incorporate the new twig into that part of the nest which needed it most. In fact, he always chose the spot which I would have chosen in his place. This may appear to be a highly self-evident and insignificant observation to many readers, but it made me strikingly aware of the fact that I had *never* seen this before in any of the large number of birds which I had observed during nest-construction. With herons and storks in particular the observer can easily become impatient when watching these birds trying to attach one twig after another in places where they cannot possibly hold.

Such intervention of insight within an instinctive behaviour pattern is probably a rare and exceptional occurrence in birds, but it is likely to be more frequent among higher mammals, thus rendering the analysis of their innate behaviour extremely difficult.

On the other hand, *variable component intercalation* would seem to be relatively common among birds.[13] One of the most conspicuous among those behavioural features which are not inherited but must be acquired individually is the animal's recognition of the species which it regards as its own; that is, the recognition of those animals towards which it will direct the instinctive behaviour patterns which are normally performed in relationship to conspecifics. This is therefore another case in which the recognition of the object appropriate to the instinctive behaviour is not itself innately determined. When the human being acts as a substitute for a conspecific, as in cases where young birds are reared by a human foster-parent, there is a resultant pronounced attachment of the young bird to the foster-parent. Heinroth has described this problem in detail in his paper on tame and shy birds. In social birds, the usual gregarious drive can be transferred to contact with human beings ('social tameness'), but even in non-social species it can occur that sexual arousal may be directed towards a human being ('sexual tameness'). It can also occur that the bird regards its human foster-parent as a conspecific which is to be repelled ('aggressive tameness'). This directional transference of instinctive responses originally adapted to relationships with conspecifics, producing a relationship with human beings, appears to be such a regular occurrence in birds reared from an early age that examples where birds reared in isolation from an extremely early stage have *not* transferred the direction of their drives would be very interesting. I do know of one such case from the Zoological Garden of Amsterdam, pointed out to me by Mr Portielje. A male South American Bittern (*Tigrosoma*) is kept there together with a female of the same species and has already bred successfully. As soon as the male sees his human foster-parent, however, he is only interested in him. Under certain conditions, the male may then even respond to the female as to a 'nest enemy'. Almost exactly the same behaviour was exhibited by a corncrake kept by Heinroth. In the presence of the human foster-parent, this male similarly showed no interest in a female of the same species, and yet he would produce a fertilized clutch when left alone with the female.

Self-conditioned behaviour patterns in birds act as acquired automatisms and they have an extremely reflex-like mode of performance. Since they are in fact only identifiable during development, such patterns are easily overlooked if they occur as variable links in a chain of instinctive behaviour patterns, as described above. They then give the illusion of variability within an instinctive behaviour pattern.

In the following account, I shall regard an instinctive behaviour pattern as an extremely rigid pattern which incorporates *absolutely* no insight components and whose variability, where such is present, is entirely dependent upon corresponding variability in the eliciting stimuli. *Wherever* acquired automatism or insight mechanisms occur as links incorporated in relatively long chains of instinctive behaviour patterns, I should like to employ the terms *conditioned component intercalation* and *insight component intercalation*. I emphasize at once that the latter is rarely encountered among birds. Even in lower animals, the apparent plasticity of their complex reflex chains is probably largely due to the incorporation of 'conditioning links'. In his experiments with bees, von Frisch has clearly demonstrated (and in fact emphasized) that variability in the behaviour of his bees always occurs at points where it has a significant biological function.

Since I therefore regard the instinctive behaviour pattern as an absolutely rigid whole, my definition contrasts with the definition of 'instinct' given by Alverdes. He apparently applies the term *instinctive behaviour pattern* exclusively to entire behavioural sequences, whether or not variable links are incorporated. Such links are frequently, though not always, present in the behavioural sequences of higher animals.

Alverdes writes:

> Some authors speak of instinctive behaviour patterns in man and animals as if they were a category apart. This view is contradicted by the observation that every activity involving insight includes a significant instinctive, drive-operated component. Conversely, no single instinctive behaviour pattern operates fully automatically. In addition to the fixed, invariable components there is always a variable component in some degree dependent upon the prevailing conditions. Each activity (A) is thus a combined function of a constant component (K) and a variable component (V) – in formal terms: $A = f(K, V)$.

I make no attempt whatsoever to deny that each activity involving insight incorporates instinctive components, even in man, since the primitive co-ordination patterns involved in gazing, locomotion, grasping, etc. are inherited features. However, I do not believe that *every* instinctive behaviour pattern includes a variable component[14] whose variability exceeds that dictated by variations in stimulation. At least, I have been unable to perceive such variability in the instinctive behaviour patterns which I have observed and I have found no reliable indications of such in the literature. Whenever an instinctive behavioural sequence did appear to possess such variability, closer analysis always narrowed this down to variations in the eliciting stimuli, to variable component intercalation systems or (as in some interesting cases to be discussed later) to *secondary insight* of the animal into its own instinctive behaviour patterns. Cases of the latter type are probably rare among birds, whereas their greater frequency in higher mammals and man is high enough to give rise to great difficulties in the interpretation of behaviour.

When estimating the equivalence or variability of stimuli, it must always be remembered that the same external factors may represent quite different stimuli for different individuals of a given species, or even for the same individual at different times, according to the physiological state of the animal concerned. Failure to appreciate this fact once led me to retract a correct observation because a later observation appeared to contradict it at first. Jackdaws possess an extremely interesting, highly-specialized instinctive behaviour pattern for the protection of companions, as I have described in another paper. It is possible to elicit this response with ease at any desired time by seizing one jackdaw among a flock of conspecifics. On doing this, a quite specific *attack call* (a loud, metallic 'rattle') is heard. In some cases, the seized bird will begin to rattle first. *Completely tame* jackdaws do not rattle when seized; they only hiss softly and exhibit bill-rattling just as they do when discomforted by another nesting jackdaw approaching to the point of physical contact. But they will immediately rattle if some other jackdaw is seized. *Completely wild* jackdaws also do not rattle when seized. Instead they utter a sound which is quite distinct from the rattle and has a more squeaking character. All jackdaws squeak when seized in the dark, so it would seem that a *severe shock* is necessary to elicit this vocalization. *Rattling* cannot be elicited in the seized bird when it is too greatly shocked to produce an attack response – such a bird will *squeak*. The rattling pattern is only elicited in seized birds which have reached a *specific degree of tameness*. If the bird is so little scared by the human being who seizes him that the stimulation arising from seizure does not exceed the general level associated with a small predator which might be successfully repelled, the seized bird will (so to speak) participate in the attack made by the others. When jackdaws are completely tame, seizure does not shock them at all.[15]

My young jackdaws from 1927 would rattle when seized, as I mentioned in a publication soon afterwards. Young jackdaws reared later on were slightly less tame and this sufficed to ensure that a seized bird would only squeak and that rattling was restricted to the watching jackdaws. But I did not at that time consider the fact that such a slight difference in tameness would entirely alter the nature of the stimulus such that it would elicit a separate response in the seized bird. For this reason, I retracted my earlier (quite correct) observations in writing my paper 'Contributions to the Study of the Ethology of Social Corvidae' (1931).

An illusion of variability in instinctive behaviour patterns can also be produced when one species has similar instinctive behaviour patterns which respond specifically to different, but nevertheless related, stimuli. For instance, the offspring of my jackdaws always ejected their faeces without an enclosing membrane. The faeces adhered to fragments of nest-material and these fragments were then unfailingly removed by the parents. Subsequent publications have documented incontestable observations of jackdaws which simply carried off the faeces in the beak, but in such cases an enclosing faecal membrane was always present. In view of this, I am inclined to believe that *Coleus* has developed the removal of fouled nest-material as a special response to cope with the production of naked faecal pellets by the young. This latter phenomenon is doubtless common and it is hardly a pathological feature.

This kind of variability in instinctive behaviour patterns produced in response to different specific stimuli in an obviously functional manner is not the type of variability envisaged by Alverdes when he wrote:

> The presence of a variable component is a feature which must be regarded as typical of both the intelligent activities of higher vertebrates and their instinctive behaviour patterns as well. For example, the variable component is to be found in a bird engaged in nest-building in the search for a nest-site made by the individual and the manner in which the bird grasps twigs, stalks and feathers and incorporates them into the nest structure in an adaptive manner with co-ordinated and purposive bodily motions. The variable component can in this case be identified from the fact that older birds build more skilfully than young birds. None of the activities performed during nest-building are fully automatic ('reflex') and none are purely intelligent activities (where V would be greater than K). In all cases, genuine instinctive behaviour patterns are involved, in which K is greater than V. The internal instinctive factor K in all such cases forms the invariable basis providing the 'biological sense' of the entire activity performed by the animal. The purposive individual activities represented by V are rooted in this foundation. Exactly the same considerations apply to the instinctive behaviour patterns of insects, spiders, etc. However rigid and invariable such activities may appear to be, the formula $A = f(K, V)$ always applies.

It has already been stated that it is only possible to speak of variability within instinctive behaviour patterns where such is the result of variations in the eliciting factors. The flaw in the formula $A = f(K, V)$ obviously arises from the fact that the V in instinctive behaviour patterns is directly equated with the V of intelligent activities. Since V in the latter case describes insight components intimately incorporated in the framework of the activities concerned, the consequence of regarding V as the same in both types of behaviour is the automatic assumption that purposive factors must operate in the instinctive activities of animals. Even if one assumes that practice, experience, tradition and so on can exert a modifying influence on genuine instinctive behaviour patterns, it is still not permissible to equate such variability with the type of modifying influence exerted in acquired behaviour patterns by insight into the goal to be attained.

As far as the examples provided for variability in innate behaviour patterns are concerned, it must be stated that the determination of the nest-site is definitely not a result of a 'choice' made by the bird. Even quite intelligent birds will characteristically respond in a purely reflex manner to the determinative stimuli associated with the nest-site. In fact, there is little that is 'unpredictable' about the choice of a nest-site. On the contrary, the stimuli determining such choice are so open to analysis that anyone who is reasonably well acquainted with the peculiarities of a given bird genus is often able to successfully manipulate these stimuli at will. My free-flying raven pair, for example, promptly built a nest at the spot where I wanted to have it, and the raven happens to be a species which is otherwise particularly unpredictable. In fact, one can maintain that the raven, among those birds which build nests, is the species least governed by instinctive factors. The large parrot species, which are their only rivals in intelligence, do not carry nest-material to build a nest.

In other features too, there is little to be seen in the way of variability in choice of nest-site, particularly as regards differential suitability of the chosen site. Verwey has reliably demonstrated with his observations on herons that these birds will remain attached to specific nest-sites which provide no support for the nest-material which they collect, with the result that the collected material continually falls away.

An additional observation, supporting the assumption that specific stimuli (probably visual) emanating from the given site induce the bird to start constructing its nest at that particular site, is the following: In many cases where several pairs of birds are confined together in an aviary, conspicuous attempts are made by *several* pairs to obtain one specific nest-site. This is even observed when the different pairs do not belong to the same species, as long as their nesting habits are roughly similar. This effect forces professional canary-breeders to provide their stock birds with a large number of superfluous prospective nest-sites in order to bring about at least some reduction in fights over nest-sites.

The same opinion is expressed by Sunkel, who states in his paper on the significance of visual impressions in nest-site choice in birds:

> Of course, the choice of a nest-site is also based upon visual perception. In general, birds select sites for nest-construction which are so typical that the ornithologist can immediately recognize which sites in the field might be occupied by any given bird species. The suitability of any given site, which is perceived through the eyes of the bird, will virtually force the bird to build its nest at that site at the beginning of the reproductive season. It therefore occurs that birds cannot resist the nest-building instinct elicited by the optical stimulus, even at those sites where the nests are repeatedly destroyed. Such

observations have been made in different cases on storks, kestrels, jackdaws, crows and many song-birds.

Turning to the 'co-ordinated and purposive' motor patterns seen in birds during nest-building, it must at once be stated that such patterns are *in all cases* either inherited as such or developed as a result of well co-ordinated conditioning mechanisms. These patterns are *never* variable and they are only appropriate in the narrowly-defined situation for which they were developed. In the first case, the situation would be characteristic of the species; in the second it would be characteristic of the individual. Genuinely variable intelligent activities adapting to a really novel situation always appear extremely *clumsy*, both in man and in animals. This is the case, for example, when we attempt to write with the left hand. Köhler gives an exact description of this effect in chimpanzees.

It has often been observed that older birds build better nests than younger ones, but the older bird can assuredly build more skilfully even after being compelled to pass through all previous breeding seasons *without* building a nest or rearing young and subsequently building its nest under virtually the same circumstances as a young bird. I have observed this in three Bullfinch pairs, one of which lived in a small household cage and did not breed whilst the other two constructed extremely rudimentary nests in a large aviary. In the following year, all three pairs occupied the same aviary and *all three* built incomparably better nests than those built by the two pairs which had bred the previous year.[16] I could find no difference between the nests; in fact, I no longer knew which of the birds had previously occupied the household cage.

I know of no case where the increase in nest-building skill with advancing age has been investigated in the field. In captive birds, it frequently occurs that a considerable improvement in a bird's general physical condition accompanies the attainment of sexual maturity and coincides with the onset of the first breeding season. The effect of this is often such that the benefits are first obvious in the second breeding season. It is rare that one observes captive birds whose first breeding season progresses as well as the second. This applies from Völkle's breeding stock of Golden Eagles down to any given pair of canaries. This phenomenon is doubtless based upon the increased reliability of performance of the relevant instinctive behaviour patterns, resulting from the improvement in physical condition. It is not based upon learning taking place during nest-construction or care of the young. I am not maintaining that ravens, crows or other highly-specialized birds cannot achieve considerable improvement in their nest-building skills on the basis of their learning abilities. But I am utterly convinced that all of the cases so far published, demonstrating that a particular bird built better in a later breeding season than in a previous one (the latter most probably being the uncertain first breeding season in all cases), can be ascribed to general improvements in physical condition. If a genuine learning process could be demonstrated in one given case of nest-building, it would be interesting to determine whether the improvement regularly takes place at the same point in the behaviour chain and thus exhibits the character of a variable component intercalation mechanism. I would be very grateful for any information about relevant observations.

In reiteration: My concept of an instinctive behaviour pattern is that of a behavioural sequence based upon inherited pathways laid down in the central nervous system, such that the pattern is just as invariable as its histological foundations or any named morphological character.[17] This type of definition, akin to that given by Ziegler, shows sufficiently clearly that instinctive behaviour patterns are distinct from reflexes (although not entirely separate from the latter) because of their greater complexity and because participation of the whole animal and not just a single organ is involved.

Whenever a chain of instinctive behaviour patterns exhibits variability, this is most likely to be due to the incorporation of self-conditioning behaviour components in long chains of fixed, reflex-like responses (i.e. ignoring those cases where varying intensity, direction, composition and so forth are present in the determinant stimuli). Such variable component links appear to occur frequently in the most advanced vertebrates and they often attain a high degree of complexity in the latter. This renders the analysis of the instinctive behaviour patterns of mammals extremely difficult or even impossible, because of their high level of intelligence.

In lower animals, the entire 'learning ability' is represented by a restricted number of conditioning-component patterns incorporated in extremely complex reflex chains. It has already been pointed out in the discussion of variable component intercalation that the sites of incorporation are often of obvious 'adaptive' importance (in the phylogenetic sense).

II Birds as experimental animals

It is no chance phenomenon that birds have been selected for investigations of species-specific fixed action patterns. Birds are by far the most convenient experimental animals for the analysis of instinctive behaviour patterns, particularly with regard to their relationships with conditioned behaviour patterns and insight behaviour.

The limited number of plastic sites present in the long reflex chains of the lower animals often possess too much the obvious appearance of a special adaptation to be equated with learning ability in higher animals. In addition, such plasticity breaks down when the animals are driven to associate factors which bear no relationship to one another in the natural environment. For instance, Armbruster was unable to convincingly condition bees to sound stimuli, which of course have no relationship to the bees' food in the natural environment. On the other hand, von Frisch successfully conditioned bees to distinguish flower shapes of quite similar appearance. The lower animals in all probability entirely lack insight behaviour.[18]

Fish, amphibians and reptiles probably possess far better powers of learning and association. Fish can in fact learn to associate factors which bear no relationship in the natural environment, when required to do so in an experiment. So we are quite justified in regarding these abilities as homologous with those of the highest evolved animals.

In all of these animals, however, the fixed instinctive behaviour patterns still bear little relationship to the variable activities. The two types of behaviour are indeed easy to distinguish and analyse, but this does not bring us much closer to an understanding of the extremely complex conditioned component (or insight component) intercalation mechanisms of the highest animals, including man himself.

Birds in fact occupy an ideal position, so to speak, on the phylogenetic ladder. The reflex chains[19] of birds incorporate a sufficiently low number of variable component links for the latter to be distinctly recognizable, while the links themselves have an influence on the overall behaviour great enough to approximate to that seen in mammals. This provides us with a certain insight into the origin of such behaviour as found in the highest mammals and man. We can only hope to gain an understanding of the more complex system by analysing the simpler form, although much of the behaviour of mammals, by virtue of its exceedingly complex structure, will never be open to analysis even with this new addition to the methods available to us.

A quite separate supplementary factor facilitating the study of the behaviour of birds is the great correspondence between the sensory capacities of the observer and those of the observed animal. Birds, like men, are largely optically-orienting animals. In both, the major function of hearing is the perception of the vocalizations of conspecifics rather than scanning for signs of approaching danger. This may appear to be a commonplace, but when dealing with an animal whose sensory capacities are greatly different from those of the human observer one is made extremely conscious of the large number of favourable circumstances which are normally taken for granted in psychological studies of birds. Even with owls, which are probably very dependent upon their sense of hearing, the probability of being unable to decide upon an interpretation of particular observations is much greater than with other birds, as Heinroth points out in his book *Die Vögel Mitteleuropas* ('The Birds of Central Europe'). If an animal such as a bat is taken, where every single sense functions differently to those which we possess and where an additional sense which we lack and cannot really imagine is present in the flight membranes,[20] then each and every sense must be tested under experimental conditions precluding the operation of the other senses in order to determine reliably which sense the animal is employing to solve a given problem or to perform any general activity. Since a task such as the attainment of a physical goal can present varying degrees of difficulty to different senses, the solution of this question is utterly vital for the estimation of the intelligence of any such animal. For instance, I was once almost convinced that my bats had 'understood' that I always placed the mealworm jar in a particular box, until I discovered that they could hear the larvae crawling around the jar inside the box. I usually value *chance observations in the natural environment* far beyond any other observations, since they are least likely to be harnessed to sources of error. However, in cases where it is necessary to conduct specific experiments under set conditions to determine which sense a given animal is employing, such chance observations are worthless.

In my attempt to supplement the results of experiments 'which nature may be said to make' (Selous), without increasing the number of sources of error through alterations in the natural environment or by producing pathological effects of captivity, I was drawn to the performance of experiments with tame, free-flying birds.

Without going into the technique of accustoming various birds to a free-flying existence, I intend simply to mention that this method of study proved to be extremely profitable with an amazingly large number of birds of the species which were kept in this way.

Most birds provide particularly suitable objects for the study of the whole range of behaviour patterns related to interaction with conspecifics, especially where social species are concerned, since there is the possibility of transferring the direction of interaction to encompass human beings (as described above). In almost every individually-reared bird, it is conveniently possible for the human foster-parent to elicit and closely observe those instinctive behaviour patterns which would be directed towards conspecifics had the bird been reared by its natural parents. It is obvious that many instinctive behaviour patterns of such socially, sexually or aggressively tame hand-reared birds (Heinroth) must of necessity remain incomprehensible to the observer when activities or motor patterns are concerned which are intended to elicit specific responses in a conspecific partner. It is frequently possible, however, to recognize such 'releasing' patterns from the fact that the bird adopts postures or performs movements which differ extremely from normally encountered patterns. This applies particularly to the releasing instinctive behaviour patterns of courtship, but it also applies to the postures and motor patterns of young birds begging for food and to the many displays found in social birds. In many cases, special organs have been developed in association with these signals. This is so often the case that the mere appearance of these organs (especially where elongated feathers, inflatable organs, coloured gape patterns, etc., are involved) provides good reason to suppose that they are present for some rôle in a given species-specific 'ceremony'. But even birds which lack such signal-organs are capable of producing such a bizarre visual effect by means of special postures and by ruffling specific areas of the plumage that it is easy to understand that their appearance under these conditions can elicit responses in conspecifics which do not respond to the image otherwise presented by the displaying bird. For example, a male raven is able to produce such an unusual visual effect by extending the head, lowering the wings, maximally ruffling the crown plumage and the abdominal feathers and extending the nictitating membrane over the eyes that an uninitiated observer has difficulty in recognizing that a photograph taken of the resultant white-eyed monster is that of a raven![21]

Such releasing patterns are particularly clearly seen in birds which have become adapted to human beings. In fact, they are then repeated more frequently than by normal, free-living birds. The human being does *not* respond when the bird 'attempts' to elicit a specific response, with the result that the bird tries over and over again. Incidentally, it is very interesting in such cases to attempt to determine what response the bird is actually 'attempting' to elicit. This can frequently lead to mistakes, however. For a long time, I regarded a motor pattern of one jackdaw as a motor pattern belonging in the context of the female copulatory invitation, when the pattern (which was in fact very similar to this courtship display) actually represented an invitation to take off and join the displaying bird in the air. The female invitation can also be employed as a normal greeting pattern.

In spite of the frequent doubt arising in attempts to interpret the performance of given instinctive behaviour patterns by hand-reared juvenile birds which have come to associate the behaviour with the human being, the occurrence of these patterns is of inestimable value in making the observer aware of the *existence* of the instinctive behaviour patterns concerned.

III Characteristics of species-specific instinctive behaviour patterns

In general, an observer with a fair amount of experience in keeping animals recognizes purely *intuitively* various instinctive behaviour patterns in animals which he keeps, and does not need to consider this any further. But, in order to lay down definite criteria for the formation of such a judgement, the observer must carry out retrospective analysis of his own perception, which is often more difficult than actually observing the animals themselves. This process of self-analysis reveals that the observer, in recognizing instinctive behaviour patterns, does not usually adhere to the two most obvious features inherent in such patterns. In the first place, the behavioural sequence is carried out in the same manner by *all* individuals of a species. Secondly, a young bird reared in isolation will nevertheless exhibit the same behaviour pattern despite the absence of a model to learn from. At least, these two criteria are not often employed in the intuitive identification of instinctive behaviour patterns.

The second characteristic is only decisive in quite specific cases; namely in those where the imitation of a particular behavioural activity lies within the range of possible activities of the bird concerned. This is seldom the case. Behavioural chains which are reasonably complex and specialized, as is often so with the instinctive behaviour patterns of birds, are in any case too complicated to be imitated by even the most intelligent bird. The fact that a bird has been reared in isolation is of no help in recognizing insight patterns 'invented' by the individual. In any case, such complex behaviour incorporates other characters which can be used as identifying criteria for instinctive control.

In those areas where imitation can actually play a part, however, rearing in isolation is extremely important in studying the possibility of innate control of displays. This is especially applicable to the vocal activities of certain birds. In this case, rearing in isolation provides extremely instructive information. In general, such information indicates that even in birds capable of vocal imitation – I do not know of any other than passerines and parrots which can do this – those vocalizations which elicit specific responses in conspecifics are *usually* innate.[22] However, examples have been discovered where the most typical of such vocalizations – the social contact call – has been shown not to be innate. For instance, von Lucanus observed that the 'Stieglitt'-call of the Goldfinch is not innately determined. It would be very interesting to show whether the response of the Goldfinch to this particular call (a social contact call) is inherited. This would mean testing a Goldfinch which cannot itself utter the 'Stieglitt'-call to see whether it is in some way influenced on hearing the unknown call. Of course, to perform this experiment, one would have to rear several Goldfinches in a group (but rigorously separated from adult conspecifics), since the result would otherwise be prejudiced by disturbances in the species-recognition behaviour of the tested birds. It is quite conceivable that a Goldfinch treated in this way and consequently possessing normal species-recognition responses while lacking only the species-specific social contact call would approach in response to the 'Stieglitt'-call of a conspecific.[23] In fact, innate knowledge of all those vocalizations *not* inherited as fully-developed patterns is present to the extent that the juvenile bird selects and imitates the species-specific call among the babble of vocalizations to be heard in its natural environment. This of course amounts to the same thing as an innate response to non-innate vocalizations.

In very intelligent birds, it can also happen that an imitated sound is endowed with the function of a social contact call through learning, sometimes without detriment to the development of the species-specific contact call. For instance, a starling reared by Heinroth copied and employed a whistle with which the foster-parents were apt to call one another. Parrots often behave very similarly. An old raven which is very attached to me and is able to speak his own name employs this 'call' when he wants to summon me to him. This especially occurs when I take him out and then stop at places where he does not dare to come down out of the air. He nevertheless calls conspecifics with the inherited, species-specific calling signals, which vary according to whether the raven is flying or perching when calling. He only employs his own name within the context of the perching social contact call and most frequently utters it (as with the latter call) soon after landing. If the raven is still in the air, and wants to draw me away from a place which he finds uncomfortable, he either remains silent and attempts to induce me to 'take off' by repeatedly flying down over my head, or he employs the brief aerial guiding call typical of ravens. After landing some distance away from the uncomfortable area, he will immediately utter his name with a mock human voice rather than give the sessile social contact call. Indeed, the exact imitation of my pronunciation of his name is extremely startling; starting with a friendly tone, he passes in a smooth transition to a commanding and eventually irate tone.

These imitated social contact calls are particularly interesting because they represent the only case in the entire animal kingdom where a vocalization atypical of the species acquires a communicative significance in the rôle of a releaser for a behaviour pattern of a companion.[24] When observing such acquired social contact calls, one has a quite definite impression that the bird has really formed a mental connection between the call and the subsequent arrival of the called object. This immediately poses the question as to how far such purposive motivation may be present in a typical innate social contact call. Bühler divides human vocal display patterns into 'expressive', 'evocative' and 'representative'. The majority of acquired vocal expressions of birds, including those which are developed by the species as a definite 'releaser'[25], would not fall into the category 'evocative' if classified according to the division in human speech, since the bird has no idea when vocalizing that a particular behaviour pattern should subsequently be performed by a conspecific. Indeed, both a hand-reared bird and an isolated adult will perform the subsequent pattern. These calls therefore belong quite definitely in the 'expressive' category within Bühler's classificatory scheme. But, extrapolating from the described observations of *acquired* social contact calls, it is perfectly possible that the *species-specific* social contact calls of the most intelligent birds may be to some degree under purposive control, i.e. there may be an awakening insight into the species-specific behaviour pattern. Thus, the vocalization concerned would also approximate to the category of voluntary evocation in Bühler's classification of human speech. One must remember, however, that the social contact call represents the simplest form of social releaser and that conscious perception of the goal and introduction of a certain amount of variability through supplementary learning is only present in the most intelligent birds, and even there it cannot be conclusively demonstrated. These two latter factors seem to me to be strongly interdependent. If, for example, the above-mentioned raven had not comprehended that conspecifics fly up in response to the typical social contact call, he could hardly have hit upon the idea of attempting to call *me* with the call which he had heard from me in the context of a social contact call. The possibility of a conditioned behaviour pattern being responsible for this effect can be excluded, since such a pattern could only have developed if I had repeatedly approached the bird in response to the call, which was definitely not the case.

Whenever the behaviour pattern which is to be elicited in a bird[26] by a vocal signal is relatively specific in nature, the pattern is probably entirely innately determined. If a bird reared in isolation utters such calls, it is of course impossible to decide at first what particular response is to be expected from conspecifics. These vocal releasers are very often combined with quite specific, visually operative motor patterns, as has been described previously.

Hand-rearing in isolation (i.e. to the exclusion of examples set by conspecifics), apart from its application in studying the innate vocal signals of birds, is only important in analysing very simple behaviour patterns; it is of no use in analysing complex and highly specialized behavioural sequences. To put it simply: when one sees a reed-warbler first soften a plant fibre in water and then wind it around a twig, it is not necessary to isolate a developing young bird in order to learn whether the behaviour pattern is inherited, since the young bird would never be in the position to copy it from an older conspecific.

The very simple behaviour patterns which I have mentioned as being imitated from the parents or from older conspecifics seem to be restricted in occurrence among birds. However, these few patterns which are determined by 'tradition' actually include behaviour patterns which one would assume to be innately determined. Apart from this, there is also the possibility (as with vocal patterns) that active displays may be inherited *in toto* and yet be determined by tradition in a closely-related species.[27] For example, a jackdaw reared in isolation shows practically no drive to flee from human beings or other large animals and it will only exhibit instinctive avoidance of contact when a very close approach is made. In fact, the escape response is elicited in young jackdaws in the field not by threat from another animal but by the sight of alarm or actual flight in the parents. However, it suffices if young jackdaws under the guidance of their parents take part just a few times in such an escape. They then develop a fear of the object provoking flight of the leaders. In fact, I am not altogether sure that a single occurrence would not suffice. Since the young birds are thus exclusively dependent upon the example of their parents in order to flee at all, a young member of the species which is hand-reared and isolated from seeing such examples consequently appears to be excessively over-confident and easily falls prey to the first cat to come along. This also explains why young jackdaws reared by tame parents are also tame, providing a clear contrast to the majority of passerines, whose offspring scarcely differ from wild captures of the same age if reared by perfectly tame parents in an aviary. Similar behaviour is found among mammals, as in the domestic cat. Kittens of very tame mothers are quite shy on their first encounters with human beings and do not absorb any part of the fearless behaviour exhibited by the mother. In most Corvids other than the jackdaw, one also finds that the instinct to flee from human beings awakens at a particular age, even when there has been no preceding unpleasant experience with man or an example provided by an escape response from other Corvids. In the virtually non-social jays, it requires great skill to be able to submerge the awakening escape instinct beneath a conditioned association between man and food. Even the crows, which are generally quite prone to become 'socially' tame, pass through a critical phase in the period immediately following keratinization of the flight feather quills. In this phase, crows can literally become hopelessly shy overnight unless one spends a great deal of time with them. The Anatidae, like the Corvidae, also show a great variability from group to group in the part played by innate or traditional factors in the escape response. Freshly-hatched ducks, for example, will instinctively crouch or attempt to flee when the brooding apparatus is opened for the first time, whereas geese of the same age exhibit no fear at all (Heinroth). Nevertheless, the long period of guidance of the young in all Anatidae has the effect that traditions transmitted by the adult birds have a fairly large influence on the tameness (or shyness) of their offspring, as is the case with all nidicolous birds.

Another very important group of behaviour patterns which (like the escape response) can be differently determined in different birds occurs within the context of migration. Sometimes, tradition plays a major part; sometimes the behaviour is determined predominantly or exclusively by fixed action patterns. Once again, one is very much dependent upon the hand-rearing of young birds in isolation and the consequent exclusion of the example provided by older conspecifics in conducting a behavioural analysis.

Quite generally, one of the few things which the young birds really learn from the adult birds in cases where the young remain with the parents for some time after fledging is *path-conditioning*. These conditioned pathways are so rigidly followed by jackdaws that one could almost speak of 'runways'. They are just as rigidly transmitted (or, more exactly, passed on as a tradition) from one generation to the next. This was very obvious when my colony of tame jackdaws had met with a mishap and I subsequently set some young birds together with the only surviving bird, an old female, in order to provide the germ of a new free-living jackdaw colony. The young jackdaws, twenty-nine of which were introduced to the adult female one-by-one in the course of two years, took over the conditioned pathways of the surviving adult so exactly that (to name one example) they still avoid those parts of the garden where our tom-cat used to hunt. The cat is long since dead and the juvenile birds have never seen him themselves. In view of the great part played by path-conditioning in birds which cannot be regarded as migratory and which are in any case occupying their breeding territory, it is not surprising that similar conditions operate along the migratory pathways followed by the birds when migrating by day in family groups. The 'knowledge' of the path to be followed is not innate; it is passed on as a tradition. Young Greylag geese, for example, will not usually migrate if there is no leader acquainted with the migratory route. In these birds, the only apparently instinctive factor is the general drive to fly large distances. They do not possess an accompanying inherited drive to hold to a particular direction, so that in hand-reared individuals the autumnal phase of migratory restlessness only results in unsystematic and non-directional wandering within a fairly small radius. I was repeatedly able to observe with free-living wild geese which flew past our house along the Danube in the Autumn that evidently young individuals which had lost contact with the flock on foggy days would remain in the area like resident birds until they were taken along with the next flock that passed by. One such bird, which I was able to observe in the vicinity of my home for several days, continually sought to attach itself to domestic geese in the area. The visibly unoriented circling of such stray geese provides a sharp contrast to the purposive flight of migratory flocks led by adult birds. It is a conspicuous feature that a stray bird of this kind, which has naturally migrated some distance previously, should remain attached to the area where it lost contact with the leader. I am not of the opinion that the geese I observed were ill, since such stray individuals were always found after a bout of heavy fog. This apparently represents a severe hazard for geese. I have observed this effect at least three or four times, so the possibility of chance occurrence can be ruled out.

The migration of cranes seems to be dominated by tradition in a manner similar to that in geese. Leaderless young storks, according to several reports, apparently have a general directive drive to the south, but they lack the 'knowledge' of the migratory avenues which are otherwise typically followed by storks.

If relevant experiments were to be carried out with hand-reared, free-living ringed birds of different species, further interesting intermediate stages would doubtless be found between the instinctively fixed migration of nocturnally or solitarily migrating birds and migration entirely dominated by tradition (as found in geese and cranes).

Of course, this interaction of fixed action patterns and tradition is nothing more than a special case of conditioned component intercalation. The only special feature is that the conditioned component of the behaviour pattern is not determined by self-conditioning; it is derived from the example set by the parents. But even in other behavioural sequences of this kind it is possible to determine the components which are not instinctively determined, by using the method discussed previously. The conditioning factor must be excluded in a hand-reared captive bird so that the conditioned component cannot be acquired by the bird in the species-specific manner, which is adapted to agree with the other purely innate links of the behavioural chain.

However, since this breakdown in the conditioning link usually leads to disruption of the behavioural chain, a prior requirement for correct evaluation of the behaviour is previous acquaintance of the observer with the complete behavioural sequence typical of the species. For instance, if one were to observe an isolated hand-reared Red-backed Shrike wiping morsels of food along the cage-wire with its beak, one would never understand the significance of the behaviour unless the bird were to chance upon the conditioning object (i.e. a suitable thorn).[28]

Performance of a given behavioural sequence is of undoubted analytical importance when demonstrated in an isolated hand-reared bird, but it is vitally necessary to be very cautious in the interpretation of *non-appearance* of a given behaviour pattern. In the first place, as has already been demonstrated, the failure of an automatism incorporated in a variable component intercalation system (which a free-living bird would normally unfailingly acquire in a manner appropriate to the behavioural sequence in the course of its development) leads to disruption of the behavioural chain and results in the non-appearance of the remaining links. Secondly, it is quite possible that a habit acquired entirely by chance may block an instinctive behaviour pattern, as is the case with the submergence of the escape drive by object-fixated drives in birds which are aggressively, socially or sexually tame. This obscuring of drives by acquired elements of behaviour appears to occur mainly among birds of extremely high intelligence, whereas the submergence of one drive by another can occur in birds at all levels.

In spring 1931, I had the unwelcome opportunity to observe with a female raven how an acquired habit (presumably largely an effect of captivity) completely prevented the elicitation of certain ritualized patterns which are apparently vitally necessary for pair-formation. The female had learnt through bitter experience at an early age to keep the physically stronger male at beak's length. As the two birds, which had in the meantime reached the age of two years, attained reproductive condition, the male began to abandon his fraternal roughness and to court the female. Whenever he approached her, adopting the species-specific courtship posture, she would in fact adopt the female greeting (and courting) posture, crouching and fluttering her wings and tail, but she would then hop away at the last minute before the male had reached her. The male at first strode patiently after her, still maintaining the courtship posture. After the process had been repeated a number of times, however, the male's pursuit became more violent. He eventually abandoned the courtship posture, and a wild chase usually ensued. If the male caught the female, an equally wild fight would develop, which only served to increase the timidity of the female. This entire series of events was doubtless due to a conditioned inhibition in the female against allowing the male to approach, i.e. he had approached her so often with aggressive intent that she had developed her response to this to such an extent that eventually *every* approach of the male released the same response. Provided that he did not approach her directly, she would often perch close to him and preen his crown feathers. She would even engage in nest-building alongside the male. But, since the ritualized courtship of ravens requires a direct approach of the male towards the female and since my female raven was absolutely unable to tolerate such approaches, an insurmountable obstacle prevented mating. I am of the opinion that this pronounced escape response of the female to the male arose largely because the birds, prior to achieving breeding condition, had lived together in one cage. In the cage the female was of course more limited in the space available to avoid the male than she would ever have been in the field. Since even free-living ravens which are closely acquainted with one another usually maintain a peck-order (to the extent that a subordinate will avoid an approaching dominant raven at pecking distance), it is possible that pair-formation can in fact only occur when the two ravens concerned first become acquainted when they are both already in full reproductive condition. When the female (vexed by the male's continual persecution) finally flew off for good, the male turned his attentions to a one-year-old female occupying another aviary together with her sibling. Until the other female flew away, I had kept the two younger birds locked up almost continuously to avoid disturbing the pair. The younger female had thus had virtually no contact with the male for several months. I liberated the two young females, which (like many birds which mature at an advanced age) had developed marked reproductive motivation in the spring despite their immaturity, so that they could join the male. When I did this, one of them – in fact the dominant female – soon responded to the courtship displays of the male without distrust and allowed him to approach closely. This was followed by a continuation of a very interesting ritualized courtship which I had never seen with the male and the initial female. The details of this need not concern us here, however.

It has already been explained how one drive can be submerged by another, taking the example of escape responses. At this point, I should like just to mention that in aggressively tame birds the submergence effect deriving from the hypertrophied aggressive drive is usually not restricted to the escape drive. Practically all drives directed towards a living object, i.e. social and reproductive drives as well, are affected. It is almost impossible to introduce a conspecific of either sex to a cage containing an aggressively tame cage-bird, which usually originates from a group with a pronounced territorial demarcation drive and is in most cases an isolated adult male. It is very often the case with older male canaries which have been kept isolated for some time that aggressive tameness is developed to such a high degree that all other drives directed towards a living object are obliterated, making it impossible for the bird to respond other than with an attack to any stimulus emanating from a living organism. Such canaries appear almost deranged because of their continual utterance of the 'tsit-tsit-tsit' attack-call.

In all of these cases, there is no actual loss of instinctive behaviour patterns as defined in Ziegler's definition of instinct, since their histological foundations (Ziegler's 'kleronomic pathways') are presumably completely developed. It is only the *elicitation*[29] of the behaviour which fails. This may be due to suppression by another innate drive which has undergone abnormal development in captivity or by a habit formed under the unnatural conditions of captivity. It may also be due to a lack of elicitation of species-specific behaviour patterns which should follow upon the acquired component of a behaviour chain with variable component intercalation, since the adequate stimulus for the elicitation of a component pattern in a behaviour chain (as in a reflex chain) is usually only provided as a result of the performance of the preceding pattern.

Since the number and importance of individually-acquired patterns incorporated in the instinctive behaviour chains of birds doubtless increases with the general development of intelligence, most care must be taken with the most intelligent birds in interpretation of the lack of appearance of specific instinctive behaviour patterns.

In my opinion, this form of disruption of instinctive behaviour patterns is also the reason for the fact that the large ravens and parrots (which can be admirably maintained in captivity and will achieve full reproductive condition) seldom fully perform the long behaviour chains involved in reproduction and parental care. Smaller, less intelligent species reproduce successfully and present no difficulties. (I find it difficult not to believe that the larger members within groups of closely-related species are always the most intelligent.)

A second form of disruption of instinctive behaviour patterns can also be found, however. Unlike all of the previously described cases, where only the elicitation of the behaviour is suppressed, this second type of disruption is caused by defective development of the pathways underlying the instinctive patterns concerned. This form of disruption is extremely common among captive birds, such cases being much more common than cases where a full complement of species-specific innate patterns is present. It is always allied with constitutional deficiency in the birds concerned. This must therefore be treated as a necessarily pathological phenomenon, whereas the first category is to be regarded as including only those cases where completely normal animals exhibit peculiar responses to stimuli which are never encountered as such under natural conditions.

In birds, all central nervous processes seem to be far more dependent upon the physical condition of the individual than is typically the case with mammals (including ourselves). When mildly ill, a jackdaw will fail to solve a very simple detour problem which it has previously solved every evening over a period of months to reach its nocturnal roost. In mammals, on the other hand, the performance of old-established conditioned behaviour of this kind is not usually disrupted even by severe illness. One only needs to think of severely ill sledge-dogs which cannot be induced to surrender their places in the team, or of similar behaviour in horses.

I am not aware of any observations on the disruption of genuine insight behaviour in sick birds which were previously in full health. Persistently ailing birds are extremely retarded in their intelligent activities (as compared to unhampered conspecifics), but this certainly does not apply to man and it is generally atypical of higher mammals. But even within the Class Aves, the difference in intelligence between sick and healthy birds is *least evident in those species with the greatest development of intelligence.*

Strangely, innate, instinctive behaviour patterns are also less dependent upon general physical condition in intelligent birds than in the more primitive species. In general, however, instinctive patterns of birds are far more dependent upon physical constitution than variable behaviour patterns. The slightest inherited or incidentally acquired physical defect suffices to cause quite severe disruption in the performance of instinctive behaviour patterns. These disturbances always result in the absence of components of such patterns and never in the performance of novel ones;[30] there is always breakdown rather than alteration. This loss of innate behaviour patterns is a completely reversible process, since birds which were once sickly and have subsequently attained full health may suddenly exhibit previously missing patterns, just as previously healthy animals may exhibit loss of behaviour patterns, which they have already performed correctly a hundred times, on falling ill.

It would seem that a quite specific order of loss of the drives governing particular behaviour patterns is evident in each individual species when physical deterioration occurs. The only bird species which I have maintained in sufficient numbers to be able to form an opinion on this matter is the jackdaw, but it is very probable that most birds behave similarly in this respect. For instance, the instinctive behaviour pattern for the softening of hard food morsels is extremely prone to disruption in jackdaws. Such disruption even occurs in birds which one would not initially regard as physically deficient. One only becomes conscious that a physical defect is in fact present after having observed jackdaws which will perform the pattern faultlessly. I only observed the complete form of this behaviour pattern, together with full performance of the similarly inherited response for the transport of birds' eggs in fairly old and extremely healthy jackdaws. The species-typical response for the protection of a companion seized by a predator or by a human being is far less prone to breakdown. At least, I was still able to elicit this latter response in markedly unhealthy jackdaws. I did once possess three extremely weak jackdaws which I had bought privately and which completely failed to exhibit the protective response. It is important to mention in this context that all three birds did exhibit the instinctive behaviour pattern concerned after they had completed their first adult moult. In Corvids, this moult is accompanied by a quite distinct amelioration in general physical condition, which is objectively demonstrated in a sharp change in mortality rate. I have never lost a Corvid through illness after attainment of the first adult moult (subsequent losses occurred through mishaps), despite the fact that jackdaws are extremely prone to succumb to infectious diseases prior to this moult. The jackdaw's instinctive behaviour pattern for concealment of food is also very resistant to breakdown. Even sick and dying birds which have lost the instinctive behaviour patterns for preening the feathers still exhibit indications of the food-concealment pattern.

I am convinced that an experiment conducted to investigate this loss of fixed action patterns through disturbances in general physical condition would provide proof of regularities far greater than can be derived from my chance observations.

In birds of high intelligence, the dependence of instinctive behaviour patterns upon physical constitution is far less rigid. In any case, I observed evidently ailing ravens and crows which were correctly performing certain Corvid instinctive patterns (e.g. food-softening) which would have long been disrupted in jackdaws of similar constitution. In my paper on jackdaws, I expressed the opinion that it is the highly-specialized instinctive behaviour patterns which are most prone to disruption. It is self-evident that the delicate and complex instinctive behaviour patterns involved in the behavioural chains of reproduction are particularly affected, as is the fact that this is the reason for the low proportion of captive birds which actually succeed in producing fledged young among those which actually *begin* to breed in captivity.

Although physically inferior specimens of intelligent species are less likely to exhibit serious disruption of instinctive behaviour patterns than those from species with a primitive level of intelligence, the latter will far more reliably provide specimens exhibiting complete development of the full complement of behaviour patterns *provided that* the specimens are in immaculate physical condition. In such species, there is a far lower probability that a behaviour chain will be disrupted by the absence of development of an acquired component or that some individually-acquired habit will intervene (as in the case of my female raven). Since this particularly applies to the long behavioural chains involved in reproduction, we can also expect a far greater reliability in elicitation, such that elicitation and response will interlock in a well-co-ordinated series of chain-reflexes like the cogs in a well-oiled clockwork mechanism. An experienced animal-keeper soon notices when the responses have been performed in the correct manner and when not. However, the actual reasons for the failure of a system to function are by no means easy to analyse.

At the risk of introducing a large number of self-evident facts, I feel that it is necessary to analyse this particular problem in some detail in order to demonstrate how far we are justified in drawing conclusions from the non-appearance of a given behaviour pattern in a captive bird.

The second condition which must be fulfilled by a behaviour pattern in order that it be regarded as instinctive is that all individuals of the species must respond in the same way. This identifying feature of instinctive behaviour patterns is relatively seldom decisive with birds. Even when only one specimen is available for observation, there is rarely any doubt as to whether or not a specific behaviour pattern should be regarded as instinctive. Other criteria are available to aid the reasonably practised observer in reaching a reliable decision long before the opportunity arises for him to confirm his observations with other birds of the same species. Nevertheless, a control of this kind derived from observations of a maximum possible number of individuals of the given species is extremely desirable in the interests of scientific criteria of reliability. But heavy dependence rests upon the observation of the given behaviour pattern in *all* members of the species when interpreting that given behaviour pattern in an extremely intelligent bird or in a mammal (such as man himself). In these cases, as will be shown later, many other identifying features of instinctive behaviour patterns are inadequate and even the test of rearing in isolation is often inapplicable.

Among the birds which I have studied closely with regard to the performance of instinctive behaviour patterns, the only species which repeatedly posed the question as to how other members of the species would behave in a particular situation was a large Greater Yellow-crested Cockatoo. This happens to be one of those species which cannot be maintained within a large free-flying group in a fairly densely-populated area. One of these birds on its own will make great demands on the patience and finances of its owner because of its unlimited destructive tendencies. With large parrots it is unavoidably necessary to allow them to fly free in order to gain an insight into their species-specific behaviour patterns. Whatever the size of the aviary which one might provide for commercially-acquired cage parrots, they will still retain a large proportion of those symptoms of captivity which I collectively denote as 'cage imbecility'. Above all, such parrots always retain some part of the psychological inhibition against taking off, which they acquire in a small cage. They continue to move largely by climbing and they fly little, i.e. changes in position which an uninhibited bird would achieve by flight are performed as far as physically possible by means of climbing. This all gives such a different character to the general behaviour of the bird that an erroneous concept of the locomotory drive and tempo typical of the free-living bird is produced. If, on the other hand, such a bird is allowed to liberate itself from its psychological cage inhibitions in a long period of free flight (months are necessary), subsequent confinement is very bad for the bird, whatever the size of the aviary may be. The parrot will either make every attempt to escape or (after recognizing the futility of such efforts) sit around in an even more listless fashion than a bird which has been taken direct from a tiny cage. In a word, the behaviour of such a bird towards its cage differs from that of a freshly-trapped specimen only in the absence of occasional frenzies of fright. For all these reasons, it is only possible to learn something worthwhile about the large parrots when they can be kept under free-flying conditions. However, one would have to maintain a large number, and one on its own can easily bring its owner to poverty or legal prosecution in a short space of time.

Although I have kept the cockatoo mentioned above (an extremely tame, completely healthy, high-spirited and mobile bird) under free-flying conditions for some time (a number of years), I still know practically nothing about the innate behavioural repertoire ('innate system of actions' – Jennings) of this species, apart from the easily-recognizable ritualized patterns. Because of the multiplicity of motor patterns exhibited by this bird, the multitude of uses found for its beak and the corresponding versatility in intelligent activities, all of the behaviour patterns of the cockatoo in fact had the appearance of genuine insight behaviour. In addition, self-conditioning ('learning by heart'), which plays such a large part in other birds, is apparently little evident in the cockatoo's behaviour. If the cockatoo was at all able to solve a given problem, the performance of the solution would be no smoother even after frequent repetition of the initial pattern. If I wanted to condition this bird to do something, he would either abruptly exhibit the requisite mental association after a small number of trials, or fail completely. The only features of conditioning present in the associations thus formed were the persistence with which these associations were conserved and the fact that they were far less open to reversal than to formation.

Since all of the bird's activities thus had the deceptive appearance of insight behaviour, even when they could be fairly reliably recognized as conditioned activities, it was reasonable to assume that this would also apply to the instinctive behaviour patterns. When observing, I was never certain of the type of behaviour pattern involved. To take one of many examples: when harvesting cherries or any other fruit, the cockatoo always behaved in a peculiar manner. He would climb out on to the thinner branches of the tree as far as his weight would comfortably allow and then bite off a twig with the fruit attached. Indeed, he would always quite purposively take the richest bunches first. The twig was then hauled in by alternate grasping with the beak and one foot, so that the cockatoo could reach the otherwise inaccessible fruit. This would actually be a remarkable achievement if it were a novel solution to the problem confronting the bird. The reader should try to visualize just what is required for the optical recognition of the relationship between the cherries and the twig and for the actual solution of pulling in the cherries with the twig. According to Köhler, such a problem provides some difficulty even for chimpanzees. Further, the bird had to recognize the inhibitory connection between the twig and the tree and to 'visualize' the twig free from the tree before purposively rupturing this connection. This requires a mental ability which one would not normally ascribe to a bird. If I were now to see a large number of cockatoos in their natural environment, all stereotypically harvesting fruit in the manner described, I would heave a sigh of relief and say: 'so I was right after all'. At the moment, however, I have no grounds to assume that the cockatoo is incapable of such a feat of intelligence when comparing this behaviour with other, quite considerable feats which must definitely be regarded as intelligent achievements. It is remarkable that in interpreting the psychological characteristics of the large parrots one is continually confronted with difficulties similar to those otherwise encountered only in the higher mammals.

In these advanced animals, one criterion of fixed action patterns which is very reliable in most other birds and in all lower mammals cannot be applied:

Whenever we observe an animal performing a behaviour pattern whose biological function is immediately obvious to us, it is often possible to identify instinctive control of the pattern from the degree of incongruity between the intelligent activities of the animal as observed in other situations and the abilities which would be required to develop the pattern by insight. Reference was made to incongruities of this type by Cuvier some time ago. When a weaver bird beginning nest-construction ties a plant fibre with a knot which is so complex that a chimpanzee cannot undo it (let alone make a similar knot itself), the instinctive character of the bird's actions is immediately obvious. The same applies when a jackdaw carrying an egg solves the problem of keeping the fluid in the vessel by keeping the opening at the top. In lower animals, this incongruity is naturally even more distinct. A hermit crab would really have to be at least as intelligent as an able monkey to carry out the extremely 'intelligent' transfer from one mollusc-shell to another on a basis of insight.

We have already seen from the observations on the cockatoo that this particular character is of no use in the recognition of instinctive behaviour patterns in very intelligent birds. It was also indicated that this is especially true of mammals. For instance, I still do not know whether a behaviour pattern which I noticed in dogs some years ago is instinctive or based on insight. Strangely enough, this pattern is one concerned with the treatment of stolen eggs, just as in the jackdaws mentioned previously. The dog takes the egg very carefully into its mouth, apparently such that it rests on the tongue and is not pressed forcibly against the teeth. The egg is then carried, without being broken, to a place where there is a flat, stony substrate, so that no part of the contents is lost when the egg is bitten. It is quite conceivable that an intelligent dog might hit upon the idea of seeking out a substrate which can be licked clean after breakage of the egg, but the fact that many dogs behave similarly in this respect makes it reasonable to assume that this is a fixed action pattern. The observation that not all dogs perform this pattern proves nothing in view of the long period of domestication of the dog.

A fourth feature of fixed, instinctive behaviour patterns arises when the behavioural sequences are performed incompletely, so that their functions are not fulfilled. The complete lack of purposive control is then obvious. The malfunctions which thus arise are in many cases the decisive characters for interpretation in the practical identification of instinctive behaviour patterns in birds.

Such imperfections in the development of innately-determined behaviour chains may appear in both juvenile birds and adult birds which are re-entering specific seasonal drive-states (though they are less pronounced in the latter). They may also occur as deficiencies persisting for life in the behaviour chains of captive adult birds.

It has already been shown that instinctive behaviour patterns directed towards particular objects which are not themselves recognized on some inherited basis are at first performed in the absence of these objects. But even in animals where the recognition of the object appropriate to a given object-fixated instinctive pattern is completely innate, i.e. where the behaviour pattern is immediately elicited on the first exposure to the object without prior experience, the pattern may be performed without the object (I use the term *in vacuo*) or with a substitute object, as is the case with prey-killing in Corvids. However, in birds of advanced intelligence and in many mammals it is possible that a purposive element is present to the extent that the behavioural sequences concerned are (so to speak) run-in by repetition, so that they are more reliably performed in a 'serious' context. This 'pre-tuning' (Groos) is generally referred to as *play* in young animals, but it must be remembered that most forms of play in young children are conscious activities and thus of an entirely different nature. I only regard playing with dolls in young girls and playful manipulation of various objects as seen in many small children as genuine 'pretuning' effects. Many young monkeys exhibit the same type of manipulation without this giving rise to increased manual skill later on. In fact, this 'mechanistic interest' gradually disappears with advancing age. None of the adult's abilities correspond to the play of the young. Frequently, one almost has the impression that the capabilities which only appear in a vague form in the young are lost at a later stage.[31]

In young birds, on the other hand, it is almost always possible to recognize from 'pre-tuning' what function the behaviour pattern concerned will have in the 'serious' context. Almost all predatory animals develop the behaviour patterns involved in predation by interaction with a substitute object. A typical example is provided by the young kitten playing with a ball of wool. My young kestrels began to 'kill' suitable objects (particularly soft materials) before they were even able to fly. They would often follow the edge of a thick carpet, quite as if there were a mouse running alongside. It is interesting that this linear structure produced the impression of movement, just as it does with us ('a line running between two points'). The kestrels followed this running line as if trying to catch it. A little later, a very tame Common Buzzard did exactly the same thing.

In observing the prey-catching instinctive patterns of predatory birds, one does not actually form the impression that the prior period of practice without the object really has a biological function. This is further supported by Heinroth's observation of a young hawk which seized a pheasant in flight on its first attempt to catch prey. The bird did this inside a room, so the conditions were rendered even more difficult. A similar sudden appearance of the prey-killing pattern appears to occur among owls; the only representative which I know well being the Tawny Owl. None of the individual isolated birds which I reared in different years exhibited any playful 'pre-tuning' of the prey-killing pattern. Nevertheless, one which had just left the nest killed and ate an adult hooded crow – a relatively large bird well able to defend itself.

This abrupt appearance of a behavioural sequence in its complete form is always a sign that no self-conditioning components are involved. The presence of acquired links in otherwise innate behaviour chains is conspicuous because of the initially incomplete (and therefore nonfunctional) performance of these chains. In such cases, the biological necessity for 'pre-tuning' is obvious and the behaviour has a far more distinct play-like character. Since this is naturally most pronounced in the most intelligent birds, this provides us with one of the few characters from which the innate behaviour patterns of birds of advanced intelligence can be more distinctly recognized than those of primitive birds. For this reason, the behaviour of a juvenile large parrot species would be very informative as a subject of study.

In the raven, many behaviour patterns which are performed with a substitute object, or in the complete absence of the object, exhibit a playful character which is just as pronounced as that of mammals (in particular that of young carnivores). It has already been stated that the instinctive pattern for food-concealment is initially performed entirely in play with all sorts of objects. I should like to provide one further example of a behaviour pattern in the raven which first appears in play, since it has an interesting parallel among the Carnivora. As soon as ravens have fledged, they can be seen to perform a quite distinct locomotor pattern with objects of medium size. The raven takes one step away from the object concerned, as if afraid of it and then grabs at the object extremely rapidly with one foot. (In fact, I usually saw the pattern performed with objects which really did frighten the birds.) The beak is *not* extended: on the contrary, the bird draws the head back between its shoulders as far as possible, as if attempting to keep a maximum distance between its face and the seized object. Frequently, the bird will not even dare to make a definite grab at the object. At such times, this behaviour is convincingly reminiscent of that exhibited by Carnivores towards a struggling prey which will not (or cannot) flee, as seen in the play of young dogs or cats. In the typical spectacle of the young dog with a wasp, the retraction of the head and the simultaneous swing of the paw is almost exactly analogous to the described pattern in the raven.

I saw this behaviour for the first time in my ravens as they followed me to a tennis-court and treated tennis-balls in the manner described. The balls, which were entirely novel objects, were frightening to the ravens and yet the birds were obviously stimulated by their rolling motion. I immediately assumed this to be the 'instinctive behaviour pattern for the conquest of resisting prey' and this later proved to be completely correct.

The ravens performed this behaviour in more complete form later on. I then noticed frequently that a raven would play with a piece of wood, cloth, or some other object, suddenly grasp it with one foot and then make a rapid strike with the other foot as well. Of course, the raven would then fall over on one side or even onto its back; but it would never let go. The raven would merely extend both legs as far as possible, as if attempting to hold the stricken object as far away from its body as possible. The head was always held right between the shoulders and the raven would often lie immobile and maintain this posture for some time. This immobility following seizure of the prey can also be seen in many other predatory animals, particularly in those which have efficient prehensile organs (e.g. in pike, grass-snakes and all predatory birds). It presumably serves the function of allowing the prey to struggle until it is forced to stay still from sheer exhaustion. The predator can then alter its grip in order to get a better hold, or finish off the prey straight away.

I have myself been able to observe the same two-footed seizure pattern and the apparently intentional toppling pattern in one specific predatory bird – the goshawk. I only ever saw this bird exhibit the pattern in a serious context, when it had been given a medium-sized animal or a particularly large piece of heart-meat. In such a case, the pattern is so self-evident in appearance that one is inclined to think that any similarly equipped bird would be driven to hit upon this method of overpowering its prey of its own accord.[32] The entire behavioural sequence would not necessarily have to be innately determined; but this is in fact likely to be the case, judging from my observations on ravens.

Behavioural sequences do exist, however, in which the links are even more self-evidently interdependent than in the case described above, and where nobody would form the impression that the sequence were innately determined *in toto* until seeing the bird perform the behaviour patterns without an object. For instance, I once had a hand-reared starling which, although it had never trapped a fly in its whole life, performed the entire fly-catching behavioural sequence *without* a fly – i.e. *in vacuo*. The starling behaved as follows: It would fly up to an elevated look-out position (usually the head of a bronze statue in our living-room), perch there and gaze upwards continuously as if searching the sky for flying insects. Suddenly, the bird's entire behaviour would indicate that it had spotted an insect. The starling would extend its body, flatten its feathers, aim upwards, take off, snap at something, return to its perch and finally perform swallowing motions. The entire process was so amazingly realistic, particularly with regard to the bird's convincing behaviour before take-off, that I always took care to see whether small flying insects which I had previously overlooked were in fact present. But there were really no insects to be seen. When observing such behaviour, one is immediately conscious of the question as to what subjective phenomena are experienced by the animal, since this behaviour is so reminiscent of that of certain human psychopaths who experience hallucinations.[33]

Innate control is not only indicated by *initial* incomplete performance of a behaviour pattern, as caused either by the awakening of the controlling drive in a young individual or the yearly resuscitation of drives in an adult bird. *Breakdown* in innate behaviour patterns in captive birds, as described previously, can also produce inconsistencies which render the behaviour of the birds so irrational that it becomes obvious that purposive control is not operative and that the behaviour is consequently innately controlled. This is particularly clear when we can observe and understand the function of the behaviour (which the bird itself does not perceive).

The complex behaviour patterns involved in reproduction are, as has already been mentioned, particularly prone to disruption. These patterns are consequently far more often to be seen in their disrupted form in our zoos than in their entire form, which leads to the ultimate goal. It would be pointless to attempt to list here all those birds which have been observed to exhibit such disruption of partially-performed behaviour chains (e.g. suspension of nest-building). Think how few of the birds which fly around in the aviaries of various zoos with nest-material in their beaks actually go on to breed successfully! It must, of course, be remembered that the behaviour patterns involved in brooding are not simple chain-reflexes. They are always assembled into so-called instinct-interlocking mechanisms (Alverdes), where each instinctive behaviour pattern in the one partner is elicited by the prior performance of a given pattern by the other partner. This can naturally give rise to a third source of error.

Breakdown in the reproductive behaviour following a promising beginning can in many cases be simply caused by unnatural surroundings or by development of one of the partners under conditions atypical for the species. In many cases, however, the failure may depend upon the breakdown of a single instinctive behaviour pattern or failure of a single inhibition, which actually amounts to the same thing in effect. As already stated, such breakdowns are to be regarded as pathological, but they can nevertheless occur in birds which are healthy enough to produce fertilized eggs. A typical case of inhibitory loss is exhibited by a pair of American Bitterns which breed every year in the Schönbrunn Zoological Garden and eat their young straight after hatching. The response represented by innate behaviour patterns which are elicited by the first sight of the young seems to be a point which is generally rather prone to be disrupted. Predatory animals in particular – from Völkle's Golden Eagles to a friend's dachshund – seem particularly liable to exhibit breakdown of the inhibition against simply devouring their young without differentiating them from other small animals.

Even though the described imperfections and breakdowns in fixed action patterns are to be regarded as pathological from a medical standpoint, one can nevertheless learn much from them and draw conclusions about the nature and constitution of the normal instinctive behaviour patterns of birds.

The last, and virtually 'classical', characteristic of instinctive behaviour patterns which I intend to describe is the invariable, fixed quality of such patterns. Even when a behavioural chain is interrupted or exhibits breakdown of one or more links, or even when the entire sequence begins at a point other than the normal initiating pattern, it is only possible to find *incomplete* sequences and never *novel* variations.

Fixed action patterns provide characteristics of species just as stable as (or even more stable than) any chosen morphological character. Heinroth has demonstrated adequately enough that they are therefore quite likely to be of taxonomic significance in his book: *Beiträge zur Biologie insbesondere Psychologie und Ethologie der Anatiden* ('Contributions to the Biology of the Anatidae, with special reference to their psychological and ethological features').

The marked constancy in form of inherited innate behaviour patterns[34] and their conspicuous invariability under varied environmental conditions (even under those conditions which bring about drastic morphological alterations) can be seen in many domestic birds. These features of their instinctive behaviour patterns (particularly with ritualized displays) often give witness to their relationship to a specific free-living form far more distinctly than morphological features.

One of the most convincing demonstrations of the great evolutionary age of many instinctive behaviour patterns is probably the fact that interspecific hybrids often exhibit specific instinctive patterns which are not reminiscent of either of the parental species and so do not form some type intermediate between the two parental types. These patterns arise in a quite different manner, by regression to a more primitive, phylogenetically older form. This effect is an old-established feature of interspecific hybrids. The following observation is drawn from Heinroth's *Vögel Mitteleuropas:* Two hybrids from a cross between a Shelduck and a Nile Goose exhibited neck-dipping and post-copulatory play corresponding to that of the Ruddy Shelduck and therefore resembling the displays of many other Anatids. The hybrids exhibited neither the diving of the Shelduck (in which both partners dive during courtship such that the male mounts the female under the water to re-emerge in the copulatory posture) nor the extremely characteristic pattern of the Nile Goose where the two partners float opposite one another in shallow water and conduct a lengthy pre-copulatory palaver. The interaction between the characteristics of the Shelduck and the Nile Goose had thus resulted in the emergence of an apparently ancestral form of behaviour which is in fact very common in Anatidae and is specifically to be found in the Ruddy Shelduck.

In my opinion, the ritualized behaviour patterns exceed the other types of instinctive behaviour pattern and many morphological characters in taxonomic value for the simple reason that it is utterly improbable that these purely 'conventional' behaviour patterns, whose development is not directly influenced by the animal's environment, might arise in similar form in two different species by parallel evolution.[35] Quite apart from this, such ritualized behaviour patterns can survive over great lengths of time just because they are not subject to the effects of environmental change.

The other instinctive behaviour patterns of a given bird species are so closely and so directly related to the environment and fit in so well with its demands that the fixity of such patterns is likely to escape conscious recognition in observations of the free-living animal. This rigid character immediately becomes obvious, however, when the instinctive behaviour patterns are elicited under conditions departing from those in the natural environment, and which therefore do not offer the normal prerequisite factors. Even in higher mammals, it is often easy to recognize instinctive behaviour patterns when elicited under abnormal conditions. Who among us has never seen how even a dog will attempt to hide a bone on a wooden floor by performing shovelling motions with the snout 'to cover the bone with earth'? To the uninitiated, the corresponding behaviour in a raven would appear to be incomparably more intelligent, for the simple reason that the performance of the instinctive pattern is not dependent upon the presence of a particular type of material – any suitable object is acceptable for covering the prey. But all we have to do is to deprive the raven of all suitable objects to see how irrational its behaviour will then seem. The raven will, for example, 'cover' the prey with a piece of paper so small that it will not only fail to conceal the prey but actually make the 'covered' object more conspicuous!

Given time, intelligent animals can adapt to the unnatural conditions presented by the human environment to the extent that they will learn where certain instinctive behaviour patterns are *not* appropriate. One will never see a reasonably old, intelligent dog attempting to bury a bone at a spot where no earth is available. But the dog will never invent a novel pattern, such as that of covering the object by bringing up some other object in its teeth. So many incidental factors are taken into consideration, although no change actually takes place in the behaviour pattern itself, that one does not actually form the impression that negative conditioning has eliminated only those individual inappropriate behaviour patterns which do not fulfil a purpose. In some cases, one is inclined to assume that the animal maintains purposive control over its actions and that it possesses a certain insight into the fundamental operation of its own instinctive behaviour patterns. Owing to this secondary insight which the animal has into its own instinctive behaviour patterns, many identifying criteria which are otherwise recognizable will of course become virtually obliterated. For example, if I were only to have an old, experienced raven for the analysis of the concealment pattern of Corvids, I would only be left with the extremely unreliable criterion that the raven is presumably scarcely able to invent such a pattern. After one has taken away the prey just a few times an old raven will subsequently conceal its prey at sites inaccessible to human beings. The raven avoids the gaze of its companions when concealing something and for this purpose seeks out only those sites which will not normally be visited by its companions. Each of my ravens had its own particular concealment area which it alone would use. When such a raven intends to visit the area concerned with a morsel in its crop, it exhibits a quite specific pattern of slipping off unnoticed so that it will not be followed. All of this is most definitely not innate; a young raven exhibits no such behaviour and jackdaws will conceal objects in full view of their companions throughout their lives even when repeatedly encountering bad experiences. My assumption that this represents a genuine case of insight behaviour in the raven is supported by the very fact that the Corvids with a more primitive level of intelligence and a correspondingly greater innate behavioural stability (assuming perfect physical condition) show none of these purposive refinements of the concealment response. I imagine that many behaviour patterns of mammals and of man himself, which are generally regarded as pure insight activities, represent originally instinctive behaviour patterns of this type, secondarily controlled by insight. Of course, this would be incomparably greater in extent than would ever be the case with a bird. Since this process eliminates the characteristic rigidity of instinctive behaviour patterns, the only identifying feature which remains for the recognition of originally innately-determined elements of behaviour (especially when the operative drive first awakens when the intelligent faculties are fully developed) is the fact that all normal individuals of the species behave similarly. Since the concept of the normal human being does not, strictly speaking, exist, even this character cannot be applied for human beings.

The rigidity of an instinctive behaviour pattern is also quite neatly demonstrated when a behaviour pattern developed in an exactly species-specific manner in a bird in perfect condition is elicited by a stimulus *other than that* which should elicit the response (i.e. by a stimulus other than the normally adequate one). Sometimes a very small number of characters of a stimulus coinciding with characters of the stimulus for which the instinctive pattern has actually been adapted in the phylogeny of the species will suffice to elicit the innate response.[36] For instance, the entire response for the protection of a companion was elicited in its specific form in my jackdaws on one occasion when they saw me carrying a wet, black swimming costume in my hand. A newly caught wild adult Whooper Swan, whose innate responses can safely be regarded as free from any re-direction towards human beings, reacted similarly by responding to a man lying stretched out in the grass just as it would have done to the similarly stretched mating posture of a female of the species (see Heinroth – *Beiträge*). In many birds, as with my kestrels, the sight of any smooth surface will just as reliably elicit bathing movements as a water surface itself. A behaviour pattern of the Blackcap which I observed years ago provides a similar example, except that the 'erroneously' elicited reflex was in this case not so irrelevant to the survival of the species. A pair of Blackcaps, taken from the same nest and reared a year previously, bred in a small box-tree in my aviary. The offspring developed well for the first few days, but I subsequently found one of them dead on the ground some distance from the nest. Shortly afterwards, I unfortunately found the other three in a similar state. At the time, I did not know what to make of the incident, but according to Heinroth (pers. comm.) it is based upon the elicitation in the parents of the instinctive behaviour pattern developed for dealing with dead young. Because of the easy availability of food, the young became overfed and did not gape for long periods. Apparently, this never occurs in the wild and so the satiated offspring in captivity will elicit a parental instinctive behaviour pattern which should be elicited by dead young.

Summary

In the above account, I have attempted to give a brief review of the characteristics which I employ in the recognition of instinctive behaviour patterns in birds. I have already mentioned that a reasonably experienced observer of animals responds more subconsciously than consciously to these characteristics. Since introspection is one of the most difficult forms of observation, I have probably overlooked many features which might determine my own judgement in many cases.

I should like to give a brief summary of the characteristics outlined above. We are justified in considering that a pattern is instinctive:

1. If the behaviour pattern is performed by a bird hand-reared in isolation. This is particularly important when the pattern could conceivably lie within the psychologically possible range of imitation open to the bird, as is the case with vocalizations or path-conditioning (including migratory pathways). In cases of behavioural sequences of reasonable complexity, it must be remembered that the bird is far more likely to invent rather than imitate the behaviour concerned.

If, on the other hand, a particular behaviour pattern is *not* performed by an isolated, hand-reared bird (e.g. in a case where the pattern may be expected to appear on the basis of field observations) we must be extremely careful in drawing our conclusions. In the first place, any physical defect (however small) is accompanied by pronounced disruption of the instinctive behavioural system. In some species, such physical defects cannot be avoided in captivity. This dependence of instinctive behaviour patterns upon physical constitution appears to be more pronounced in animals at a primitive level of intelligence than in animals of advanced intelligence. On the other hand, there is the fact that animals of advanced intelligence, which are greatly influenced by personal experience, frequently exhibit disruption of species-specific behaviour patterns due to the unnatural features of their previous life in captivity.

2. If all individual animals within a species behave similarly in a given respect. With very intelligent birds and other advanced animals, where other characteristics of instinctive behaviour patterns do not apply, this is often the only reasonably reliable character.

3. If there is an obvious incongruity between the normal intelligent capacities generally observed in the animal concerned and those which would have to be present for the completion of the given behaviour pattern. This characteristic is lacking in some instinctive behaviour patterns of highly-intelligent birds.

4. If imperfections appear in a behavioural sequence which distinctly indicate that the animal itself is not conscious of the purpose of the behaviour pattern concerned. In young individuals, instinctive behaviour patterns related to a particular object at first develop in the absence of the object and become attached to an appropriate object through secondary effects of experience. In animals of a high level of intelligence, whose behavioural chains quite frequently incorporate links which are modifiable by personal experience, the performance by young animals of sequences not achieving their actual goal is likely to be of biological significance as preparatory practice for the animal.

In captive birds, an instinctive behaviour pattern is often characterized by pathological breakdown of components in just the same way as a behaviour pattern of a young animal is characterized by incompletion.

5. If the characteristic rigidity of the instinctive behaviour pattern and its resistance to external influences is clearly demonstrated by unaltered performance of the pattern under external conditions far removed from the natural conditions under which the pattern was evolved. In such cases, the behaviour pattern may be so irrational in appearance that the complete and precise sequence of its component parts will provide a clear demonstration of its chain-reflex nature.[37]

A second case where the rigidity of the instinctive behaviour pattern is very clearly shown is that where the behaviour pattern is elicited 'erroneously' by a stimulus which does not correspond properly with that for which the pattern was adapted under natural conditions.

Animals with the most advanced intelligence capacities in many cases learn to avoid these pointless or misfiring performances in a manner which gives the impression that the animal possesses a certain insight into the purpose of its actions. The rigidity of the behaviour pattern is not altered by this modifying effect; the behaviour pattern simply ceases to appear in the described manner.

Even though the above account probably represents an extremely incomplete survey of the characteristics available for the recognition of instinctive behaviour patterns in birds, I nevertheless hope that I have shown that these instinctive patterns are to some extent open to scientific investigation, especially since we are not forced to rely on any single one of the above criteria in any actual case. On the psychological side, one conclusion which can be drawn is the following: Every experimenter should at least to some extent investigate the species-specific system of instinctive behaviour patterns of the animal species which he is studying, since many otherwise excellent investigations of animal psychology fall down because of ignorance of this behavioural repertoire.

Companions as factors in the bird's environment

The conspecific as the eliciting factor for social behaviour patterns (1935)

DEDICATED TO JAKOB VON UEXKÜLL (IN COMMEMORATION OF HIS SEVENTIETH BIRTHDAY)

I Introduction

The concept of an *object* in our environment arises from a process of compilation of stimuli emanating from one given thing, by means of which we relate the assembled stimuli to that particular source of stimulation (the 'thing'). This also involves projection of the perceived stimuli outwards into the space surrounding us, in order to localize the object. The image of the sun formed on the retina of the human eye by the lens is not perceived simply as 'light' in the same way that we perceive the image of the sun projected onto our skin with a magnifying glass as 'heat'. We in fact *see* the sun up in the sky, a long way from our bodies. This localization effect is a product of our perceptive mechanism and not an achievement dependent upon some conscious process.

Thus, in the realization of the presence of objects in our environment, we are dependent upon those senses whose perceptive mechanisms permit us to localize things in the surrounding *space*. It is only in this way that we are able to recognize the inherent spatial correspondence of the individual stimuli, which defines the concrete unity of the object and which provides the basis for Uexküll's simple definition of an object: 'An object is that which moves as a unitary whole.'

In human beings, the senses which largely localize stimuli are touch and sight. We can therefore talk of tactile space and optical space. Even with our sense of hearing, localization is far less accurate and we seldom speak of 'auditory space' with regard to human beings. I do not propose to discuss whether this is true of all animals. Owls can localize auditory stimuli at least as easily as they can optical stimuli, whilst bats are even more accurate at auditory localization.[39] It is quite possible that these animals possess an auditory space just as well defined as our optical space.

The *compilation* of stimuli perceived by the different senses as emanating from one object, to permit recognition of that object as a unitary whole, is an achievement which must be correlated with the localization of these stimuli to a common source. We can best observe the development of the object from the sum of the stimulus-data when the process occurs slowly for some reason. For example, when one is awakening from narcosis or particularly deep sleep, even well-known things may not be recognized as objects. We see lights and hear sounds, but it takes a certain amount of time (sufficient for self-observation) until one has localized the source of these stimuli and allowed them to coagulate into the concrete object.

A particular achievement of the central nervous system is evident from the fact that we can perceive an object over a wide range as being of uniform size, form and colour although it may be seen from different perspectives, from differing distances and at differing light intensities. This effect operates over a wide range of differences, despite the variation in the stimuli actually impinging upon our sense-organs. A man a little distance away is seen as being no smaller than another man close to us, despite the fact that the image of the first man formed on the retina is smaller than that formed by the second. We can perceive the right-angles of a picture hanging on a wall even when the picture is viewed obliquely, such that it forms a non-quadratic parallelogram on the retina. A sheet of white paper is recognized as being white even in dimmed light, where the quantity of light actually reflected by the paper is less than that reflected in bright sunlight by a piece of paper perceived by us as black. The details of this high degree of constancy maintained in the perception of objects in our environment are the concern of the perceptual psychologists. It suffices here to recognize that even in higher animals this constancy in the properties of objects shows far less stability in the face of alterations of the stimuli presented than is the case with human beings.

Since the variability in environments inhabited by individual animal types is accompanied by great variability in the degree of variation in the perceptual conditions, it is understandable that the stability of animal perception of objects in the face of changes is extremely variable from species to species. Frequently, the biological significance of such differences is immediately obvious. For instance, experiments performed by Bingham and Coburn demonstrated that chickens conditioned to recognize specific stimulus patterns did not recognize these patterns when they were presented upside down, whereas crows were not affected at all by spatial transposition of the conditioning objects within the environment. For a flying animal such as the crow, which frequently searches for food by circling on the wing, it is of course vitally necessary that objects which are perceived on the ground below should be recognized independently of the momentary direction of the crow's flight and the consequent orientation of the image on the retina.

For the human being, who attempts to dominate his environment and its properties by means of insight into causal relationships, correct integration of the stimuli emanating from things in the environment, in order to permit perception of objects in that environment, is the basis of all knowledge and vitally necessary.

However, for animals (particularly lowly animals), which are largely dovetailed into their environments by means of inherited instinctive behaviour patterns, recognition of objects within the environment is not an unconditional biological necessity. Insight plays absolutely no part in their innate responses to environmental stimuli. It is sufficient that an instinctively determined response, which is developed on the basis of survival value of responses relative to a particular thing, should be elicited by *one* of the stimuli emanating from this thing (so long as this stimulus characterizes that thing so clearly that erroneous elicitation of the response by another thing emitting similar stimuli does not reach a frequency sufficient to adversely influence survival value). In order to reduce the probability that the latter effect will occur, it is common that a number of stimuli are combined to provide a pattern of stimulus-data. Such a combination is nevertheless quite simple in nature and evokes a response from an 'innate releasing mechanism'. The structure of such an innate releasing mechanism must possess a certain minimum of general improbability for the same general reason that the shape of a key is made to represent a generally improbable pattern.

There is considerable difference between these *innate* mechanisms and the *acquired* mechanisms which represent the eliciting factor in conditioned reflexes and other conditioned responses. Whereas the former appear to be constructed as *simply as possible* right from the outset, the latter seem to be rendered as *complex as possible* in all animals.

If the animal is not forced in the course of conditioning experiments to select certain characters out of the many presented (as can be achieved by repeatedly altering all other characters), the animal will in general become conditioned to the totality, the 'complex quality' of all of the stimuli presented. Uexküll writes: 'Take a dog which is trained to follow the command "on the chair" with a particular chair. It can occur that the dog will not recognize all seats as chairs, *but will only recognize that particular chair in that given position.*' Effects which 'can occur' in an unintentional manner in purposive human training of the highest mammals occur as the rule in the *self-conditioning* processes of animals with lesser developed intelligence. An animal which conditions itself to a combination of characters which has led to success on one or more occasions is of course (without insight into the causal relationships) unable to determine which of the perceived characters are irrelevant trimmings and which are causally related to the success achieved. It is thus presumably biologically significant that the animal should combine *all* characters to a complex quality and select the latter as the eliciting factor for the conditioned response. The animal repeats its actions blindly, without differentiating relevant and irrelevant stimuli, producing the behaviour which previously led to success only in situations which resemble the foregoing situation in every detail. The animal can of course be conditioned to *one* character alone, e.g. the signal 'triangular', in cases where this one character is maintained constant whilst all others are varied. However, *wherever* selection of relevant stimuli occurs *without* such pressure, I am convinced that we are concerned with a first step towards 'insight into causal relationships'.[40]

Innate releasing mechanisms, in contrast to these individually conditioned mechanisms, are incorporated within the organism from the very start as ready-formed, species-specific functional layouts. It is predetermined at the outset which characters are important, and it is therefore a fully economic principle if the releasing mechanisms are linked to as few characters as possible. It is sufficient for the sea-urchin *Sphaerechinus* to respond to one single, specific chemical stimulus emanating from its chief predatory enemy, the starfish *Asterias*, as the eliciting factor for its extremely specialized fleeing and defensive response. Such elicitation of a highly-complex motor response (adapted to a quite specific biological process) by a relatively simple combination of stimuli is characteristic of the great majority of innate responses.

One would at first expect that higher animals, to which we must attribute the concrete perception of objects in their environment on grounds of general behaviour, would also perceive the stimulatory sources related to all their instinctive behaviour patterns as objects. One is particularly prone to make this supposition in cases where a conspecific represents the object in a particular response. Strange as it may seem, in many cases a cohesive identification of the conspecific as one object linking several behavioural complexes cannot be demonstrated. I think I am able to explain why the subjective identity of the conspecific as an object related to different functional systems is actually less of a biological necessity than that of other objects linked to instinctive responses.

Even in the highest vertebrates, object-directed instinctive behaviour patterns are frequently elicited by a very small selection of the stimuli emanating from the total range of perceived objects. When several functional systems have the same object, it can occur that each of these systems responds to different, distinct stimuli emanating from the same object. The innate releasing mechanism pinpoints, so to speak, a small selection of the wide range of stimuli emanating from the object concerned and responds to these selectively to produce the given behaviour. The simplicity of the innate releasing mechanisms of various instinctive behaviour patterns may produce a situation where two such mechanisms have no single eliciting stimulus element in common, despite the fact that they are related to the same object. Normally, the appropriate object will transmit all of the stimuli relevant to the two mechanisms *simultaneously*. In an experiment, however, one can trigger off the innate releasing mechanisms with two different objects and thus effect a separation of the two behavioural complexes related to *one* object. (Because of their great simplicity, such mechanisms can frequently be triggered by *artificial* presentation of appropriate stimulus combinations.) Conversely, it is possible (for the same reasons) to elicit with one given object two opposing responses which are only biologically adaptive with respect to two different objects. This is particularly frequent in cases where a conspecific is the object. In various duck species, for example, the protective response of the mother can be elicited by the distress-call of ducklings of other species. On the other hand, other parental responses are highly species-specific and are linked to quite specific coloration and marking patterns on the head and back of the ducklings. It is therefore easy to explain why a mallard duck leading her ducklings will courageously save a distress-calling Muscovy duckling from danger, only to 'treat it non-specifically' on failing to perceive the specific mallard markings on the head and back of the duckling. The lost duckling is regarded as a 'strange animal in the vicinity of the protected ducklings' and is attacked and killed.

Uniform treatment of a conspecific, as can be seen under natural conditions where the behaviour patterns are naturally elicited, need not necessarily be based upon a relationship between the responses established within the responding subject. Most frequently, such uniformity is established by purely external circumstances, based on the fact that the object of the responses – the conspecific – transmits the stimuli appropriate to different releasing mechanisms as a coherent whole. Thus, the functional plan of the species-specific instinctive behaviour patterns incorporates the biologically necessary unifying factor in the object transmitting the stimuli and not in the responding subject.

Let us consider a case where two or more instinctive behaviour patterns are developed to be elicited by the same functional object in order to fulfil their biological functions. There are two possibilities of ensuring this unity of response to the object. The first is that the object should be perceived in an objectivating manner and will thus always emerge as the same object in the subjective world of the subject in all functional contexts. The second possibility, which we will be examining in more detail, is contained in the above-mentioned unification of the various innate behaviour patterns *through the object itself*, without coalescence of the perceived factors to a unitary phenomenon in the subject's nervous system. It is self-evident that this form of unification is very dependent upon the properties of the object concerned.

When the releasing object is some given foreign structure in the environment, such as the natural prey or nest-material, the releasing mechanisms responding to the object can only be linked to characters which are as a matter of course inseparably incorporated in the object concerned. Since the number of these characters and their peculiar qualities is usually restricted, the biologically necessary general improbability (p. 103) of the innate releasing mechanisms constructed around them is subjected to a fairly tight upper limit, to the extent that the complexity (signal coverage) of these releasing mechanisms cannot be extended beyond a certain limit. Consequently, the probability of chance misdirected elicitation cannot be reduced below a certain level. This limitation on the range of performance of instinctive responses and their innate releasing mechanisms presumably considerably increases the survival value of subjective, concrete perception of objects.

The situation is quite different when the common object of two or more instinctive behavioural sequences is an individual of the same species as the responding subject. Since the specific morphological plan of a species and the specific structural plan of its innate behaviour patterns are constituents of a single, indivisible functional plan, the releasing mechanisms of the subject can in this case be evolved in parallel with the corresponding characters of the object. Thus, the general improbability involved can be increased indefinitely to virtually exclude misdirected elicitation of the behaviour pattern. I have previously applied the term *releasers* to those characters exhibited by an individual of a given animal species which activate existing releasing mechanisms in conspecifics and elicit certain chains of instinctive behaviour patterns. These characters may be either morphological structures or certain conspicuous behaviour patterns; usually they are a combination of both. The development of a releaser incorporates a compromise between two biological requirements: maximum simplicity and maximum general improbability. An exclamation which is often made by an uninitiated observer seeing the tail-fan of a peacock, the display plumage of a Golden Pheasant or the colourful gape-pattern of a young Hawfinch is: 'How unusual!' This expression of naïve astonishment in fact hits the nail right on the head. The gape of a nestling Hawfinch, for example, is so 'unusually colourful' for the very reason that it combines with the instinctive gaping pattern to act as a key to the species-specific feeding response of the parents. The biological significance of the colourful pattern lies in the prevention of 'mistaken' elicitation by coincidentally similar stimuli from another source. Sometimes, the improbability of such a releaser is not quite sufficient: certain African Estrildine finches have offspring whose gape-pattern is almost as highly specialized as that of the Hawfinch. Nevertheless, these birds are parasitized by a brood-parasite which has 'imitated' the key by developing almost identical head and gape coloration in its young.[41]

It is possible to continue specialization of the eliciting characters of the object and those of the existing releasing mechanisms in the subject to an almost unlimited extent. Unitary response to a conspecific, which acts as the object of different instinctive behaviour patterns, can be ensured in this way almost as thoroughly as through subjective recognition of the unity of the conspecific in the environment by the individual performing the instinctive behaviour patterns. It is obviously easier for animals of such limited intelligence as the birds to achieve extremely advanced specialization of releaser and releasing mechanisms than to develop a subjective identity of the object of instinctive responses, which will apply to all functional systems.

It is now possible to understand why an inanimate or foreign object of a species-specific behaviour pattern is more often responded to as a concrete unity than is a conspecific. This can be roughly summarized as follows: A stick used in nest-construction does not possess a sufficient number of conspicuous characters to allow the development of a sufficient number of adequately improbable releasing mechanisms to fulfil the requirements of elicitation of the number of different behaviour patterns involved in the total pattern of nest-building. For this reason, recognition of nest-material is acquired by instinct-conditioning intercalation and thus exhibits considerable constancy throughout all of the relevant functional systems. The nestling of a bird species can incorporate characters of unlimited complexity, for which an unlimited number of releasing mechanisms exist in the adult animal: a characteristically patterned gape for the feeding response, a characteristic, conspicuously coloured wreath of feathers around the anus for the faecal-removal response and a specific call (which indicates that the nestling is cold) to elicit the brooding response.

Under the natural environmental conditions of the species, releaser and innate, instinctive response ensure unitary response to a conspecific, although the latter does not represent a unitary object in the bird's environment. A unity of this kind can perhaps be demonstrated only in the acquired or 'intelligent' (i.e. non-innate) behavioural responses of the highest evolved animals. The identity of the object is lost, however, as soon as the physiological/instinctive relationship of the subject to the perceived object is altered. In human beings, intelligent reflection can enable us to calculate the effects of our own instinctive relationships within the framework of our total behaviour and thus prevent their projection outwards, with consequent alteration of the character of environmental objects. It can also occur, despite such insight gained through reflection, that an instinctive response to a purely subjective alteration of an environmental object may break through, e.g. when we kick a door against which we have stumbled in the dark, even though we 'know better' before the action has reached completion. But if an environmental object may appear to a *human being* simultaneously as constant (from the standpoint of experience) and labile (from the standpoint of objective codes of conduct), then in *animals*, even the highest, we shall not need to consider an introspectively-determined dichotomy between receptor and motor processes.

For most birds, we can confidently assume that the conspecific represents, with each functional system (*Funktionskreis*, in Uexküll's terminology) in which it appears as a reciprocating object, a separate object in the environment of the subject. The peculiar rôle which the conspecific thus plays in the bird's environment has been neatly described by J. von Uexküll as that of a 'companion'. By 'companion', we of course understand a fellow human being to whom we are bound only by the links of a single functional system, which themselves have little to do with higher emotional impulses, as is the case with a drinking or (at the outside) a hunting companion.[42]

The 'companion' in the bird's environment is interesting not only from the standpoint of environmental research, as is to be conducted here, but also because of its special sociological importance, which I believe merits closer investigation.

I owe to the personal encouragement of Professor Dr Jakob von Uexküll the courage necessary to at least attempt to set out the exceedingly complex matter contained in the following passages.

II Observational techniques and principles

1. OBSERVATION

The factual data upon which all of the following investigations are based is derived almost entirely from chance observation. I kept various bird species in an environment as close as possible to their natural habitats, for the purpose of general biological and specifically ethological observations. The birds were to a large extent allowed complete freedom. I was predominantly interested in colonially-breeding species such as the jackdaw, Night Heron and the Little Egret, since I had set myself the task of investigating their sociology as far as possible. Since the structure of these bird societies can be explained almost exclusively from the instinctive framework of the species concerned, such behaviour provided the natural starting-point of my investigations.

When one attempts to induce an animal species in captivity to unravel its entire life-cycle, to perform all of its chains of instinctive behaviour patterns, one usually obtains so many glimpses into the functioning of the instinctive framework of the species that subsequent experimentation need not be based on blind interference with environmental factors. This effect is enhanced by the behavioural deficiencies and pathological effects resulting from captivity. I have already discussed the methodology of investigation of instinctive systems in an earlier publication. In addition, in the course of several years of keeping animals for this express purpose, many unexpected secondary results emerged, particularly when simultaneous maintenance of several species helped to produce a continual supply of new situations. In this way, behavioural responses were frequently observed as a result of influences which I did not intentionally introduce, but which appeared as the only alteration in the usual natural habitat and must thus have acted as the eliciting cause of the behavioural response. A chance experiment of this kind has the advantage that it is recorded by a really impartial observer. In the light of the fine differentiation of many animal behaviour patterns, particularly where motor display patterns and vocalizations are concerned, it is extremely valuable when the observer is demonstrably completely free of any hypothesis.

Almost all of the observations relevant to the companion context considered here are derived from such observations and involuntary experiments, which were obtained as secondary results of the ethological investigations mentioned. These observations were not systematically obtained; they assembled of their own accord over the years. This explains their fragmentary nature, which could only be supplemented to a very small extent by subsequent experiments conducted on the basis of the evidence already available.

I see no obstacle to the scientific evaluation of these observations either in their fragmentary nature or in their broad temporal spread. The fact that these are secondary results of investigations which were actually constructed for other ends similarly presents no obstacle. In any case, I hope that by this early presentation of my (perhaps erroneous) opinions I will gain information about pertinent observations made by other students of animals.

2. THE EXPERIMENT

Claparède championed experimentation rather than observation, stating that the latter renders study dependent upon chance and makes inordinate demands upon time. This can be countered with the statement that an experiment conducted in the absence of knowledge of the natural behaviour pattern is in most cases completely valueless. It is my opinion that comparative psychology should be regarded and conducted as a *biological* science, even if this should result in temporary rejection of many generally recognized treatises on animal psychology. But, before the instinctive framework of an animal species and its functioning under the natural conditions of life in which the species evolved this framework are known (at least in broad outlines), even experiments which are directed purely at intelligent behaviour *tell us nothing about learning ability and intelligence in animals*. In the behaviour of animals, it is never certain how much must be ascribed to inherited instinctive behaviour patterns and how much to learning and intelligence. Without very exact knowledge of the instinctual plan of an animal, one cannot possibly gauge *how difficult a task the test problem represents for the animal*. There is always the possibility that the species investigated possesses an inherited *instinctive behaviour pattern* which happens to correspond to the experimental situation and is elicited accordingly. In such a case, a behaviour pattern which has no bearing on intelligence could be identified as a feat of intelligence.[43]

For example, in Hempelmann's *Tierpsychologie* and in Bierens de Haan's publication 'Der Stieglitz als Schöpfer' (*Journal für Ornithologie*, 1933), the experimental observation of grasping food with the feet is described first for tits and then for the goldfinch. In both cases, this is contrasted with the behaviour of other birds which do not do this, in such a way that the uninitiated reader automatically gains the impression that the tit and the goldfinch exhibit highly intelligent behaviour, in contrast to the species with which they are compared. In neither of these two cases is there mention of the fact that this motor co-ordination is an innately, reflexly-determined[44] instinctive pattern and has just as little to do with intelligence as the fact that we human beings protect the cornea of our eyes from disruptive desiccation by regular blinking of the eyelids. A *learning effect* can only be determined to the extent that the goldfinch and the tit learn to *employ* the behaviour pattern concerned in a novel, unnatural situation. A fixed instinctive behaviour pattern can under certain circumstances be employed for novel purposes like an immutable tool. There is some difference between construction of a novel tool through insight and conditioned use of an inherited tool.

But it is not just necessary to be familiar with the innately-controlled natural behaviour of an animal species *before* passing on to the experiment. It is also necessary to have a general acquaintance with the individually-variable behaviour patterns when attempting to conduct intensive analysis of one given pattern. To take one example: one source of error, which receives absolutely no consideration in many labyrinth and puzzle-box experiments, is based on the fact that any state of panic reduces intelligent abilities to almost nothing, particularly in the highest animals. If I should subject a bird of advanced intelligence (which is by nature easily aroused) to the slightest fear-inducing situation in a detour experiment, its feats of intelligence will at once remain far behind those of a much less intelligent animal which does not respond timidly to the same alteration of its environment. Tame and shy specimens of the same species will therefore give quite different results. The outcome is just as erroneous as if one were to attempt to estimate the intelligence of *Homo sapiens* on the basis of the behaviour of a mass of people at the fire at the Viennese *Ringtheater* in 1881. The panic-stricken people failed to solve the detour problem posed by the fact that the doors of the theatre opened inwards and not outwards. But in order to assess such fine details in overall psychology, and in the case of advanced animals in individual psychology, and to exclude such mistakes from the experimental results, *it is essential to conduct an extensive period of general observation, which must precede the performance of experiments.* He who maintains that he has no time for such observation, which is at first not directed at a particular goal, should leave animal psychology well alone.

Probably, the only subsidiary branch of psychology in which one can dispense with a knowledge of instinctual behaviour patterns in animal experiments is that of perceptual psychology. In this field, the response to the proffered stimulus is relatively unimportant, and it is only decisive to the extent that it represents a characteristic and objectively identifiable reaction to a particular perceptive experience.

3. EVALUATION OF ANOTHER AUTHOR'S OBSERVATIONS

A great disadvantage of pure observation in its narrowest sense, as is freely admitted here, derives from the difficulty in communicating these observations to others. Whereas the experiment achieves a high degree of objectivity through the repeatability of the experimental conditions, the same does not apply to pure observation.

The great difficulty in the investigation and description of the behaviour patterns of higher animals stems from the fact that the observer himself is a subject who is too similar to the object of his observations to allow of genuine objectivity. The 'most objective' observer of higher animals cannot avoid allowing himself to be repeatedly drawn into making analogies with his own subjective experiences. Our language itself *forces* us to employ 'experience terminology', which is derived from our own emotional framework, and to speak of 'fright'-postures, 'angry' displays and the like. With lower animals, which are taxonomically less related to us, it is easier to steer clear of such involuntary analogies. No observer has ever compared the attack response of a soldier termite to anger, nor the defensive response of a sea-urchin to fear.

However, it would be pointlessly doctrinaire to insist on the removal of all expressions used in the description of human emotions from the description of behaviour of higher animals. It is only necessary that these expressions should *always be used in the same sense*. Expressions which had to be *invented* for the description of behaviour patterns in the biological investigation of lower animals (since colloquial terms were lacking) are quite generally used by the rule of priority in the same, narrowly-defined sense as that intended by the inventor. On the other hand, those expressions which were already present as colloquial terms to describe our own emotions have carried their ambiguity in the colloquial context over into the scientific literature. The converse has also occurred. The word 'instinct' has practically lost its scientific usefulness because it has been taken up in colloquial language.

A further source of error in the use of another's observations stems from the fact that one worker will employ expressions concerning human emotions only where he can see genuine homologies, whereas another worker will use them where only superficial analogies are present. We might say on observing the flight response of a shrimp after brushing the tentacles of an anemone with its antennae: 'Now it is frightened', or on observing a maturing young male bird, which cannot yet utter the complete courtship call: 'Now he wants to say something, but he cannot do so yet.' But these statements are consciously entered in inverted commas. Many observers, including those schooled in psychology, *write* such statements down without any indication whether they are meant to be within inverted commas or not.

However, quite apart from these purely linguistic difficulties in transmitting observations, the greatest source of error in the evaluation of others' observations derives from the fact that any two observers watching the same thing do not *observe* the same thing – in other words, everyone lives in his own personal environment. Above all, the fact that a particular observer has *not* seen something in a particular animal does not justify a negative pronouncement. Thus, if in the following account I use (in addition to my own observations) only those results derived from workers operating from a standpoint close to (or identical with) my own, this is not the product of narrow-mindedness and it is certainly not a device to exclude facts which are in opposition to my hypotheses. I do this simply because it is only possible to read between the lines and to exert criticism similar to self-criticism in such cases, because only then is it possible to really understand the written accounts. In doing this, it is very useful if one has a close personal acquaintanceship with the author. For example, if I should read through some observation made by my friend Horst Siewert, then I can imagine what the observed animal really did and come quite close to the actual event. When reading the observations set out in Lloyd Morgan's books, on the other hand, I can only imagine very vaguely what really happened. Of course, this says nothing about the value of the observations made by different authors, but it would be very misleading to evaluate all in the same way.

For these reasons, in the following account I shall supplement my own observations only with those of the limited number of genuine animal observers. Among these, I make particular reference to my paternal friend Heinroth. His opinions and judgements concerning animal psychology were in such close agreement with mine before I had even met him that it is no longer possible to determine how much of his intellectual wealth I have inherited. (Luckily it is completely unnecessary to do so.) In view of this, I hope I shall be forgiven if I occasionally introduce his ideas without particular reference to his name.

4. THE OBSERVATIONAL MATERIAL

The material for the following analysis of observations and experiments (both intentional and unintentional) was provided by the following birds, which I have maintained over the years in a *free-flying* state: 15 Little Egrets, 32 Night Herons, 3 Squacco Herons, 6 White and 3 Black Storks, many mallard, many domestic ducks (*Hochbrutente* – call-duck), many domesticated Muscovy duck, 2 Carolina Wood-duck, 2 Greylag geese, 2 Common Buzzards and 1 Honey Buzzard, 1 Imperial Eagle, 7 Cormorants, 9 Kestrels, approximately one dozen Golden Pheasants, 1 Great Black-backed Gull, 2 Common Terns, 2 large Greater Yellow-crested Cockatoos, 1 Amazon Parrot, 7 Black-capped Parakeets, 20 ravens, 4 Hooded and 1 Carrion Crow, 7 magpies, more than 100 jackdaws, 2 jays, 2 Alpine Choughs, 2 Grey Cardinals and 3 Bullfinches. (I have also given the number of individuals investigated, since I regard it as very important to keep as many individuals of a given species as possible, in order to avoid unjustified generalizations. Of course, I did not keep all of the individuals at once – particularly with those species which were well represented; the maintenance experiments were generally continued over several years.) It is virtually unnecessary to detail those birds observed in closer confinement, since observations taken from such birds are used little in this account.

It is my agreeable duty at this point to thank all of those who have supported me by providing me with live birds. Above all, my heartfelt thanks go to the directors of the Zoological Gardens in Vienna (Schönbrunn) and Berlin. In addition, I should like to thank Dr Ernst Schüz (Rossitten), Herr Frommhold (Essen) and Fräulein Sylvia and Oberst August von Spieß (Hermanstadt; Sibiu). The latter provided me with an overwhelming present of 12 Little Egrets.

As can be seen, domesticated forms were largely excluded from the birds chosen, and observations on such forms have been analysed with extreme care, for the following reason: In my opinion, the main emphasis in the analysis of animal psychology should initially be placed on study of innate, instinctive behaviour rather than on variable behaviour patterns representing acquired products of intelligent processes, largely because (as has already been explained) one can never judge the extent of learning and intelligent abilities of an animal without prior knowledge of its instinctive properties. Nobody will doubt the importance of analysing inherited behaviour patterns. Detailed study of the instinctive behaviour patterns of domesticated animals in conjunction with comparison with the corresponding wild forms (particularly in birds) has shown that domestication produces *disruptive mutations* of fixed action patterns similar to those evident in morphology. Any attempt to study the instinctive behaviour patterns of domesticated animals always strikes me as being closely similar to an attempt to conduct an investigation of the structural coloration of birds' feathers using a white Pekin drake. In fact, one would be better off in the latter case, since one can *see* the disruptive effects (an advantage which one does not enjoy in the study of behaviour patterns). In this context, one must remember that in birds by far the major part of the total behaviour is instinctively determined. Of course, it is possible with some skill and a certain biological acumen to reconstruct the instinctive behaviour of the wild form from that of the domesticated form, in the same way that one can calculate the wild coloration of a duck from the colour patterns of various mottled domestic ducks which have white patches on *different* parts of their bodies. Brückner has achieved such a feat amazingly well in his thesis *Untersuchungen zur Tierpsychologie, insbesondere zur Auflösung der Familie*, which is entirely based upon the observation of domestic chickens. He does indeed say in his introduction: 'Even if the primary social instincts of these animals are disrupted, this does not prevent us from taking the present existing relationships as object of an investigation.' However, in the remainder of the text he takes great pains to avoid including abnormal responses in his analysis, with a great measure of success. To take one error from this otherwise exceptionally good article, in order to illustrate the typical result of neglect of the manner in which disruptive mutations of instinctive behavioural chains can emerge and be 'Mendelled around', I should like to quote the following passage from Brückner: 'Not all broody hens lead with equal proficiency. There are hens which are famous for their magnificent leadership and which are loaned out for this purpose, but there are others which perform this important task in a barely satisfactory form. This is a question of temperament and personality.' This can be answered with the observation that one might well describe variously effective disruptive mutations of species-specific instinctive behaviour patterns as 'variations in personality', but certainly not as variations in temperament. In any case, it can be demonstrated that *every* healthy wild hen (Burmese Jungle-Fowl or Golden Pheasant) of course measures up to the *ideal* broody hen. An adult animal which performs its task only incompletely does *not* occur in the normal range of variation of a wild form, or at least will occur only as a single mutant individual which will not leave descendants.

Variability in domesticated forms obeys its own particular laws, which are quite distinct from those governing the wild form. What factors in the hereditary complement of a species such as the mallard have maintained the beautifully and so exactly determined rich details of the plumage for centuries, only to disappear through domestication, rapidly producing such wild mutations of coloration? We do not know the answer. Thus, if we investigate the underlying principles of the behaviour patterns in domesticated animals, we are examining principles which are overlain, and partially reduced, in a completely aberrant manner. The resultant picture is admittedly not chaotic, but enormous complication is introduced by the fusion of two entirely different sets of principles (one of which lies close to the border of pathology), and the minor advantages of examining domestic animals – mainly represented by ease of acquisition and maintenance – are more than counteracted. Before we study variations, we must be acquainted with the invariable basic pattern. Even though I may greatly admire and revere the work of Katz, Schjelderup-Ebbe and (more recently) Brückner on domestic chickens, I nevertheless feel justified in maintaining that the generally important results of these investigations would have been greater in extent if a non-domesticated form had been chosen as the experimental animal, for example a Greylag goose instead of a domestic goose. This particularly applies to Brückner's article on the fragmentation of the chicken family, mentioned above. The form of fragmentation described by Brückner is a particularly good example of something which *only occurs in the domesticated chicken.* Only the domestic hen chases away her chicks by pecking at them, since only the domestic hen begins to lay a fresh clutch before the chicks have left of their own accord. In this case, the domestic chicks have the normal instinctive behaviour patterns of the wild type, which do not harmonize with the present uninterrupted egg-laying process and consequent reduction of maternal care produced in the hen by artificial breeding.

I have never observed pecking at the offspring in any single non-domesticated bird species, and I have been able to demonstrate in various cases in the literature, where the relevant details are given, that fragmentation of the family is determined by the young themselves. Just as the onset of parental instinctive behaviour patterns corresponds to the onset of infantile behaviour and interlocks with the latter in accordance with the plan of the species-specific instinctive behavioural system, there is a harmonized phase of extinction of their behaviour (provided that Man has not made an egg-laying machine out of the adult animal through planned breeding). Whenever we can see friction in the performance of interlocking parental and infantile behaviour, there is an indication that a pathological effect is present.[45]

5. THE FUNDAMENTAL APPROACH TO THE INSTINCT PROBLEM

Since the opinions and hypotheses presented here are constructed on a quite specific approach to the question of instinctive behaviour, it seems advisable to briefly outline this fundamental approach.

It is a generally held biological belief that instinctive behaviour is – so to speak – a phylogenetic predecessor of those behaviour patterns which we describe as 'learned' or determined by 'insight' (W. Köhler). Lloyd Morgan, in his book *Instinct and Experience*, describes how acquired behaviour could arise by a gradual process from instinctive behaviour, from graded modification of the existing instinctive framework by experience to produce behaviour better adapted to the goal. This view is shared by a large number of German animal psychologists. It is a thankless task to oppose a generally-accepted opinion, but I regard it as my duty to resist the fetters of authority.

In my opinion, an instinctive behaviour pattern is something fundamentally different from all other behaviour patterns of animals, be they simple conditioned reflexes, complex conditioned behaviour or the highest feats of intelligence based on insight. I cannot perceive any line of separation between pure instinctive patterns and chain reflexes[46] composed of unconditioned reflexes, though I must emphasize that I do not accept a strict mechanistic explanation based on a theory of nervous pathways even for a pure reflex. I do *not* regard the instinctive behaviour pattern as *homologous* with all acquired or insight-based behaviour patterns, however great the functional analogies may be in individual cases. Neither do I believe in the existence of genetic intermediates between the two types of behaviour.

I originally developed this view in a completely naïve fashion, regarding it as a self-evident fact and believing this to be obvious to every practical student of animals. In my earlier works, written in a state of restricted knowledge, I included this view as an obvious statement without particularly emphasizing it.

I am quite aware that these assertions, temporarily set down as a working hypothesis, cannot be proved at the present stage of our knowledge. Therefore, before I introduce the facts which agree with my views (and lend them a good degree of *probability*), I should like to demonstrate that the prevailing concept of the modifiability of instinctive behaviour patterns by experience is purely dogmatic in nature and that the amount of supporting factual evidence is far more limited.

Morgan quotes the acquisition of flight by learning in young swallows as a classical example of the adaptive modification of an instinctive framework by the addition of experience. I have serious objections to this view of the gradual progress in flying ability of a juvenile bird as proof of the modifiability of an instinctive behaviour pattern under the influence of experience.[47]

In the first place, it is essential to exclude quite definitely the possibility of a *maturation process* in all cases where the term *learning process* is used. Exactly like a maturing organ, the developing behaviour patterns of a young animal may begin to function before the maturation processes are completed. The development of an instinctive response and the organs necessary for the performance of the response need not necessarily exhibit temporal conjunction. When the development of the behaviour pattern is in advance of that of the organ, the effect is easy to identify: All ducklings have disproportionately small and entirely useless wings. Nevertheless, in the course of the fighting response (which can be elicited a few days after hatching) they exhibit exactly the same co-ordination of wing-movements as is evident in the adults of the species, which flail at their aggressors with bent wings. This, despite the fact that the wings of a duckling are so disproportionately short that they cannot reach the opponent at all in the innately-determined, instinctive fighting posture adapted to the dimensions of the adult bird. If, on the other hand, the development of the organ is completed earlier than that of the instinct which governs the use of the organ, the state of affairs is not so obvious. In many birds (including swallows) and particularly in large birds such as storks, eagles and the like, the organs of flight are functionally complete before co-ordination of the flight movements is complete. Now if the maturation of this co-ordination is in the process of catching up with the advanced development of the organ concerned, the behaviour of the young animal has just the same appearance as a learning effect, whereas in reality an internal maturation process is continuing along an exactly predetermined pathway. We shall return later to discuss a few genuine learning processes which accompany these maturation processes.

The American worker Carmichael maintained amphibian embryos under continual narcosis in weak chloretone solutions, and this – strangely enough – completely inhibited all movements, whilst permitting bodily development. When he allowed these embryos to 'awake' at late stages of development, there proved to be no difference between the swimming movements of these animals and those of normal control animals, which had 'practised' these movements for some days. Of course, we cannot perform this experiment with young birds, but some facts do indicate that a similar state of affairs exists. For example, juvenile Ringed Doves (*Columba palumbus*) leave the nest at a very early stage, when the primaries are very short and still vascularized. In the hole-nesting Rock Dove, the young birds are obviously far less threatened than those of its open-nesting relative the Ringed Dove, and flight away from the nest does not appear until the primaries and secondaries are fully-grown and keratinized. Although Rock Doves at this later stage still have no practice at flying, they show no difference in flying ability from Ringed Doves of equal age which have been flying for several days.

My second objection is that neither Lloyd Morgan, nor any of the biologists following him, have considered the undeniable existence of a phenomenon which I described in an earlier paper as '*instinct-conditioning intercalation*'. It is a property of many behaviour patterns of birds and other animals that a functionally unitary behavioural sequence may include innate and individually-acquired links in abrupt alternation. Neglect of the intercalated conditioned behavioural link automatically leads one to ascribe variability to a purely instinctive behaviour pattern which is quite free of variation.

Instinct-conditioning intercalation involves a chain of several unconditioned reflexes, between which one or more conditioned reflexes are inserted (depending on the complexity of the behavioural sequence). As in the case of the simplest conditioned reflexes, the acquired components of behaviour with instinct-conditioning intercalation quite often represent the *eliciting factor in the response*, taking the form of a group of characters which are united to form one schema by the responding subject. This *acquired* releasing mechanism when subsequently activated sets the remaining, purely reflex, behavioural actions in motion.

O. Koehler describes Pavlov's nomenclature as 'dilution of the reflex concept'. I am in full agreement with this view, and I should like to suggest the introduction of the term reflex-conditioning intercalation to replace the term 'conditioned reflex', in order to express the distinct duplicity of the factors involved.

An example of instinct-conditioning intercalation is provided by the impaling response exhibited by the Red-backed Shrike (*Lanius collurio*) with captured prey. In young birds of this species, not every component of the entire behavioural sequence of impaling the prey morsel is innate. Firstly, instinctive recognition of the necessary *thorn* on which the prey should be impaled is lacking.[48] On the other hand, the entire motor co-ordination pattern of impaling *is* innate, as is the recognition of the fact that impaling must be carried out on a solid object. A hand-reared Red-backed Shrike soon begins to perform impaling motions with a morsel in its beak, but at first performs these on randomly-selected sites within the cage. Even if a thorn (or a nail) suitable for impaling is present, this object is ignored. The young bird presses small pieces of meat held in the beak backwards and forwards along perches and twigs in the cage and now and again performs peculiar brief tugging movements, which are provided in the instinctive chain of behaviour patterns for the impaling response. These tugging actions are particularly evident when the bird meets with resistance when wiping the morsel along an object. This special adaptive feature has the result that tugging immediately appears when the bird *by chance* hooks the morsel on to a nail or a thorn. The bird subsequently *learns* very rapidly to recognize thorns as the object of the otherwise entirely innate chain of instinctive patterns. Through such a process (which is definitely to be equated with *learning*), what we may call 'gaps' in innate behavioural chains are filled out in many instances. These 'gaps' are links in behavioural sequences provided as restricted 'learning tendencies' rather than as an equivalent link in a chain of completely innate segments.

If we compare between different bird species instinct-conditioning intercalation effects which serve analogous functions, it often emerges that in one species a particular link in the behavioural sequence concerned is innate, whereas the functionally analogous link in another species is individually acquired. This mutually *substitutive* occurrence of links which are sometimes innately determined and sometimes represented by a specific ability to acquire a given behaviour pattern is particularly evident in 'learning' to fly in young birds. There it often emerges that in one species the ability to gauge distances reliably and to arrive accurately at landing-sites is completely innate, and is present immediately after emergence from the nest without any prior practice, whereas this ability must be acquired in a drawn-out learning process in another species. In many cases, the operation of biological compulsion can be demonstrated: Young Reed-Warblers, Roseate Starlings and some rock-dwelling birds, in all of which misjudging the landing-site would have severe consequences most particularly exhibit an entirely innate ability to localize a landing goal in space. Young ravens, on the other hand, have both the ability and the time in their (so to speak) progressive emergence from the nest to acquire the requisite skills individually. In very many of these processes of acquisition, an important part is played by what we generally describe as *play*. Where one can observe playful activity in the development of a behaviour pattern in a young animal, it is reasonable to assume that conditioning links are incorporated in the behavioural sequence.

The distinctly *substitutive* occurrence of innate behaviour patterns on the one hand and acquired behaviour patterns on the other can be regarded as a case of *analogous functions* in Werner's terminology, and this renders the existence of transitions between the two very unlikely.

The instinctive behavioural chains of animals with primitive powers of intelligence doubtless contain fewer acquired links than those of more advanced organisms. If the question should be raised as to the manner in which the phylogeny of the behaviour of the two groups should be interpreted, I would present the following hypothesis: Instinctive behaviour on the one hand and acquired behaviour involving insight on the other are not successive steps in a process, either ontogenetically or phylogenetically; they represent two divergent developments. Wherever one of the two behavioural types experiences a particularly high degree of specialization, the other type is extensively excluded. In those cases where instinctive behaviour patterns are progressively excluded in the wake of higher developments of learning ability and insight, I do not believe that gradual transition of one into the other has been demonstrated and I do not think that such a transition is possible. In my opinion, fixed, instinctive components in a behavioural sequence do not become more labile and increasingly modifiable by experience with increasing development of learning ability and intelligence. It is far more likely that these instinctive patterns drop out completely, one by one, to be replaced by acquired or insight-controlled behaviour patterns.[49] In addition, novel variable links can be incorporated between the retained instinctive links of a behavioural sequence. As with all phylogenetic hypotheses, we can only indicate the validity of this hypothesis by setting out a series of types. But the fact that we are at all *able* to construct the beginnings of such series in the light of our present fragmentary knowledge of the instinctive behaviour patterns of higher animals is itself illuminating. No such factual evidence can be produced to support the prevalent view that instinctive behaviour patterns can be gradually rendered more subject to experience. As a first attempt at constructing such a series, we can compare the food-concealment responses of two Corvid species. When a jackdaw has a morsel in its beak and is motivated to conceal it, it responds regardless of the general environmental stimuli and usually hides the morsel in the deepest and darkest of the corners or holes present. The jackdaw is unable to learn by experience that the adaptive purpose of the concealment response disappears if another jackdaw is allowed to watch the process of concealment. Similarly, this bird never appreciates that certain localities are inaccessible to human friends and that stolen objects of value would be safely concealed in such places. The raven, which is closely related, but nevertheless greatly excels in learning ability and intelligence, learns even as a very young bird that concealment only has a purpose when no other animal is watching and that human beings, because of their inability to fly, cannot reach all of the available sites. Thus, in the raven – in contrast to the jackdaw – the timing and the site of concealment is not instinctively determined. However, in every other respect the concealment response of the raven is just as fixed as that of the jackdaw. The motor co-ordination patterns themselves are never altered and do not differ from those of the jackdaw. Experimental alteration of the natural conditions for the behavioural sequence shows that even in ravens there is an obvious and complete absence of actual recognition of purpose, which is characteristic of inherited instinctive behaviour patterns.

I imagine that with more detailed analysis of functional organization of species-specific instinctive behaviour patterns of many more species (particularly closely-related forms), much better and much clearer serial reconstructions will be possible.

In the earlier works which are extensively referred to in this article, I adopted Ziegler's definition of instinctive behaviour. I am aware that this definition, based on a theory of nervous pathways and centres, does not agree with certain overall regulatory phenomena. I remind the reader of Bethe's experiments, which demonstrate extensive regulatory properties for innate behavioural co-ordination of walking patterns in widely-different animals. In strongly emphasizing the fixity of instinctive behaviour patterns at that time, I was referring to their absolute impermeability to influences stemming from experience and intelligence. I emphasized that Bethe's experiments in fact demonstrate quite clearly that the regulatory properties of instinctive behaviour represent a form of plasticity which has nothing to do with *learning* and *experience*. Wherever Bethe's experiments demonstrated regulatory properties in the locomotor co-ordination of an animal, the regulator effects were *immediately* evident in full strength after the operation! The contention that instinctive behaviour patterns are immutable in the face of experience is in fact supported rather than questioned by Berthe's results. Such patterns are neatly covered by Driesch's definition of instinctive behaviour: 'Instinct is a response that is complete from the outset.'[50]

Of course, the negative observation that the innate behaviour pattern is not modified by learning is difficult to prove and does not represent a definition. However, the only course open to me is to supplement this negative statement with another in order to restrict, at least to some extent, the concept of the instinctive behaviour pattern as I understand it.

Edward C. Tolman says in his truly excellent book *Purposive Behaviour in Animals and Men:* 'Wherever learning ability occurs in relation to a particular goal (and where, excluding the very simplest tropisms and reflexes, is that not the case?), we have the objective appearance and definition of that which we appropriately designate as *purpose.*'

This objective definition of purpose in the terminology of behavioural theory is extremely valuable. In my opinion, it is possible to assemble all *non-instinctive* behaviour, be it conditioned or insight-determined, under the heading of 'purposive behaviour'. Instinctive behaviour patterns, however, completely lack purposive adaptability ('docility relative to some end') of this immediately obvious kind, so that this 'behaviouristic' definition of purpose can be used as a negative statement about the instinctive behaviour pattern. Nevertheless, the opinion that 'learning ability relative to a particular goal' is lacking only in the simplest tropisms and reflexes must be vigorously opposed. It must be emphasized that chains of instinctive behaviour patterns are in general fixed to an extent corresponding to their complexity. The simple detour responses of a starfish or a *Paramecium* are much more prone to exhibit plasticity than the comb-building responses of the bee. Uexküll once said: 'The amoeba is less of a machine than the horse.' In the same way, we can say that the highly-specialized parental care patterns of a female dog are much purer and show their reflex nature much more clearly than the simple, but conditioned, salivating 'reflex' of a Pavlovian dog. We can find both unconditioned and conditioned reflexes among the more highly-specialized innate behaviour patterns. But it is possible to state that in the highest specialized behaviour chains, as are found (for example) in the colony-constructing insects, unconditioned reflexes[51] predominate.

In the less complex instinctive behaviour patterns of higher vertebrates, we scarcely ever find instinctive behavioural chains of any length; usually we find highly complex instinct-conditioning intercalation. The fact that such behaviour chains are frequently not open to the limited number of analytical methods available, and can hardly ever be fully analysed, provides no grounds for not carrying out a *conceptual* separation between acquired and reflex components.

We can apply a second characteristic of purposive behaviour, as defined by Tolman, in order to specify this concept of the instinctive behaviour pattern. Tolman writes: 'Eliciting stimuli are not in themselves sufficient; certain additional supporting factors are necessary. Animal behaviour cannot be performed in a vacuum. A certain supplementary "sustenance" is required to maintain it. A rat cannot "run along a passage" without an actual floor to support its feet and real walls, between which it orients itself.'

There is no more impressive characteristic of the instinctive behaviour pattern than the fact that it will be performed '*in vacuo*' if the species-specific eliciting factor is absent. If a particular instinctive pattern is never elicited under conditions of captivity, because of the absence of the adequate stimulus, the threshold for this stimulus is remarkably lowered.[52] This process can sometimes go so far that the instinctive pattern concerned will eventually 'fire off' *without* any demonstrable stimulus. It is as if the perpetually latent behaviour pattern itself eventually becomes an internal stimulus. I am reminded of the example of the starling described in my earlier article, where the behavioural sequence of hunting a fly was performed *in toto* without the object. One would think that the presence of flies would be the most important and indispensable 'behaviour support' (Tolman) for the performance of fly-hunting behaviour. This would indeed be the case if some 'purpose' relevant to the animal subject were present in the performance of this behaviour. The animal does not obey the slightest purposive impulse, however; it follows the 'blind plan' (Uexküll) of its instinctive behaviour patterns. The developmental level and the biological significance of this plan have as little to do with the psychological capabilities of the animal as the biologically-adaptive structure of its body.

This delineation of my concept of the species-specific instinctive behaviour pattern, largely constructed from negative observations, is not intended as an explicit definition. The views expressed here have already proved themselves to some degree as working hypotheses in the analysis of animal behaviour. At least they stand up to examination on the basis of the facts so far available better than the prevalent view of instinctive behaviour open to adaptive modification by experience. How far they will stand up to further factual evidence cannot be predicted at our present minimal state of knowledge about the instinctive behaviour of higher animals.

III Imprinting of the object of species-specific instinctive behaviour patterns

The *acquired* component of instinct-conditioning intercalation patterns is very often the *object* of the innate behaviour, as I attempted to demonstrate with a number of examples in the paper already quoted Although the acquisition of this object is generally equivalent to a conditioning process, a basically different process of acquisition takes place with a particular group of innate, instinctive behaviour patterns lacking an innately-determined object. In my opinion, this process cannot be equated with learning – it is the acquisition of the object of *instinctive behaviour patterns oriented towards conspecifics.*

To the uninitiated, it is often surprising (even incredible) that a bird does not recognize conspecifics innately and purely 'instinctively' in all situations and respond accordingly. Very few birds behave in this way, however. *In contrast to all mammals investigated in this respect,*[53] *isolated, hand-reared individuals of most bird species do not recognize conspecifics as their own kind when introduced to them later*, i.e. behaviour patterns which would normally be elicited by conspecifics are not elicited. On the contrary, young birds of most species will direct the instinctive behaviour patterns adapted to conspecifics *towards a human being* if they are reared in human care, isolated from their kind.

This behaviour appears so bizarre – so 'deranged' – to the observer, that any individual observer who encounters this phenomenon when hand-rearing young birds is at first prone to regard it as a *pathological* process, explained as a 'confinement psychosis' or the like. Only when one repeatedly encounters this behaviour even with completely healthy specimens of extremely varied bird species, and observes that it occurs in animals reared in complete freedom, does one gradually realize that a controlled process is involved and that *the object of instinctive behaviour patterns oriented towards conspecifics is not innately determined* in most bird species. Instead, recognition of the object is acquired during the individual's life-time by means of a process which is so peculiar that it merits detailed consideration.

If an egg of the curlew (*Numenius*) or a Godwit (*Limosa*) is hatched in an incubator and the young bird is taken into human care straight after hatching, it will be seen that the bird will not seek contact with human beings as 'foster-parents'. It will flee as soon as it sees a human being, and one will not be able to observe any part of the instinctive behaviour patterns directed towards the parents, except perhaps by using finely-adjusted dummy-experiments. (Unfortunately, nobody has so far conducted such experiments.) In these two species, as with many nidifugous types (which leave the egg at a very advanced stage of development) these instinctive behaviour patterns can *only* be elicited by conspecific adults. Translated into the terminology of environmentalist studies, this means: The young bird possesses an innate 'schema' of the adult. The image of the adult animal is defined by so many innately-recognized characters that the infantile instinctive behaviour patterns only respond 'species-specifically' to adults of the same species. We can sometimes determine quite accurately how many characters are involved, in cases where *imitation* of these characters successfully elicits infantile instinctive behaviour patterns.

If, instead of taking a curlew, we should take a Greylag gosling into our care, after it has grown for several days in the custody of its parents, we should find the same effect. In this case too, it is impossible for a human being to elicit any infantile instinctive behaviour patterns. *The result is quite different, however, when a Greylag gosling is taken into human care directly after hatching.* Then, all of the instinctive behaviour patterns directed towards the parents are at once elicited by the human foster-parent. In fact, it is necessary to employ specific safety precautions to induce young Greylag goslings which have been *artificially* brooded in an incubator to follow a mother Greylag goose which is leading other goslings. The artificially-hatched goslings *must not be allowed full sight of a human being* between hatching and transfer to the mother goose, since otherwise the following drive will immediately become attached to human beings. Heinroth described this process quite exactly in his work *Beiträge zur Biologie, insbesondere Psychologie und Ethologie der Anatiden:*

> I frequently had to attempt to introduce goslings hatched in an incubator to a pair already leading very young goslings. This involves a number of difficulties, which are in fact quite characteristic of the entire psychological and instinctive behaviour of these birds. If the lid of an incubator is opened after young ducklings have just hatched from the egg and dried out, they will at first crouch quite still and then dash away with lightning speed when the attempt is made to pick them up. In the process, they often jump to the ground and rapidly crawl under nearby objects, so that it is often a difficult task to catch the tiny things. Young goslings behave quite differently. Without any display of fear, they stare calmly at human beings and do not resist handling. If one spends just a little time with them, it is not so easy to get rid of them afterwards. They pipe piteously if left behind and soon follow reliably. It has happened to me that such a gosling, a few hours after removal from the incubator, was content as long as it could settle under the chair on which I sat. If such a gosling

> is carried to a goose family accompanied by goslings of the same age, the result is usually the following: The approaching human being is watched suspiciously by the adult male and female, and both attempt to take to the water with their offspring as fast as possible. If one approaches so fast that the goslings do not have sufficient time to flee, the parents naturally angrily turn to defence. One can then rapidly deposit the orphan gosling among them and retreat hastily. Aroused as they are, the parents naturally regard the tiny newcomer as their own offspring at first and will attempt to defend it as soon as they see and hear it in the human hand. But the worst is to come: *The young gosling shows no inclination to regard the two adults as conspecifics:* The gosling runs off, piping, and attaches itself to the first human being that happens to come past; it regards the human being as its parent.

Heinroth continues to explain that a gosling can be successfully foisted onto a goose family if it is placed in a sack straight after removal from the incubator, so that it does not see human beings at all. He rightly expresses the opinion that the freshly-hatched goslings look on the first living thing that they see in the first light of the world 'with the intention of exactly imprinting this image, since – as has already been mentioned – these delicate woolly things do not appear to recognize their parents as conspecifics in a purely instinctive manner'.[54]

I have described this behaviour of the Greylag gosling because it provides a virtually classic example of the manner in which a *single experience* imprints the relevant object of the infantile instinctive behaviour patterns in a young bird which does *not* recognize this object instinctively. *This object can only be imprinted during a quite definite period in the bird's life.* A further important feature is the fact that the Greylag gosling obviously 'expects' this experience during a receptive period, i.e. *there is an innate drive to fill this gap in the instinctive framework.* It should also be emphasized that the genus *Anser* represents an extreme, to the extent that so *few* characters of the adult companion are innately determined in the freshly-hatched bird. Apart from an instinctive response to the species-specific alarm-call, no instinctive response to any character present in the parents can be demonstrated. In particular, the instinctive response to the summoning call of the parents evident in so many small nidifugous birds is lacking.

The process of imprinting the object of otherwise innately-determined instinctive behaviour patterns directed at conspecifics is *markedly different to the acquisition of the object of other instinctive behaviour patterns* whose releasing schemata are not innately determined, but acquired like conditioned reflexes. Whereas in the latter case the acquisition process is presumably equivalent to self-conditioning – a learning process – *the process of imprinting of the object of instinctive behaviour patterns oriented towards conspecifics possesses a series of features which are basically different from learning*. There is no equivalent process in the behaviour of any other animals, particularly mammals, but at this point it should be pointed out that there are certain analogies to pathological fixations of the drive object in *human* psychology.

The following factors distinguish the process of imprinting from typical *learning*:

In the first place, the described acquisition of the object of instinctive behaviour patterns can only take place during a narrowly-defined period of time in the individual's life. Thus, the process of imprinting of the object depends upon a *quite definite physiological developmental condition* in the young bird.

Secondly, imprinted recognition of the object of instinctive behaviour patterns directed towards conspecifics, following the *expiration* of the physiological imprinting period determined for the species, has *exactly the same appearance as corresponding innate behaviour – the recognition response cannot be 'forgotten'*! The possibility of 'forgetting' is, as Bühler particularly points out, a *basic feature of all learning processes*. Of course, it is not yet permissible, in the light of the relative novelty of all observations on this process, to make a final statement about the permanence of these acquired objects. The justification for such a pronouncement is based upon the fact, observed in many cases, that birds which have been hand-reared by a human being and have come to direct their conspecific-oriented instinctive patterns towards the human frame do not alter their behaviour in the slightest, even when kept together with conspecifics and away from human beings for many years. This latter measure is just as unsuccessful in bringing such birds to recognize conspecifics as beings of their own kind as the attempt to induce a bird captured as an adult to recognize a human being as conspecific. (Behaviour directed towards substitute objects, which provides an apparent exception, will be discussed later.)

These two facts, the determination of later behaviour by an external influence (derived from a conspecific) during a specific ontogenetic period and the irreversibility of this determination process, provide a remarkable analogy between the developmental processes of instinctive behavioural systems and processes which have been identified in morphological development.

If, at a certain time in development, cell material is taken from the ectoderm of the posterior abdominal region of a frog embryo – where it would normally form a piece of abdominal skin during further embryonic development – and grafted onto the posterior end of the outer surface of the neural groove, it will form a constituent part of the spinal cord in accordance with its position. The cells are thus influenced by the organizational determinant of the local environment – an effect which Spemann terms 'induction'. The possibility of induction forces us to distinguish between 'prospective potency' and 'prospective significance' of a given cell. The ectoderm cells of the frog embryo transplanted in Spemann's experiment had the prospective significance of a region of abdominal skin; in the absence of experimental manipulation, these cells could *only* have had that fate. At the same time, however, these cells still had the normally latent ability to develop into a region of the spinal cord. Thus, their prospective potency was greater than their prospective significance. If a similar experiment is carried out at a later stage of development, or if the abdominal ectoderm cells (now determined as spinal cord cells) are *transplanted back* to their site of origin, the prospective potency is found to be identical to the prospective significance. When this second transplantation is carried out, those cells which would have otherwise become spinal cord constituents do so even at the new site, and the transplantation experiment produces a 'monster'. The prospective significance of the cells is thus 'determined' under the influence of the local environment, by induction. In other words, the cell material does not possess inherited 'knowledge' of its fate – this is determined by the site it occupies. The local environment imprints the final organ character of the cells. After completion of this determination process, which takes place at a particular period in development, the tissue can no longer 'forget' its determined fate. The presumptive spinal cord cells transplanted back to the abdominal epithelium can no longer be 're-determined' to form this type of epithelium! In some animals, such as the tunicates, even in the two-cell stage of the developing egg the behaviour of each of the cells is already completely determined. Here, there is almost no induction. One cell of an artificially-constricted two-cell stage will literally develop to form half an organism. Individual cell-groups in later stages of development, if isolated, will only produce the same organs or organ components which they would have produced on combination with neighbouring cells. These cells are therefore not influenced by the local environment. Each cell has exact inherited 'knowledge' of its functions and the mosaic of such exactly-balanced and co-ordinated parts produces a unitary organism without further influence exerted between the cells. Such embryos are consequently referred to as 'mosaic embryos', in contrast to the 'regulative embryos' described earlier.

The terms 'mosaic' and 'regulative' could well be applied to instinctive systems, and it would be quite fitting to use the term *inductive determination* for instinctive behaviour patterns whose object is not innately determined in the animal, but imprinted by the environment (particularly by conspecifics). The functional plan of the instinctive system of an animal and the functional morphological plan are in many ways analogous.

When a young jackdaw at about fourteen days of age directs its infantile instinctive behaviour patterns towards its parents, these patterns have the prospective significance of direction towards the parents, towards conspecifics. However, at this time the species-specific instinctive behaviour patterns oriented towards conspecifics have a much broader prospective potency in the choice of object. The parents, which already function as this object for the young jackdaw, can still be replaced as such. A young jackdaw taken from the nest, is at first shy towards human beings and will crouch in their presence, indicating that the young bird is already familiar with the sight of the parents. Despite this, the parental companion – the object of the infantile instinctive behaviour patterns – can be functionally replaced in another respect. Within the space of a few hours, the jackdaw will gape towards a human being; after about twenty days the young bird has fledged and will direct its aerial following drive towards human beings and can no longer be functionally re-oriented 'to jackdaw'. A young jackdaw left with its parents until reaching the same age is no longer open to imprinting on human beings. The prospective significance and prospective potency of the object are now coincident.

We must therefore distinguish between two phases in the developmental period of instinctive patterns inherited without an object: an initial, usually very short, phase during which the bird seeks out the object of the innate behaviour and a second, longer, phase during which an eliciting object for the instinctive behaviour has already been found, but in which 'change of determination' is still possible. In some birds, as in the nidifugous types already mentioned, where there is determination through a single impression, the second phase is extremely short. The entire psychological development which the nidicolous birds undergo during the long nest-phase is, so to speak, compressed into the few hours which a nidifugous bird remains in the nest. The shortest imprinting phase for infantile instinctive behaviour patterns lacking an innately-determined object (which at the same time occurs at the shortest interval after hatching from the egg) is found in nidifugous types from extremely diverse groups, as has already been mentioned. From my own experience, I can verify that young mallard, pheasant and partridge which have followed their mothers for only a few hours can no longer re-orient their following drive to human beings. Consequently, one can only rear these birds properly when they are artificially incubated; otherwise, human beings will elicit a flight response which is so strong that the young birds may stop feeding and succumb. I regard it as quite possible that a *single* elicitation of the following response can bring about complete imprinting to the mother. I am particularly convinced that this is the case with the partridge, since I have attempted to rear partridge chicks turned up by farmers whilst reaping. These chicks could not even stand continuously and were forced to squat after each short burst of running. This stage, with which I am well acquainted, lasts for only a few hours after the chicks have dried, and the hen leads the chicks only a few metres during this phase. Nevertheless, the partridge chicks always succumbed, because they persistently fled or crouched immediately they were brought out into the light to feed. They first began to feed after they had become too emaciated to survive. If partridges are artificially incubated, on the other hand, they are at once tame towards the human 'foster-parent' and can be reared without difficulty.

The time of inductive determination, imprinting of the object of instinctive behaviour patterns oriented towards conspecifics, is in most cases not so easy to determine as with the infantile instinctive behaviour of the Greylag goose or the partridge. There are two reasons for this:

In the first place, identification of the imprinting period may be hindered by facilitation of imprinting to the species-specific object by a large number of innately-recognized characters, which serve to prevent false imprinting. The innate positive response of the young bird to these characters has the effect that a juvenile which has previously responded to a species-atypical object can be re-orientated to conspecifics at a time when the reverse process is no longer possible. For instance, the instinctive response of a young Golden Pheasant to the summoning call of the hen (i.e. definite running towards the source of sound) – something which is certainly *not* evident in the Greylag gosling – permits re-orientation of a 'humanized' pheasant at the developmental stage where 'transplantation' from hen pheasant to human being would be unsuccessful. It would be interesting to perform experiments employing people talented with the mimicry of bird-calls – a group to which I most definitely do not belong.

A second obstacle to the exact determination of the time of imprinting is derived from occasional temporal overlap of the time of object-imprinting of one instinctive behaviour pattern with that of another. This remarkable form of behaviour is apparently not uncommon in nidicolous types. Particularly with passerines, I have observed that specimens which are fostered at a relatively late age will still respond to a human being with their infantile instinctive behaviour patterns, and yet will later direct their equally object-less instinctive behaviour patterns towards conspecifics. This struck me most forcibly when I once had several young jackdaws of the same age, three of which had been taken as naked young and six of which were obtained shortly before emergence from the nest. All of these jackdaws were tame towards me whilst they were still gaping for food from me. However, after extinction of the infantile instinctive behaviour patterns the late-reared young rapidly became shy towards me, as if object-imprinting of the sexual instincts had taken place in the jackdaws *before* final determination of the infantile instinctive behaviour patterns. Determination does not seem to be quite final at this time, since I was able to record re-orientation 'to jackdaw' in some cases – something which never happened at a later stage.

Imprinting of different conspecific-oriented functional systems to the relevant object occurs at different points of time in individual ontogeny. This is very important in the present context: It provides the basis for the fact that the different functional systems can be voluntarily or involuntarily imprinted on different objects under conditions of captivity. For instance, I once possessed a young jackdaw reared in complete isolation, in which all behaviour patterns related to conspecifics were imprinted to human beings, with the exception of two behavioural complexes: flying in the company of a flock, and feeding and care of young conspecifics. The former behaviour was imprinted on Hooded Crows at the time of activation of the gregarious drive, these being the first flying Corvids with which the jackdaw became acquainted. This jackdaw even continued to fly persistently with free-living Hooded Crows when the attic which the bird occupied was used as a home-base for a whole flock of other jackdaws, which were not considered as flight-companions. Each morning, after I had liberated the birds, this particular jackdaw would climb high into the air and set off in search of her crow flight-companions, which were always found with great accuracy. At the time of rearing the young, however, this jackdaw abruptly adopted a freshly-fledged young jackdaw, which was guided and fed in a completely species-typical manner. It is in fact obvious that the object of instinctive parental behaviour patterns *must* be innately determined. This object cannot be acquired previously by imprinting, since the jackdaw's own offspring are, of course, the first that it sees. Thus, in this jackdaw's environment, the human being was a parental and sexual companion, the Hooded Crow a flying companion and the young jackdaw an infant companion!

Imprinting which determines the object of instinctive behaviour patterns related to conspecifics in the young bird frequently results through the influence of parents and siblings, yet it must nevertheless determine the behaviour of the young bird to *all* conspecifics. Thus, in the *imprinted* schema of the conspecific, as with the innate equivalent, only *supra-individual* species-characteristic characters may be derived from the image of parents and siblings, to be permanently imprinted. It is amazing enough that this should succeed in normal, species-typical imprinting, but it is astounding that a bird reared by, and imprinted to, a human being should direct its behaviour patterns not towards *one* human but towards the species *Homo sapiens*. A jackdaw for which the human has replaced the parental companion, and which has consequently become completely 'humanized', will thus direct its awakening sexual instincts not specifically towards its former parental companion, but (with the complete unpredictability of falling in love) towards any *one* relatively unfamiliar human being. The sex is unimportant, *but the object will quite definitely be human*. It would seem that the former parental companion is simply not considered as a possible 'mate'. But how does such a bird recognize our conspecifics as 'human beings'? A whole range of extremely interesting questions await solution![55]

In conclusion we must consider *which* conspecific acts as the source for the stimuli which determine the inductive establishment of the object of an instinctive behavioural chain.

In cases where imprinting of the object occurs long before the appearance of the instinctive behaviour pattern, imprinting must of course be induced by a conspecific which is involved with the imprinted bird in a different functional phase as that for which object-imprinting is induced. For instance, imprinting of the instinctive sexual behaviour patterns in jackdaws is almost certainly determined by the parental companion. At least, young jackdaws will become sexually imprinted on the human being even when reared in the company of several sibling companions, as long as the human foster-parent gives the bird sufficient attention to occupy the rôle of a full parental companion. Many other birds kept by Heinroth similarly showed sexual imprinting on human beings even though reared with several siblings (e.g. owls, ravens, partridges and many more).

On the other hand, there are birds for which the *siblings* determine later sexual behaviour. The mallard ducks mentioned on p. 137, which were intensively fostered in my care, proved to be completely normal sexually, whereas a Muscovy drake reared with them was imprinted 'to mallard'. Since this mixed-species company of siblings remained as a group until the next Spring, I am unable to give any answer to the question 'when' regarding imprinting. However, I intend to conduct experiments to investigate this with the easily-reared genera *Cairina* and *Anas* in the near future.

Birds reared in complete isolation from their own kind frequently become imprinted to human beings in all their instinctive behaviour, even where object-imprinting is normally determined by the sibling companion. Since the human being, as will be discussed later, never acts as a sibling companion, the imprinting process does not seem to be unconditionally bound to a particular type of companion.

IV The innate companion-schema[56]

It was demonstrated on p. 103, in the first section, that the innate releasing mechanisms (schemata) of many object-oriented instinctive behaviour patterns often lack any kind of correlation, since each responds to separate characteristics of the object and functions in a manner completely independent of the responses of other mechanisms. It was shown that this independence is particularly evident in those instinctive patterns whose object is a *conspecific*.

Through imprinting, the bird acquires a schema of the conspecific animal which is distinctly constructed and differs from the innate releasing mechanisms of species-specific instinctive behaviour patterns (as does a schema acquired by learning) in the incorporation of a wide range of characters.

Under natural conditions, an innate schema and an acquired schema of a conspecific form a functional unit.[57] Only under experimental conditions can the instinctive behaviour patterns which are directed towards a conspecific be oriented towards different objects. In the free-living species, it is the imprinted companion-schema which unites the instinctive patterns (representing unconditioned reflexes) elicited by innate schemata to form a functional entity. The complex form of the acquired companion-schema fits into the single, mutually-independent releasing mechanisms of individual instinctive behaviour patterns. This process resembles that of children linking the pre-punched holes in a lacing puzzle. The holes have no actual connection with one another and they are only integrated through a functional plan whose entirety was determined by a separate agency. We may only speak of a unitary, innate schema of a companion on the condition that a species-specific imprinting process integrates the mutually independent characters of the companion to produce one schema. Conversely, the development of the imprinted companion-schema is dependent upon innate patterns. It is the response of one or more instinctive behaviour patterns possessing innate releasing mechanisms which directs the imprinting process towards a particular object. I call to mind the young partridges, in which a single response of the following pattern, for which the instinctively-recognized guiding call of the mother acts as releaser, irreversibly attaches imprinting to the complex form of the maternal companion. The components of the innate schema provide, so to speak, a frame for points fixed independently of one another in space. The imprinted schema is somehow 'pressed into' this framework, as Uexküll puts it.

Just like the instinctive behaviour pattern and the acquired behaviour pattern, innate and acquired schemata form functional units whose components fit together discretely and are subjectively unconnected. Such schemata occur substitutively in different species, and this substitutive occurrence can go so far in different birds that in one the imprinted schema of the conspecific can be reduced to the point of disappearance whilst in another the same occurs with the innate schema.

In the first case, the instinctive behaviour patterns form a 'mosaic' which is only brought to function as a unity through the releasers of the object, as discussed on p. 106. In this extreme case, the conspecific represents a separate environmental 'thing', a separate companion in the functional system of each individual instinctive pattern which is related to a particular releasing mechanism. Strictly speaking, it is not permissible to refer to 'parental companion' and the like with such birds, as will be done in the following discussion, since the parental bird can represent in the environment of the offspring a 'feeding companion', a 'warming companion', a 'guiding companion' and various other autonomous entities. An extreme 'mosaic-type' with this kind of orientation of behaviour patterns towards conspecifics is represented by the young curlew discussed on p. 124.

On the other hand, in cases where blind response to species-specific releasers is not so decisive, where the imprinted (or simply acquired) schema of the conspecific is better represented, the conspecific is doubtless a more unitary environmental object for the animal concerned. This particularly applies where personal recognition of a particular, individual companion plays a part. In Greylag geese, where siblings remain together for years although they almost never mate with one another, and in other Anatids, where the offspring maintain an amicable relationship to their parents for several years and resume the old family connections in the Autumn even when they have reared their own offspring, the environmental picture of the conspecific similarly does not exactly correspond to the concept of 'companion' – though for opposite reasons.

It is amazing that within one Class of vertebrates we should find behaviour which on the one hand corresponds to the fixed instinctive behaviour of lower invertebrates and on the other may include patterns reminiscent of corresponding behaviour in human beings. It is this very fact which makes observation of birds so valuable.

The extreme 'mosaic-type' of the curlew and the extreme 'regulative-type' of the Greylag goose represent extreme cases, between which there is every conceivable intermediate stage of substitutive occurrence of innate and acquired conspecific schemata. When imprinting takes place in a species-atypical manner, extremely instructive behavioural inconsistencies in the relationship to the companion often emerge. This produces contradictory behavioural tendencies, which demonstrate, through their utterly non-adaptive appearance, the manner in which abnormal imprinting disrupts the species-typical functional plan of innate and acquired companion-schemata. A jackdaw which is imprinted on the human being for all behaviour patterns with an acquired object is friendly to the human foster-parent and hostile to other jackdaws, but it will still respond with the species-typical defence response when the human foster-parent seizes one of the other jackdaws. Conversely, lack of a species-specific releaser in an acquired, species-atypical companion can cause a similar disruption of the adaptiveness of the behaviour. A female White Stork in the Schönbrunn menagerie was mated with a male Black Stork and built a nest with him every year. The greeting ceremonies associated with nest-entry differ somewhat between the two species. In the White Stork, the well-known beak-rattle is used, whereas the Black Stork employs a peculiar hissing call. This dissimilarity had the result that the pair in question, which had been mated for years, repeatedly exhibited mistrust and fear during the greeting ceremony. The female White Stork, in particular, often seemed to be on the point of attacking the male partner when he was not prepared to rattle. Heinroth describes similar inconsistencies in a mixed pair containing a Rock Dove and a Wood Pigeon.

It would be wrong to maintain that species such as the Greylag goose, which innately recognize very few of the characters of the parental companion, have no innate schema. In such forms, the schema is simply *extremely broad* because of the *small number* of characters involved. For the freshly-hatched Greylag goose, which initially lacks an object for its following instinct, not *every* object can become a guiding companion. The relevant object must possess certain characteristics which are necessary for the elicitation of following – above all, it must move. The object need not necessarily be alive, since cases are known where very young Greylag goslings attempted to attach themselves to *boats*. Thus it would seem that the size of the companion is a further 'signal' that is not specified in the innate schema.

Experiments should be conducted in which the size of the object and various form characteristics would be varied to determine the range of elicitation of the following drive of Greylag goslings.

A budgerigar (*Melopsittacus*) with which I conducted experiments in the spring of 1933 exhibited extremely interesting behaviour, which departed radically from the described behaviour of young Greylag goslings. The bird was removed from the nest at the age of one week and reared in isolation. Until fledging, it was kept in a non-transparent container, so that it saw little of the fostering human beings. After fledging, the bird was kept in a cage in which a white-and-blue celluloid sphere was mounted on a vertical spring so that it began to move slightly even when the bird was simply climbing or flying within the cage. I had not designed this experimental set-up on the basis of pure imagination. Some time previously, I had seen a talking budgerigar (reared in isolation by the Viennese budgerigar-breeder Grasl) which courted a dummy budgerigar attached to the cage roof. Since I was convinced that the innate companion-schema of the budgerigar was too rudimentary in signal-composition to really necessitate such an exact copy of a conspecific, I selected for the first experiment a celluloid ball as an object of maximal simplicity. My intention of re-directing the instinctive behaviour patterns of the budgerigar was an unqualified success. The bird rapidly began to remain close to the vibrating sphere and would only settle down to rest in its vicinity. Soon afterwards, the budgerigar performed the instinctive behaviour patterns of 'social preening' on the sphere, using the same movements as parrots mutually combing their plumage. The motor co-ordinations of preening small feathers were performed in great detail, although the sphere was of course completely bald. In addition, after 'preening' the sphere for some time, the budgerigar would present its own neck for preening with the invitation gesture well known in parrots. Sometimes, the bird almost succeeded in allowing itself to be preened by the vibrating sphere.

One observation indicates that the *spatial* schema of a conspecific is indeed innately determined in more of its components than that of the Greylag goose: The budgerigar treated the sphere in all respects as if it were the *head* of a conspecific. All of the social activities it exhibited were such as are directed against the head of a conspecific in the normal performance of instinctive behaviour of *Melopsittacus*. When the sphere was attached to different parts of the cage-wire, so that the budgerigar could approach unhindered and settle on the wire at any preferred height relative to the sphere, the bird always hung on with the sphere at head height. However, when the sphere was attached to a horizontal perch so that the squatting height of the budgerigar was predetermined at a level below head height, the bird behaved uncertainly and exhibited clear signs of 'embarrassment'. When the sphere was detached from its stem and simply thrown to the floor of the cage, the budgerigar responded with prolonged and silent crouching on its perch in exactly the same manner as a budgerigar reacting to the death of a conspecific cage-mate.

The only instinctive behaviour pattern normally directed towards the body of the conspecific, rather than its head, which I observed in this budgerigar, was the following: Courting male budgerigars commonly grasp at the body (usually the rump) of the female whilst dancing up and down and chattering in front of her. When my budgerigar was older and had begun to court the sphere, I observed this movement, particularly when I had arranged the sphere so that it vibrated on the upward-pointing spring. When the budgerigar fondled the sphere, his foot often swung at the spring with what I interpreted as the same movement as that described above for a normal male budgerigar. When the spring was otherwise arranged, the budgerigar often swung his foot beneath the sphere into thin air! Unfortunately, this bird died before reaching sexual maturity.

According to these observations, the innate schema of the conspecific in the budgerigar appears to incorporate spatial subdivision into head and rump. This subdivision is lacking in the innate schema of the maternal companion in young Greylag geese. The reason for this is probably the division in direction of the adult budgerigar's actions to separate body areas of its companion, which is itself lacking in the young Greylag goose.

The innate schema of the 'flying companion' of a jackdaw is more restricted than that of the Greylag goose parental companion or the budgerigar social companion – the ability to fly, the black coloration and probably the general Corvid form are incorporated as signals along with other features. In this schema, which 'should' actually characterize the jackdaw to serve the adaptive function of the instinctive behaviour patterns, a Hooded Crow can be 'squeezed in' (to use Uexküll's expression) by imprinting of additional signals. A human being cannot fit into this schema, because too many of the necessary signals are lacking. The absence of *single* signals incorporated in the innate schema need not always prevent the occupation of the schema by an inadequate imprinting object, as we shall see. The interplay between the innate schema and imprinting is particularly clear in cases where so few quite characteristic signals of the companion are innately incorporated that they can be artificially provided as appropriate stimuli in an experiment.

Whenever one rears young mallard from the egg and attempts to elicit as many of their infantile instinctive behaviour patterns as possible, one gains the impression that the attempt is not completely successful. Heinroth has said of artificially-incubated mallard ducklings: 'If one has several, their need of companionship and social contact is largely satisfied. They scarcely miss the guiding adult and do not attach themselves to the human being. They are not in fact shy; they of course eat out of the hand, but they do not take to handling since they always preserve a certain independence.' Thus, mallard chicks and Greylag goslings provide the greatest imaginable contrast within a single group in their behaviour towards human beings. My experiences were at first identical to those described by Heinroth. I had also made the experience that mallard and even races of the domestic duck close to the wild-type (such as the so-called 'Hochbrutenten', which have more than a second part of wild-type blood in their veins) will not accept a non-conspecific foster-parent. The chicks of heavily-built domestic ducks will very easily transfer their infantile drives (particularly the following drive) to human beings or that familiar substitutive mother, the domestic hen. This loss of specificity of the stimulus response is an effect of domestication, however, as can be found with various other domestic animals. Unfortunately, conclusions are often drawn from these thoroughly uncharacteristic, debilitated instincts of domestic animals to give a completely unqualified picture of 'the instinct'. Mallard ducklings which possess the full instinctive behavioural complement will not even respond to foster-parents from different duck genera with their infantile instinctive patterns. For example, if they are incubated by a Muscovy Duck *Cairina moschata*, they will lose their foster-mother whilst she is still sitting on the nest, because they simply run away. This, despite the fact that *Cairina* can be easily hybridized with *Anas*! On the other hand, young mallard will apparently follow a fostering domestic duck without difficulty, although the foster-image departs just as much from that of a mallard mother as that of *Cairina*. However, the domestic and the wild duck still share the patterns of communication, particularly the summoning call, which have been little altered in the course of domestication. Whereas a *Cairina* mother has only a faint quacking call, and is in any case silent for the most part, both mallard and domestic duck call almost uninterruptedly whilst guiding their chicks. After these experiences with the replacement of the mallard maternal companion by a human being, a Muscovy Duck or a domestic duck, I was convinced that the innate maternal call must be the decisive signal which is lacking for the mallard ducklings when reared with substitute mothers other than the domestic duck. Since one can fairly easily mimic the guiding call of the mallard mother, I decided to attempt to prove this experimentally. Accordingly, in the early summer of 1933, I took three mallard ducklings hatched under a 'Hochbrutente' and six hybrid ducklings of the same age obtained by a mallard × 'Hochbrutente' cross and reared by their mother (a pure-bred mallard). The ducklings were taken straight after hatching and I gave them a lot of attention, including repeated imitations of the guiding call, even whilst they were still drying out. In the following days, which were luckily coincident with the Whitsun holiday, I spent all my time quacking. This experiment, which was self-denial in the truest sense of the word, was not without success. The first time I placed the ducklings at liberty on the meadow and moved away quacking, they immediately began to utter the 'lost piping' which is present in some form or other in almost all nidifugous birds. Just like a proper duck mother, I went back to the ducklings in response to the piping and repeated the slow withdrawal with renewed quacking.[58] This time, the entire procession set off and followed close behind me. From then onwards, the ducklings followed me almost as eagerly and reliably as they would have done with a genuine mother. The progress of the experiment demonstrated that the maternal call is very probably the decisive character of the maternal companion for the mallard and that the appearance of the companion is individually imprinted: At first, I was unable to stop quacking, since the ducklings would otherwise soon start to utter lost piping. Only when the ducklings were older was I recognized as the maternal companion even whilst I was silent.

Thus, one can extricate in individual cases the property which the maternal companion must without exception possess, and which of her characters are imprinted on the young bird only in the course of ontogeny.

V The parental companion

Since the infantile instinctive behaviour patterns of many birds in fact provide the best examples of object-imprinting and the innate releasing mechanism, I have already had to mention some features of the behaviour of young birds towards the parent, which will be referred to again in the following chapter. We must still consider what functions the parental companion performs in the functional network of the young bird – how this companion is reflected in the young bird's environment.

1. THE INNATE SCHEMA OF THE PARENTAL COMPANION

The innate recognition of the character combinations of the conspecific parent by the young bird, or the latter's response to these combinations with a specific behaviour pattern, vary greatly from species to species with respect to the number of characters involved. This variation produces in some species a generalized innate schema of the parental companion and in others a more restricted, specifically 'structured' schema. These variations frequently occur within one and the same group of birds in a thoroughly unpredictable manner. However, taking the Class Aves as a whole we can state one generally applicable (if apparently self-evident) rule: The innate schema of the parental companion is simpler, the earlier the developmental stage at which the bird leaves the egg. It is obvious that in a freshly-hatched passerine, which emerges from the egg with its eyes and ears closed, a complex schema is prohibited by the rudimentary powers of the sense organs, whilst the long nest-phase permits such birds to acquire much through imprinting at their leisure. In contrast, a freshly-hatched plover possesses functional sense organs which permit response to quite complex innate schemata, whereas the bird's mental development, which occupies weeks or months in nidicolous types, is compressed into a matter of hours. The Greylag goose demonstrates that the latter condition is no barrier to imprinting characters of the parental companion, but overall imprinting plays a greater part in nidicolous than in nidifugous types. For this reason, the following discussion of the various innate parental schema in young birds will not be based on the Zoological classification. The few birds about which anything is known in this context will be arranged according to the developmental condition exhibited at hatching. We shall begin with the highest specialized nidicolous types – the passerines.

A freshly-hatched passerine is an utterly helpless creature. It has been demonstrated that the senses functional on emergence are hearing, gravity perception and temperature perception. It is obvious that only an extremely simple schema of the parental companion is possible with these three sensory input sources.

The greatest discriminatory ability for detecting stimulus differences is doubtless contained in the auditory sense. Even though the only behavioural response of the nestlings – gaping – can usually be elicited by stimuli other than the voice of the parents or a suitable imitation, it can be demonstrated that lower (indeed much lower) stimulus intensities are required to elicit the gaping response with the parental enticement call. I was able to verify this with freshly-hatched jackdaws left in the parental nest in spring 1933, since the feeding call of the adult jackdaws is easy to imitate. Heinroth made the same observation on blind raven nestlings at an age of about nine days. The nestlings were more easily provoked to gape by low-pitched notes (corresponding to the voice of the parent ravens) than by high-pitched notes. In this case too, the response was not specific, i.e. the response was given to species-atypical stimuli at higher intensities, but the behaviour was not so easily elicited as with the appropriate stimulus emanating from the parent.

The responses to rocking stimulation which we can observe in some very young passerines are extremely characteristic. This response exhibits an interesting adaptation to the nest-site characteristic of the species. Nestlings of hole-nesting species, which normally have a completely stable cradle (e.g. the true tits), are *alarmed* when the artificial nest in which they are reared is accidentally shaken and will abruptly stop gaping if in the process of feeding at the time. In species whose nests are typically built on swaying branches, or suspended from them, any mechanical disturbance (provided that it does not exceed a certain level) will in fact elicit the gaping reflex. According to Steinfalt's observations, this is particularly true of the Penduline Tit.

Finally, it is sometimes possible to observe that the nestlings begin to gape when the mother departs from the nest (after warming them) so gently that no stimulation results. Since the nestlings only begin to become restless and eventually start to gape after some time has elapsed, it would appear that in this case the cooling stimulus represents the eliciting factor. It must be added that young passerines remarkably do *not* possess a behavioural releaser to stimulate the parents to warm them. If young and still completely naked passerines are insufficiently warmed, they apparently do not perceive the gradually operating *persistent* cooling accompanied by 'creeping stimulation'. They do not begin to wail like undercooled herons, birds of prey and presumably all nidifugous types. Instead, like tropical reptiles which have suffered cooling, all of their movements become slower and slower. But they still gape with slow-motion actions even when they are quite cold to the touch. They can be revived by renewed warming even after spending a considerable time in complete torpor, as I was able to determine after failure of a warming device for young House Sparrows. Thus, the persistent warming function of the parent would seem to elicit no response from these almost poikilothermous nestlings. Perhaps any cooling stimulus which they perceive when the persistently warming mother *suddenly* rises from the nest acts as a parental companion signal, since this stimulus naturally precedes any feeding bout.

It is obvious that no very highly-developed and characteristic schemata can be constructed from the few signals which the described sense-organs are capable of perceiving. The situation is entirely different where the *visual sense* begins to play a part, as with older passerine nestlings and with young herons and other birds which are in fact nidicolous, but emerge from the egg with a functional visual sense. Even in the realm of the visual sense, the overall image of the parental companion is not innately determined; as with all other senses, only a relatively simple schema constructed from a few signals is innately incorporated.

Not all 'innate' behaviour need appear directly after hatching. An example is provided by the completely instinctive response of the Night Heron to the species-specific greeting ceremony. This will be dealt with later (p. 106) as a good example of a 'releaser'. As with many instinctive behavioural releasers which include an optical stimulus along with other stimuli, this releaser includes *morphological signal-organs*. In *Nycticorax*, this organ consists of a crest of greatly elongated, black erectile feathers on the head, together with three protruding, snow-white neck feathers which can be reared or spread sideways. The expression 'ornamental feathers' applied to this organ is utterly misleading, since they are not ornamental but serve to elicit a quite specific, species-typical response from a conspecific. In the greeting ceremony, the beak is directed downwards and the head is presented towards the greeted conspecific with the black crest ruffled and the three white feathers extending out of this crown. Seen from the front, the black disc with the three divergent upward-pointing feathers really does look like a signal. The instinctive response which is elicited by this species-specific signal is not a behaviour pattern but an inhibition. The *defensive response*, which is very strong in herons and is usually elicited by any approach made by a conspecific, *is inhibited*. It is very characteristic for the heron group that a special morphological organ had to be developed in order to enable the partners of a nest to mutually suppress their defensive responses. Verwey of course observed with the Heron that courting males which had already stood for days on their nests calling for a female were unable to suppress their reflex defensive responses when a female arrived and so attacked the bride they were 'longing for'. It is interesting with respect to the described defence-inhibition ceremony in *Nycticorax* that a South American relative *Cochlearius*, which is a true night heron despite its aberrant beak-form reminiscent of *Balaeniceps*, has the same instinctive behaviour pattern with a *different signal-organ*. In this heron, the three white 'ornamental feathers' are lacking. This is offset by greater elongation and lateral extension of the black feather-crest on the head. Thus, in this species, the conspecific to be pacified is presented not with a black disc and three white lines, but with a large black, acute-angled triangle standing on its sharpest corner. *Therefore, the ceremony is more archaic than the constituent signal-organ.*[59]

The response to this species-specific instinctive behavioural releaser is of course innately determined, but it first appears some time after the hatching of the young Night Heron. Indeed, it emerges a few days after appearance of the defensive response! At first, the young herons do not normally *see* their parents, since the latter continue to warm their offspring (as long as they are quite small and in need of warmth) just as persistently as they did the eggs. The male and female continue to relieve one another in the same way. As with pigeons and other birds, nest-relief usually takes the form of the relieving partner joining the other in the nest and slowly driving the latter off the eggs (or offspring), so that any eggs or hatched offspring cannot be seen in the process. Naturally, the hatched offspring do not see their parents either. However, with one of my free-breeding Night Heron pairs, nest-relief did not always occur in the normal manner because the nest-site lay very close to the feeding-site. The partner on active brooding service at any time would often leave the nest briefly during feeding to fetch a fish. When this partner returned to the nest, the young nestlings regularly adopted the defensive posture and even attempted to attack the returning parent (though the latter was little disturbed by this). For a young Night Heron which is still under brooding-care, the parental companion is not an animal which approaches on the wing, but one which broods continually, since the nestling presumably perceives no change at nest-relief. It is not a 'normal' occurrence for the parent to fly back to the uncovered nest.

With somewhat older juvenile Night Herons, the response to the greeting ceremony of the parents emerges almost simultaneously with the performance of this instinctive behaviour pattern by the young bird towards its parents. From this time onwards, the greeting ceremony appears to be the most important signal by which the schema of the parent bird is characterized for the young bird. The fact that specific motions combine with the form in the stimulus pattern is doubtless related to the fact that in birds (particularly in the less intelligent species) form perception is far less developed than perception of movement. In order to permit operation of the form as a specific stimulus, signal-organs as characteristic and relatively simple as those described must be developed.

The highest development of the innate schema is found in certain nidifugous types. In these types – as was initially described for curlews – so many of the characteristic signals of the conspecific are innately recognized that the conspecific parental companion is almost unequivocally determined. For these reasons, the response to the parental companion cannot be elicited by substitute stimuli. Since such young birds are scarcely open to experimentation, we know virtually nothing about the nature of the innately-recognized characters of the parental companion. Experiments with dummy birds or fostering birds which are closely related and very similar to the species under investigation might provide information on this topic.[60]

In some cases, as in that of the mallard already discussed, the parental companion is characterized for the young bird by the summoning call or the guiding call. In general, *innate companion-signals appear to be more often incorporated acoustically* than in any other sensory modality. It appears to happen particularly often that an innate response to an acoustic stimulus directs imprinting of the instinctive behaviour patterns related to the parental companion towards the species-specific object. This would explain the described behaviour of young mallard to fostering duck species.

Finally, it must be mentioned that certain signals occurring in the innate schema of the parental companion, which are *not* incorporated during complementation of the schema in the course of imprinting to an abnormal (i.e. non-conspecific) parental companion, be it a fostering bird or human being, will not in fact prevent imprinting to the parental companion, but will often indicate their presence in a quite characteristic manner. This applies, for example, to the *warning call* of the parental companion, recognition of which is innately determined in the young bird. Lack of this call in the human parental companion has a result which will be discussed later.

2. PERSONAL RECOGNITION OF THE PARENTAL COMPANION

In very many cases, it is not at all obvious to what extent young birds recognize their parents individually. For individually nesting species, it is of course entirely unnecessary for the nestlings to respond only to their own parents with their infantile instinctive behaviour patterns, since they suffer no harm when (for example) they beg in vain for food from a stranger which flies past. Such behaviour would be much more likely to cause harm in colonially-breeding species, if the nestlings did not go on the defensive against alien (and possibly hostile) adults. Young Night Herons do in fact exhibit such defence, but on the other hand they apparently do not recognize their parents in the proper sense of the word. The parent is in fact largely characterized by a signal within the *innate schema*, in that the parent enters *its own* nest with the greeting ceremony described earlier. If an adult does this under experimental conditions *without* the ceremony (which never occurs normally) it is met with an angry defensive response like any alien Night Heron. The adult bird does not represent an animal of a particular appearance for its offspring – it is an animal which settles on the edge of the nest and behaves in a particular way. In summer 1933, my hand-reared, extremely tame Night Herons from 1931, bred in complete freedom in a beech-tree in my garden. The tamest pair had two offspring, which had just begun to take part in the defensive response of the parents as I first climbed the tree and looked into the nest. The mother was so little afraid of me that she had no hesitation in flying into my face or onto my head and pecking angrily. It was not at all easy to fight off this fearless bird from my precarious perch on a sloping beech branch, without a proper hold for my hands. The male, who was somewhat shyer, did protect the nest, but did not proceed to attack like the female. I wanted to be alone with the young birds, since I wanted to see whether they would be provoked to adopt a defensive posture by the anger-call of the parents and would behave quite differently towards me without the influence of the parents. So I waited until the female was alone in the nest, the male having flown off to the meadows bordering the

Danube. I climbed into the tree, seized the female as she approached angrily and threw her as hard as I could down from the tree. The Night Heron is such a poor climber that the female was fairly exhausted when she approached me again to renew the attack. I threw the poor thing down again, and this time she was forced to rest for a few minutes, which left me some time to examine the nestlings undisturbed. The nestlings did in fact threaten me and stabbed at my hand; but when I left my hand motionless in the nest between the young birds, they soon became calm enough to pick up and eat pieces of fish alongside this human hand in the nest. At this point, the male suddenly returned from the meadow and immediately came up to the nest from the other side exhibiting the threat posture. As he was about half a yard away, the two nestlings turned their backs on me to face their father and reciprocated his threat posture. As he clambered onto the edge of the nest in order to attack me, still showing the threat posture, both nestlings stabbed at his face and uttered the species-specific fear croak. They took the threat posture directed against me as a threat against *them*. For these young birds, the parental companion is strictly the bird which mounts the nest with the species-specific greeting ceremony, which they reciprocate. The angrily threatening father was a stranger to them.

Thus, a very young Night Heron 'recognizes' its mother 'from' her warming properties, whilst a somewhat older nestling 'recognizes' her 'from' the greeting ceremony. After fledging, the young birds do not recognize their parents at all – they will beg food in like manner from any approaching adult bird. Still later, they learn to recognize the territory belonging to the parents within the colony and the parents are 'recognized from' the sites on which they perch. Begging and performance of the greeting ceremony is probably elicited in the young bird by a *very small number*[61] of characters of the greeting parent. It is possible that the nestling would still greet and beg if we should present him with the three ruffled neck feathers as an optical stimulus and the greeting call as an acoustic stimulus, without any other form of stimulation. But it is the 'without any other form of stimulation' which prevents the possible performance of dummy experiments to investigate the characters which are the prominent eliciting factors in most cases. Only in a few special cases is the situation so favourable that we are able to isolate the characters relevant to a behavioural response. In these cases, there is an astounding *paucity* of innately-recognized characters. The word 'astounding' illustrates very well how much we are inclined to imagine that the environment of an animal is too like our own. In fact, it is *not* astounding that the environment of a bird should be less diverse than that of mankind.

However, the few characters of the parental companion which are instinctively recognized by the young bird are so incorporated into the functional plan of the instincts that they are sufficient to characterize the parent unequivocally under natural conditions. The adult Night Heron approaching the nest on the wing with the described greeting gestures and calls is, of course, always one of the actual parents, as is the creature which first encounters a tiny Greylag gosling and directs the 'imprinting drive' of the gosling towards itself. It seems as if *economy*[62] is essential in the incorporation of *innately-recognized* characters.

At this point, it is worth mentioning a number of observations, partly belonging in the category of pathology, which demonstrate that unequivocal personal recognition of the parental companion can also be determined in the young bird by the fact that the characters incorporated in its innate parental companion schema undergo *one process of development* (a temporal pattern) which runs parallel to an equivalent development of the innate behaviour patterns of the parent incorporating these characters. In other words, the co-ordination of the instinctive behaviour patterns of parent and young undergo a regular process of development, such that releaser and released undergo changes which are always parallel, thus avoiding any disruption of their co-ordination. In this way, the young bird's innate schema of the parental companion also changes. This process of change is *independent* of the changes of the characters emanating from the parental companion, but normally *corresponds* to the latter developmental process. Thus, the parental companion is characterized for such young birds by the correspondence between the *stage* of the instinctive behaviour patterns of parental care and the developmental stage of the instinctive patterns and the innate parental schema in the young birds themselves.

This 'signal of the equivalent stage' appears to play a considerable part in individual characterization of the parent animal in certain duck species. For 'Hochbrutenten' and Muscovy ducks (i.e. two types not extensively altered by domestication) and similarly for the Eider duck (a wild form) relevant observations have been recorded. 'Hochbrutenten' and Muscovy ducks have roughly the following behaviour in this context: If it occurs that two clutches hatch on the same day and the families encounter one another on their first excursions on the pond, it frequently – nay regularly – happens that the two flocks of ducklings unite. This fusion is brought about by the ducklings very much against the wishes of the mothers, which at first fight bitterly and only gradually become accustomed to one another. Fusion begins when ducklings of one flock follow the leading mother of the other. The ducklings' own mother is forced to attach herself to the unified flock of ducklings, come what may, and she regularly succeeds after a number of fights with the other mother. The young Muscovy ducklings in particular seem to recognize their mother in their first days of life exclusively from the fact that her parental instinctive behaviour patterns are at the same developmental level as their own infantile instinctive behaviour patterns. The development of the instinctive behaviour patterns forms a temporal pattern, comparable to a melody, and for a juvenile *Cairina* the most important character of the mother is the fact that she follows the melody in time with the duckling. It is only under quite specific – and very informative – circumstances that a juvenile *Cairina* will follow the guidance of a strange mother whose offspring are of a *different* age.

If any young bird, including the *Cairina* duckling, is retarded in morphological development, the development of its instinctive behaviour patterns exhibits retardation corresponding to the inhibition of morphological development. For example, in many nidifugous birds the drive to crawl *under* the mother to sleep is extinguished almost simultaneously with the emergence of the dorsal plumage. This coincidence occurs whether or not the development takes place in a physically-perfect duckling or in a weakly individual, where it occurs at twice the age. The mother herself loses the drive to warm the ducklings at a certain time after the hatching of the offspring, and this change is largely independent of the developmental stage reached by the latter. This means that she will indeed warm under-developed ducklings to some extent, but not appreciably longer than normal ducklings. However, if just a few weakly individuals appear among a number of normal ducklings in a clutch, the instinctive behaviour patterns of the mother certainly do not adapt to the weaker offspring. The latter usually succumb at the time when nocturnal warming ceases. In the Summer of 1933, I twice observed a situation where two weakly *Cairina* ducklings left their mother of their own accord and transferred to another *Cairina* female leading young, whose offspring were *younger* and whose maternal instinctive behaviour patterns still corresponded to the physical developmental level of the weakly individuals. Thus, a duckling of this kind regards a leading mother duck with *corresponding* instinctive behaviour patterns as its own mother, without any appreciable degree of personal recognition. The duckling very probably recognizes its own species, of course, since it will never follow a duck of another species.

Naturally, the above account does not demonstrate whether the *wild form* of the Muscovy duck behaves in this way. The Muscovy duck is a type in which personal recognition between individuals plays a very small part in behaviour, in contrast to almost all other Anatids. Since *Cairina* is also the only Anatid known to be completely lacking pair-formation, it is an obvious step to try to link the weak ability for personal recognition with this feature. Even if two-mother flocks can be found in mallard, and even if such flocks are almost the rule in large flocks of Eider duck, I greatly doubt that in offspring of these species the character of 'equivalent development of interlocking instinctive behaviour patterns' dominates over the individual characteristics of the mother (as it does in *Cairina*) when there is conflict between the two.

In chickens, equivalent age of interlocking instinctive behaviour patterns is also an unconditional requirement for any exchange of the maternal companion. However, the acquired recognition of the individual characteristics of the mother becomes so important at such an early stage that errors of recognition apparently never occur without experimental interference (Brückner).

In contrast to those cases where the *characters* of the *innate schema* actually act to characterize the parental companion by virtue of their spatial or temporal arrangement (thus rendering personal recognition unnecessary), there are many bird species in which genuine personal recognition of the parents by the offspring is present. This applies to the great majority of nidifugous birds and to those nidicolous types which are guided by their parents after fledging. The young bird must acquire recognition of a sufficient number of characters of the parental animal to individually characterize the latter unequivocally. Whereas the innate schema, as described, incorporates a relatively small number of characters, the acquired schema of the individually recognized parental companion is typified by such an *abundance* of characters that it would appear improbable that the bird perceives and 'registers' each individual character as a separate entity. The bird nevertheless responds unambiguously to the loss or alteration of any single one of these many characters. But this is not the place to go into detail about the concepts of 'complex qualities' used to explain this and similar phenomena.

We shall, however, go into detail about the manner in which personal recognition of a companion develops in different birds, although this is not only relevant to the parental companion – it is equally applicable for any other type of companion.

Relatively unintelligent birds, whether young birds which are still of restricted intelligence or species which never advance any further, generally only recognize a human being, who has been incorporated into one of their functional systems through imprinting, when he corresponds in all (or almost all) of the characters in the image that they usually associate with him. If this human companion wears a hat or takes off his coat, they are afraid of him. This well-known phenomenon can be explained in various ways. One explanation is based on the postulation of so-called *complex qualities*, i.e. the assumption that the animal does not divide the total impression which it perceives into single, separable properties. According to this assumption, any alteration of just one property occurring in the perceptual environment (Uexküll) of the animal is equivalent to complete alteration of the total quality of the impression. Another possible explanation is the following: Many birds, including some very intelligent birds, characteristically react to any alteration of *known* objects with extreme fright. For instance, erection of a wood-pile in the localized home-range of a raven suffices to vex the bird for a whole day. If the same wood-pile is erected far from this home-range, where the raven will similarly see it for the first time, the bird shows no tendency to be afraid and will even settle on it. With less intelligent birds, the same behaviour is even more pronounced. Some small birds suffer wild panic when the litter-tray of the cage is filled with earth for once instead of with the usual sand. But if these birds are taken to an entirely new environment, where the same earth is a component feature, they are far less afraid of it. In the same way, a bird could be just as afraid of the unfamiliar hat on the head of a familiar human being and yet still essentially recognize the human companion. Of course, the massive elicitation of the escape drive completely obscures any other response.

This assumption is further supported by the fact that these less intelligent birds in fact adjust to a slight alteration in the appearance of the human foster-parent more rapidly than to complete replacement of the latter.

However, the concept of a complex quality in perception is itself supported by the fact that many unintelligent birds respond to the familiar and unaltered human companion as to a novel object when seen against an unusual background, *which may in itself be familiar*. At the present time I have a cow-bird which is far from shy; it lives in a cage between my writing table and a bay window alcove which I scarcely ever frequent. As long as I am moving around in the room, the bird is calm and tame, but it begins to flutter as soon as it sees me between itself and the bay window. The concept of complex quality in animal perception (Volkelt) explains this in that the bird does not separate the perceived retinal image into object and background (the highest degree of non-separation) and will perceive a well-known human being differently against different backgrounds even when each background is itself well known.

With very intelligent birds, the entire process of recognition appears to differ from that in unintelligent birds. I experienced the following with my ravens: These birds were generally indifferent to alterations in my clothing and would even accompany me on the wing when I went on skiing excursions, although I presented a quite different image with the long boards on my feet. The only feature of my clothing which affected them was evidenced by their reluctance to perch on my arm when it was covered with a strange sleeve. I once had a motor cycle accident in which I broke my jaw, and for some time I had to wear a dressing which incorporated a spring on each side to pull the jaw upwards. As I came to the ravens for the first time with this weird helmet on my head, they panicked and rushed against the cage-wire on the side opposite me. But as soon as I uttered a few words they suddenly stopped their frenzy and looked at me motionless for some moments. This they did over their shoulders, since they had crammed themselves into the hindmost corner of the cage, facing away from me. Then, their sleeked feathers abruptly loosened and all three ravens shook themselves simultaneously, in the manner generally typical for birds relaxing after a great fright. After this, all approached me just as usual and began to peck at my shoes as they had always done. Thus, these birds had obviously taken me for a stranger and had then abruptly subtracted the head dressing as something not belonging to my person. So far, I have only experienced the like with dogs and children. These ravens had therefore recognized me at first sight and had only been afraid of the head dressing – otherwise they would not have been able to relax so completely a moment later.

3. THE FUNCTIONS OF THE PARENTAL COMPANION

In this discussion, having set out from an environmentalist standpoint, we are not concerned with *all* of the functions of the companion which have any relationship to the bird under investigation, but only with those which always produce a specific *behavioural response* (i.e. a 'counter-function' in Uexküll's terminology). Wherever this is not the case, we are not justified in maintaining that the considered function of the companion actually appears in the bird's environment. On the other hand, we will regard stimuli emanating from the companion and responded to by our investigated bird as 'functions' of the companion, *even when* these do not impress us at all as activities of the companion, because the latter behaves quite passively during the whole process and possesses no knowledge (not even innate) of its own eliciting function. For example, as one 'function' of the juvenile companion we will be dealing with the mere presence of offspring in the nest-cup as the factor eliciting the parental drive to warm them.

(*a*) *Elicitation of the begging response*

The offspring of all nidicolous and some nidifugous birds respond to certain stimuli emanating from the parents with behaviour patterns which themselves, within the framework of the species-specific instinctive behavioural system, have the function of eliciting the feeding response in the parents – a typical example of instinct interlocking. These elicitatory behaviour patterns of the offspring can be summarized as *begging responses*, and we shall need to determine the stimuli which elicit them. It has already been mentioned on p. 140 how certain acoustic and tactile stimuli can elicit begging responses in young birds which are still blind. It should be mentioned at this point that optical stimuli which are unrelated to visual image-formation can also produce this effect in a typical manner. In many nidicolous young birds, as long as they are still warmed continuously by the parents, any intensification of the illumination indicates that the parent animal has risen from the nest and that feeding will soon take place. Conversely, many young hole-nesting birds will begin to beg when the light intensity decreases. For these birds, overshadowing of the nest entrance indicates the arrival of a parent bird and this evokes a corresponding instinctive response.

In many birds the offspring will perform the begging responses when its hunger exceeds a certain level, even if the parent does not see this at all. Particularly prominent in this are growing young herons, which will perform their begging movements and calls for hours even if the parents are nowhere near the nest. Nevertheless, the observer notices a considerable magnification of these activities when the young bird sights the returning parent. Even with nidicolous birds, which do not continuously beg '*in vacuo*' like the Night Heron, the mere sight of the adult bird arriving often suffices for elicitation of begging.

A particular releasing pattern of the parents which acts to produce readiness to accept food in the young is found in the form of a greeting ceremony – the well-known beak-rattling – in the White Stork and some closely-related species. Night Herons show similar behaviour. The young do in fact beg when the parents appear on the horizon, but they nevertheless quickly utter the greeting call when the adult lands on the nest, before greedily grasping the latter's beak. In this case, too, one has the impression that reciprocal performance of a 'bothersome ritual' suffices.

Even birds which do not possess a greeting ceremony often possess instinctive behaviour patterns (very often specific calls) which will elicit the begging response of the offspring if the response does not otherwise appear. In the jackdaw, this call is interesting in that one can understand the evolution of the function, as I have reported in an earlier paper: The normal summoning call is in fact altered by the physical fact that the jackdaw does not open its beak when calling, since it would otherwise lose the crop contents. However, this form of the call is instinctively determined, since it is also produced when the crop is demonstrably empty. Warblers elicit the gaping response of their offspring by springing onto their backs. I once saw a female Blackcap

(*Sylvia atricapilla*), which had returned to her offspring to find them asleep, awakening her young – but not with a call. The young birds were perching close together in a row along a branch, and the female jumped *along* the row from back to back. She made use of her wings, however, so that she only touched the young birds lightly with her feet. I believe that this interesting instinctive behaviour pattern is present in very many passerines, for the very reason that with a great number of species (e.g. the jay, *Garrulus glandarius*) the young birds can be provoked to gape with a sharp prod on the head and upper back, even when other stimuli fail.

At this point, a few words must be said about the gaping response of passerines. Opening the beak wide, which originally took place simply for the purpose of food intake, has (as has already been indicated) adopted a new function and thus a quite distinct significance during evolution of the group. In a typical case of instinct interlocking between young bird and parent, this instinctive pattern predominantly serves to elicit the feeding response of the adult. We shall need to return to this feature in the chapter on the juvenile companion.

However easily a human being can replace the feeding adult bird in this special form of food provision, the instinctive behaviour pattern of the young bird remains completely resistant to adaptation to changed conditions. A human being functioning as the parental companion is thus frequently able to determine the exact details of the parental behaviour instinctively expected by the young bird. It is extremely conspicuous, and in some cases irritatingly so, that a young fledgling whilst gaping will not move even a fraction of a centimetre to approach the feeding human being. In most cases, such a bird will not even take off and approach before gaping. The schema of the parental companion *can* fly, of course. Only the young of a few bird species, without exception types which exhibit persistent following of the parent after emergence, will gradually learn to approach the human foster-parent after fledging. But even these fledglings will normally approach *only when they are not in the act of gaping*. If I should enter a room where a freshly-fledged young jackdaw is perching on some object inaccessible to me, the bird will at first gape towards me from its perch. As long as the fledgling continues to gape, it is prevented from flying down towards me. Only when the gaping response tires is it possible that the bird will fly towards me in the interval between two bouts of gaping. Hand-reared jackdaws which have already been flying for some time will not respond to the sight of the human foster-parent with immediate gaping, but will fly towards him. This adaptive behaviour can be promptly converted to a non-functional response, however, if such birds are induced to gape *from a short distance away*. In this situation, even an older juvenile will be unable to fly over the short stretch involved. It is as if gaping itself were to bind the young bird to the spot, and at such short distances the bird is in fact more likely to respond with gaping than with approach on seeing the parental companion.

Apart from preventing movement from one spot, the gaping response of young passerines can similarly block other responses in certain circumstances. Young birds which are about to begin independent eating initially eat only in a more-or-less playful manner *when they are almost entirely satiated.* As soon as they become slightly more hungry, they will begin to gape. Only when such birds are alone and unaffected by stimuli which might elicit gaping do they succeed in eating independently when quite hungry. The following can happen with such birds in human care: The 'foster-parent' is away for several days and returns to find the birds apparently healthy and adequately nourished, having been dependent upon independent eating in the keeper's absence. The young birds beg from the keeper after his return, but he regards it as unnecessary to continue to feed them, since they have demonstrably fed independently for several days. But if the human foster-parent is forced to stay continuously in the same room with the young birds, their incessant begging becomes vexatious. More than that, one is amazed to see the birds becoming weak and exhibiting all the signs of nutritional deficiency after a few hours. The sight of the foster-parent actually causes persistent elicitation of the gaping instinct, and this so completely blocks the bird's ability to eat independently that such a bird, which has already eaten alone for several days in the absence of the keeper, will literally starve rather than eat in his presence.

(*b*) *Elicitation of food-uptake*

Even in species where the innate schema of the parental companion generally incorporates few signals, the species-specific form of food-uptake from the parents is completely innately determined. In no single bird species is there the slightest degree of modification of the instinctive behaviour patterns of food-uptake in the young bird by individual experience. This absence of adaptability in the instinctive patterns of food-uptake presents great difficulties in the rearing of many species.

There is some variation in bird species which exhibit direct regurgitation of food into the crop of the young, according to whether the beak of the young bird is thrust into that of the parent during regurgitation (as in pigeons, etc.), or the beak of the young bird encompasses that of the adult (e.g. in herons, etc.).

Even when reared by a human substitute parent, young pigeons attempt to bore their beaks into crevices of any kind. Swallowing actions are only elicited in such birds when the base of the beak is successfully subjected to tactile stimulation from all sides, as when one encompasses the beak with the finger-tips. But it is by no means simple to elicit swallowing actions from the young bird by all-round stimulation of the base of the beak and to simultaneously pump a paste of swollen seeds into its crop. The sucking effect of the swallowing actions is in fact so weak that feeding is not successful without extra pressure supplied by the food-providing companion. The young pigeon is completely incapable of food-uptake unless all of these conditions (partially reflex-eliciting and partially mechanical in nature) are fulfilled. The simplest device to aid artificial hand-rearing is to take the swollen seeds in the mouth and to surround the young bird's beak with the lips. The young bird immediately begins to perform swallowing actions and the food can be easily injected into its throat. As long as the young pigeons are still blind, any tactile stimulus elicits head movements which are directed towards pressing the beak into crevices. Guided only by the tactile sense, they reliably find such crevices (e.g. those between the fingers of a human hand). Birds which have remained with their parents accurately attempt, at a later age, to bore their beaks into the angle of the parent's beaks, obviously employing optical orientation. However, the recognition of this source for tapping food is probably acquired rather than innately determined, since such birds reared by human beings aim just as accurately at the human foster-parent's mouth.

Just like pigeons, a young cormorant pushes its beak into the parent's beak, but in this case the young bird appears to grasp the food actively. Although actual regurgitation under pressure on the part of the parent bird does not occur, the reflex for grasping and swallowing food is still unconditionally dependent upon pushing the beak into the throat of the parent. 'Since one cannot easily feed them from the throat', as Heinroth puts it, forced feeding must be employed in artificial rearing of very young cormorants, i.e. the fish must be pressed into the throat with mild pressure. This is not at all easy, since the young bird does not keep still but continually carries out greedy seeking motions, since it 'wants to push its beak into a throat'! With such young birds, where hand-rearing takes weeks, one is particularly struck by the absolute lack of adaptive modification in their instinctive behaviour patterns!

In contrast to birds of that kind, young birds of species in which parental feeding by regurgitation is conducted with the offspring closing its beak round the *outside* of the parent's beak are easy to feed artifically. In this case, the young bird attempts to grasp *something* (i.e. the parent's beak) and when this something is not the intermediate goal (the parental beak) but the food itself, then the latter is simply swallowed. But one only has to hold onto the food to demonstrate that the grasping action of the young bird is not 'meant' for the food, but for the parent's beak.

When this is done, the juvenile scarcely attempts to pull the food away – it performs with the grasped object begging motions and calls (i.e. the characteristic 'milking' actions) which normally elicit regurgitation by the adult.

Thus, one can easily feed young parrots from a spoon, since they grip one edge and 'milk' it just like the parental beak and one can then pour in the food by tipping the spoon. At an advanced stage of development, they 'apparently' eat independently from the stationary spoon, but they nevertheless perform begging motions and the appropriate calls throughout food-uptake. They therefore behave as if they believed that if the spoon were not continually prompted to supply food with the described releasers it would stop feeding, which is exactly what the parent bird would do. The young birds 'feel' that they are 'being fed', and if this were not so they *would not be able* to take up food independently in the manner described, since the actual drive to eat independently does not appear until later.

The young of Heron species behave similarly in human care. They also grasp the proffered food with the action actually intended for grasping the parental beak and naturally swallow the – in fact unexpected – food which has been taken into the beak. But there is a difference here in that when reared by the parents the young birds will pick up fallen animal food from the edge of the nest. Regurgitation into the throat in Herons is probably derived from *external regurgitation*. In the Night Heron (*Nycticorax*), the food is regurgitated in front of the very small nestlings by the parent, which props its open beak on the nest. The nestlings peck at and grasp the parent's beak before the food has been regurgitated, so that the semi-digested paste more or less dribbles between their jaws. But even after the first few days, they optically track, pick up and eat the scattered food-morsels in the nest-cup. During regurgitation, the adult bird often re-ingests a part of the regurgitated food, possibly in order to keep it warm, or perhaps even to keep back morsels which are too big. Storks do exactly the same. Whereas this form of feeding is persistently retained in storks, with the young continually begging with an upwards-pointing beak and waiting for regurgitation, young Night Herons soon begin to stretch towards the arriving parent bird as soon as it lands on the edge of the nest and to grasp the base of the parental beak. The young bird starts to perform swallowing actions at once and the whole sequence gives the impression of the nestling trying to swallow the parent's beak.

(*c*) *Elicitation of the following response*

One function of the parental companion, which is almost as important as feeding and which in some cases makes the latter unnecessary, is *guiding* of the progeny. The drive to follow a guiding parental companion is found in varying degrees of development with the great majority of nidifugous birds and with a number of nidicolous types. Nidifugous birds lacking the following drive are the Megapodidae, the gulls and the terns, in which the young are provided with food just like the nidicolous types, although they are far from tied to the nest. Only a relatively few nidicolous species exhibit following of the parents by the fledged young in the nidifugous manner. These few particularly include extremely intelligent types, e.g. the large Corvids, some parrots, etc.

We have already considered the following drive of some *nidifugous* birds in the section on imprinting, covering, for example, the linking of elicitation of following to specific stimuli emanating from the parent in some species and demonstrating how such stimuli can be replaced by arbitrary stimuli in other species. I intend here to discuss only a number of peculiarities of the following drive, which show quite plainly how the young bird is sometimes adapted to quite specific characteristics of the parent – characteristics which are in fact less conspicuous than summoning or warning calls (which the young bird also recognizes innately), but which must remain just as constant if the complementary function of the parental companion oriented towards the following drive of the young bird is to remain unimpaired.

We have already considered the fact that the guiding companion of the young mallard must quack continuously in order to fulfil its rôle. This continuous vocalization is found in very many guiding nidifugous types. However, the parental companion must also keep moving whilst the young are awake and active. If the human parental companion should stop still for some time, the stream of chicks will pass by and keep on going. The ducklings first become uneasy and begin to pipe when they have gone on for some yards. If the human companion stays still and begins to call, the ducklings do not run back immediately – they remain some time in a group and continue to pipe with their heads stretched high. Engelmann has described in some detail similar behaviour in domestic chicks. One of the young birds will then abruptly run a short distance in the approximate direction of the call, stop again and pipe in the same posture. A second duckling will follow and approach the human companion a little closer, and this will be followed by others until the avalanche breaks loose and the entire flock comes back at full pelt. Having arrived back at the 'parent', they will all begin to quack vigorously, using the 'conversation call' which can always be heard when a group of mallard has been disrupted and has eventually re-congregated.

It is just as difficult to induce a flock of mallard ducklings to bear off at an acute angle to their previous direction as it is to bring them to a sudden halt. If the human companion leading the flock should take a sharp turn, the ducklings carry straight on and begin to 'cry' after progressing a few yards, just as when the leader stops still. They turn back in the same manner and thus eventually take the sharp corner as well, to go in the direction intended. I cannot say for sure whether the mother mallard possesses specific signal-calls evoking a specific response from the young for this particular situation, but I believe this to be unlikely, since I have never seen such behaviour in many years of observation of mallard. Just like the mother hen, the mallard mother possesses a motor pattern which communicates the intended course to the young. This is a forwards nodding movement of the head, like that performed by many birds when they attempt to optically fixate an object when moving along. But this directional signal-nodding movement is characteristically faster and more extensive than is usual for the speed of the moving duck. Cichlid fish which lead their young have developed a corresponding movement, through an amazing process of convergence. Exaggerated swimming intention movements, which also give the impression that the parent intends to depart at top speed in the direction shown (although the fish in fact progresses quite slowly), are employed to signal the intended swimming direction to the school of young fish. However, all of these directional changes accomplished through signalling by the parental animal are restricted to obtuse angle turns.

Acute angle turns and abrupt halting normally only occur in the mother mallard when she has perceived a frightening stimulus ahead. In this case, however, she utters a warning call, which a human being unfortunately cannot imitate. The only other case where the mallard mother interrupts her continual, appetitive forward motion occurs when she chances upon a spot which is so rich in food that she can initially find enough to eat without moving. But the young follow suit and do not move away from the mother. Of course, under natural conditions this all seems self-evident. It is only when one attempts to act as a replacement mother for the young birds that one realizes how delicately the behaviour patterns of mother and young are mutually adapted, how slight a change will suffice to disrupt the functional pattern of species-specific instincts, and how sharply demarcated are the instinctive behaviour patterns of the young adapted to the conspecific parent.

The function of the guiding parental companion is quite different in the few *nidicolous* types in which guiding of the young takes place after fledging. This nidifugous kind of guidance of the young occurs among our endemic Corvids in the jackdaw, less markedly in the raven and even less developed in the Hooded Crow and the magpie. The rook appears to behave similarly to the jackdaw. The behaviour of the guiding

Corvids appears to be equalled in the parrots, which are similarly very intelligent, but here no reliable information is available.

The following drive of young jackdaws and ravens does not appear until some time after abandonment of the nest, when the young birds are physically capable of following their parents everywhere. Corvid following is further distinguished from the following drive of nidifugous types in that it is composed of a *terrestrial following drive* and an *aerial following drive*, which are remarkably independent of one another.

If one takes a jackdaw which has been artificially reared from an early age (i.e. for which the foster-parent represents the parental companion in every respect) and brings it out into the open at the time when its following drive is awakening, it will usually attach this to the human companion. On the other hand, one can also induce a humanized jackdaw of this kind to fly after an adult jackdaw or the like. It will then only respond to the human foster-parent as the parental companion in feeding behaviour.

However, if one should wish to study the following drive of a humanized young jackdaw by eliciting this oneself, roughly the following tactics are necessary: The bird should be set down on an object of only medium height in the open and one should then give attention to the bird until it has overcome its fright response to the change of environment. If the bird is placed on the ground, it will often fly off prematurely through fear of the ground; if the perch is too high, the procedure described below will not work. After calming (and perhaps even feeding) the bird for a while, one must suddenly leap up and run away from the bird at top speed. The abrupt rising movement is very important for elicitation of the following instinct, and the jackdaw must therefore be perched at a height below the experimenter's head. If this has all been done correctly and the jackdaw is in good, healthy condition, the bird is sure to fly after the human companion. The jackdaw usually lands on the human companion, but it will sometimes carry on flying. If this occurs, the jackdaw is not able to turn back properly and to return to its keeper. It will usually land some way up in a tree, and it is not at all easy to bring the bird back down. One cannot elicit the following instinct by standing near to the tree and suddenly running away. Apparently, elicitation of take-off is dependent upon the guiding companion being at the same level as the bird and probably rising somewhat higher on take-off. Luckily, young jackdaws respond well to imitations of the jackdaw summoning call, and they will eventually follow their ears to fly straight back to the human companion – something which young nidifugous birds cannot do, since chicks only find the hen uttering the summoning call after a long process of zig-zag acoustic scanning of the mother's position. The chicks must be approached so that they are separated by a gap only a few yards, over which they can regain contact rapidly. But jackdaw parents also do this, at least in the initial period after the young birds have learned to follow in the air. If the young birds should lose contact with the parents (normally flying ahead), either because of overtaking the parents or flying too far behind, the parents re-establish contact by flying slowly, continually looking back over their shoulders and waiting for their offspring, or by overtaking any young which are ahead and thus resuming leadership of the flight. The following drive of such young jackdaws, which have not been flying with the parents very long, behaves rather like an elastic band stretched between parents and young: the greater the separation, the stronger the pull – up to a certain point, where the band breaks away and must be reconstituted.

After a while, this purely reflex, optically-elicited following behaviour recedes in the face of the contact drive of the young elicited by auditory stimuli. The young birds can then localize the source of parental or human calling with amazing accuracy. In accordance with this ability, they can separate themselves farther from the parental companion without danger of losing contact. Such 'Stimmfühlung' (vocal contact), as Heinroth calls this acoustic cohesion, in fact often plays a very large part in the behaviour.

Optical elicitation of take-off by the take-off of the leader remains as the only persistent optical effect. The strongest stimulus available to the human jackdaw father for inducing the young jackdaw to follow in the air consists of a fast run towards the bird followed by a sharp turn and a fast trot in the opposite direction. I found out this method of elicitation with my first young jackdaw and I was surprised, and not a little pleased, when I later observed that the same method of elicitation is a regular species-specific instinctive behaviour pattern of adult jackdaws. It is advisable for experimental purposes, however, to use this method only with young birds which have already flown after the keeper in the open for some time. When this is tried as the primary method of elicitation with a very young jackdaw, one frequently elicits escape responses instead of aerial following.

The sharp separation between the aerial following drive and the terrestrial following drive is very distinct. The most conspicuous feature is the initial great difference in intensity of the two drives. On foot, a young jackdaw which follows the parental companion 'as if on a lead' in the air, will only show an approximate tendency to maintain the same direction. As with young nidifugous birds, terrestrial following by the young jackdaw only operates persistently when the human parental companion exactly maintains the same speed as a food-seeking parent jackdaw on the move. If this speed is exceeded, the jackdaw will be left behind and will then either fly up *or not follow at all.* In fact, it is quite possible to employ 'creeping stimulation' and steal away from the jackdaw *without* eliciting the aerial following response. Under normal conditions, it is apparently unimportant if a young jackdaw seeking food on the ground should be separated some distance from the guiding adult. The take-off of the parent will in any case bring the young bird into the air, even if it has strayed forty to a hundred yards away in the meantime. This distance is of no account in the air.

In the summer of 1933, I experimented with a young starling and discovered that with this species the difference in intensity between the terrestrial and aerial following drives is even greater. This is in exact correspondence with the relatively greater flight speed and lesser walking pace of the starling. When searching for some time for food, the birds cannot separate enough to stop them from re-congregating in the air a few seconds after take-off.

This is even more pronounced with some small passerines which virtually ignore both the parental companion and other conspecifics whilst moving around on foot and yet stay close 'as if on a lead' on the wing. Heinroth has described this effect in a hand-reared yellow-hammer.

(*d*) *The warning call*

One very important function of the parental companion is that of warning the young birds of danger when necessary. The response to the warning calls and movements of the parents is presumably innate in all existing species,[63] perhaps with the exception of the Megapodidae, which do not have a parental companion.

Warning consists of calls and/or movements and *always* represents a genuine instinctive behaviour pattern, the performance of which involves no purposive altruistic impulse on the part of the warning companion – the bird will warn even when it is alone. It is also worth mentioning that the presence of specific warning calls by no means allows one to conclude that warning plays a particularly large part in the life of the species concerned, and the converse conclusion is especially unjustifiable. Of all the species I have worked with, the young of the jackdaw are most dependent upon the warning function of the adults, since they exhibit the escape response far less readily in response to the stimuli emanating from a potential predator than to the fright and subsequent flight of adult conspecifics. In spite of this, the jackdaw has no actual warning call. It is virtually characteristic that the jackdaw, which in fact responds extremely sensitively to the slightest display movements of conspecifics, should not have needed to develop an actual warning call. Warning of the young by the parents occurs in a similar manner in other nidicolous types, such as the raven, where the parents guide for some time. The raven does actually possess a warning call, but this apparently plays a less important part in the elicitation of the young bird's escape response than the accompanying display movements. To my annoyance, I never succeeded in eliciting the escape response of my young ravens by imitation of the warning call, although in certain circumstances it would have been desirable to be able to do this. Ravens typically persistently avoid places where they have once been frightened. They were always greatly alarmed by unfamiliar people who had approached quite closely unnoticed. In order to avoid the development of a localized fear response, which often severely disrupted my experiments, I would have dearly liked to induce my ravens to fly off with the imitation warning call as soon as I saw someone approaching unnoticed by the ravens. I never succeeded in this until on one occasion I happened to move off rapidly at the same time as uttering the warning call, since I had seen an undesired acquaintance approaching. The ravens actively joined in with the calling and fled most rapidly – much more rapidly than I did myself. Thus, they were not simply flying after me.

The response to parental warning is much more reflex-like in young nidifugous birds. In these birds, too, the response is completely innate, even in the Greylag goose, which otherwise exhibits so few purely instinctive responses to parental stimuli and does not even respond to the summoning call instinctively. It would be entirely wrong to assume that a young bird of this kind has some kind of understanding that an acquainted being is attempting to warn it against a third party. In many cases, one has the very strong impression that the young bird is fleeing *from the warning call.* One only has to consider the fact that a freshly-dried chick (e.g. of the Golden Pheasant) *cannot be induced to flee* by a stimulus of any kind *except* the warning call of the mother. I once saw a still-wet Golden Pheasant chick run a good yard away and hide in a dark corner as I removed its older sibling from the nest and frightened the hen into uttering the warning call. In the resting state, this fleeing chick was scarcely able to hold its head up; it could hardly stand and could not even walk slowly. It was only whipped up to perform a short, but rapid, spurt by the immense stimulatory effect of the elicitatory warning call. Such tiny chicks always flee *away from* the hen in response to warning and thus usually take cover separately a long way away. They remain in hiding until the customary summoning call of the mother is heard. If she is persistently alarmed, so that she will not summon the chicks, they will remain in hiding until she does.

Chicks of many gallinaceous birds which are somewhat larger and already flee up into trees show two distinctly separate responses to *two different warning calls* of the parents. The Burmese Jungle Fowl and the domestic chicken also have two separate warning calls for an aerial predator and a terrestrial predator. In the literature, however, one encounters the opinion that these two calls represent intergrading displays based on the same stimulatory effect. I take this fact as justification for once again expanding on well-known phenomena. *Gallus bankiva* responds to an aerial predatory bird with a drawn-out sonant R-call, which is usually represented as 'Rräh' ('Rreh'). This call can be elicited by a flying predatory bird, but it can also appear in response to any other unknown, large flying bird, even when it does not look at all like a raptor. I have heard the 'Rräh' of the local domestic cocks in response to the sight of my flying storks, Greylag geese, cormorants, ravens and other large birds, i.e. including some which are far from raptorial in appearance. Such elicitation of the 'Rräh'-call was much more reliable and not subject to habituation in response to short-necked, rapidly-flying birds: Both my Parakeet and domestic pigeons (when diving steeply downwards) regularly provoked the village chickens to utter this warning call against predatory birds.

If the same stimulatory effect which gives rise to the 'Rräh'-call reaches a higher pitch, the escape response occurs *downwards*, i.e. towards the darkness on the ground and thus possibly into cover. 'Rräh-elicitation' is automatically coupled with glancing upwards.

Apart from this warning and escape response, domestic chickens and many gallinaceous birds possess a second response. If a domestic hen or Burmese Jungle Fowl should sight a *non-flying*, but nevertheless apparently dangerous animal, an entirely different call is heard. The chicken at first calls 'Gockokokok', with heavy emphasis on the second syllable. If the stimulation persists and reassurance does not set in immediately, single 'gok'-sounds are uttered at regular intervals of just under one second. These can increase in frequency or, less often, fuse into a very loud, disyllabic 'Gokóhk', according to the strength of stimulation. Strangely, this warning call is uttered by the hen after laying and is best known in most hens for this association than for its warning effect.

The question arises as to why the warning call, of all things, should be elicited straight after laying. This behaviour at first appears so unadaptive, even biologically deleterious, that one is not inclined to believe that the wild form exhibits the same effect. However, we encounter the same behaviour in the blackbird[64] at dusk. Everyone knows the characteristic loud warning vocalization that the blackbirds utter before going to sleep. Heinroth has expressed the opinion that this loud warning vocalization is always uttered far away from the actual sleeping site, and that the blackbirds fly some distance in silence after calling. This assumption was completely backed up by observation. I assumed that a similar biological function operates in hens, and this was likewise verified by an enquiry which Heinroth kindly conducted at my request. Burmese Jungle Fowl, and even hybrids between these birds and domestic chickens, stealthily and silently leave the nest after laying and *fly* as far away as possible. Only then do they burst forth into the typical *egg-laying cackle.* Since the drive to fly away usually breaks down in domestic hens, which are more or less incapable of flight, they usually utter the egg-laying cackle in a completely non-adaptive manner right by the nest. Thus, cackling in the laying hen is also basically coupled with take-off motivation.

During the normal warning response against terrestrial predators, one can usually identify take-off motivation in the chicken in the course of the regular gock-gock-gock, etc. The chicken flattens its feathers and performs vigorous head movements (which can be unquestionably taken as aiming movements for flight) in time with the individual call-notes. If the stimulation of *this* motivational state is increased to the point of locomotor elicitation, the bird will eventually take off whilst uttering a 'Gogóhk' call and soar up, repeating this latter call in a rapid sequence. During this response, the chicken gazes alertly downwards to the ground.

This warning response to terrestrial predators is disrupted in the domesticated chicken by the suppression of flight in heavy breeds, but one can have no doubts as to its function after seeing the corresponding behaviour pattern in the Golden Pheasant.

As is apparently the case with many gallinaceous birds, one finds two warning responses in the Golden Pheasant. In accordance with the higher-pitched voice of *Chrysolophus*, the aerial warning call sounds more like 'Rrihh'. A mild 'Grix-grix-grix' corresponds to the terrestrial warning call of the Burmese Jungle Fowl. This call is uttered at somewhat greater intervals than the corresponding call of the domestic chicken, and can culminate in a disyllabic 'Girrih', which is intercalated between long sequences of the monosyllabic call, as is the disyllabic call of the chicken. But this call is heard more rarely than that of the chicken, since take-off occurs much more readily in the Golden Pheasant than in the chicken. In addition, take-off of the Golden Pheasant – in contrast to that of the domestic chicken – usually occurs in silence. The Golden Pheasant usually takes refuge in a tree without fleeing any further.

In the context of these deliberations, it is important that this dichotomy of the response is at first absent in young domestic chicks and Golden Pheasant chicks. The latter take cover immediately in response to both warning calls. They only begin to respond to the terrestrial warning with take-off motivation from the time when they can fly a little and are starting to take refuge in trees with the mother at night.

Comparative considerations permit the statement that this behaviour pattern, innately determined in mother and offspring and extremely decisive in the preservation of the species, gives a much greater impression of reflex-like, involuntary processes in gallinaceous birds than the alerting and flight of young jackdaws and other Corvids elicited only by the behaviour of the parents.

It must also be remembered that young Corvids look around them when the parents are alarmed (i.e. they look out for a predator) and thus behave as if they innately 'know' that the leaders are warning them of a third party. This is in contrast to young nidifugous chicks, which by all appearances flee directly *away from* the warning call of the mother, In any case, from the outset one does not expect that these freshly-hatched young birds will possess such highly-developed behaviour patterns as a physically-mature young Corvid. The larger domestic chicks, when able to fly, behave like the young Corvids in response to utterance of the terrestrial warning call – they look all around them for the expected predator.

I should like to include a few words about *behaviour associated with disruption of the warning function.*

Since the warning calls and displays of individual species usually elicit the appropriate response of the young birds in their specific form, a human being taking the place of the parents is usually unable to supply adequate substitute stimuli.

However, in many animals we find the tendency to exhibit elicitation of responses, which are actually predetermined for a specific mode of elicitation, by another stimulus when the adequate stimulus is persistently withheld. The stimulus threshold of the reflexes or reflex chains concerned is continually lowered until the entire response eventually occurs *without* the influence of recognizable external stimuli. It is just as if a species-specific instinctive behaviour pattern which is not elicited for some time because of lack of external stimulation were eventually to become an *internal* source of stimulation.

My interpretation of this fact is greatly at variance with that of Groos, who regards all such vacuum activities as 'play' and assumes their biological value to be that of 'behavioural practice' (running-in). I am of the opinion that only a relatively small number of vacuum activities earn this interpretation and that very few of them are really *distinct from the response in the vital situation.* The play-fighting of two young pups is of course quite different from serious dog-fights and is not just a lesser degree of the latter behaviour. The main difference is represented by the *maintenance of all social inhibitions in play*, even in the most boisterous and passionate types of play. Above all, the inhibition against genuine biting is maintained, whereas in serious fights, even in the slightest dispute, this inhibition is at once completely lifted. Similar differences between play and 'serious' behaviour can be demonstrated in many cases of animal play, and in my opinion the use of the word 'play' should be restricted to such cases. I am convinced that the central nervous processes of the animal involved in the vacuum response are in many cases little different from those in normal elicitation, particularly in those animals of restricted intelligence. I am sure that a young rabbit which suddenly begins to perform headlong doubling movements on an open meadow, and abruptly heads for the nearest cover to hide, experiences exactly the same 'fear' as if it were actually chased by a goshawk. A goat kid, on the other hand, behaves quite differently during analogous 'escape'-play. After a few doubling and leaping movements, it stops still and induces the chasing sibling to continue pursuit. This is genuine play. As with everything, the boundaries are hard to define; but we will surely not go far wrong in assuming that *in birds* the majority of vacuum activities are *identical* with serious behaviour *as regards the internal nervous processes* and therefore cannot be regarded as play. Selous, without doubt a reliable observer and interpreter of animal behaviour, denies that play occurs in birds at all. There are of course exceptions and intermediates – occurring, as may be expected, among the large Corvids and parrots.

When the warning function of the parental animal is lacking, many young birds exhibit peculiar vacuum behaviour of the escape and alerting responses which should actually be elicited by parental warning.[65] Thus, one often finds an extremely peculiar fickleness and *extreme readiness to perform escape responses*, entirely independent of the tameness of the bird towards its keeper. It is as if the bird *waits* for the persistently absent predator to put in an appearance at last. Heinroth likes to express this motivational state of tame young birds in a joking form of conscious anthropomorphization: 'When is the old devil going to turn up?' The tendency to exhibit inordinate fear in response to inconsiderable minor sources of stimulation contrasts conspicuously with the trust exhibited towards the keeper. One could even say that there is a *converse relationship* with the trust shown. I am primarily driven to consider the described fickleness and readiness to panic as an effect of absence of the parental warning function through the fact that this behaviour is in fact most clearly observed in species where the parents and offspring exhibit a particularly long and intimate mutual relationship, and in which the offspring exhibit *little sign of innate recognition of a potential predator*. Such young birds, which prove to be particularly trusting towards a human companion, are members of species (e.g. the jackdaw) in which elicitation of the escape drive is not brought about by the sight of an instinctively-recognized predator, but by sympathetic induction based on the warning and escape response of the guiding parents. It is in birds of this kind that we see a 'vacuum' response to a non-existent stimulus. According to Heinroth, we find this described effect extremely plainly in cranes, Greylag geese and ravens. My own observations have confirmed this for the latter species. I should like to add the jackdaw to this list. (I have already reported the entirely baseless occurrence of alarms and consequent panics which are particularly prone to occur in leaderless young birds of this species.

In jackdaws, the effect can briefly be explained from the fact that the escape responses of one bird elicit even greater responses in the next bird. The slightest alarm in one individual will lead to flight of the entire flock, as long as enough individuals are present for the intensification of alarm at each transference of stimulation from one bird to another to produce an adequate magnification effect. In a flock of normally-reared birds, each individual will respond to the display of fear in a companion with a response which is, at the *maximum*, of the same intensity. The avalanche intensification of panic in the young birds is based on an *abnormal lowering of the threshold value of the warning stimulus*. It is possible that a similar effect is involved in the elevated panic tendencies of some ungulates.

(*e*) *Defence*

Defence of the young by parent birds appears to be 'granted recognition' by the young in many cases. Brückner observed with gallinaceous birds that on approach of a predator which elicited the defence response of the mother hen the young would collect *behind* the hen – in the 'danger shadows', as Brückner called it. I observed the same with Night Herons and Little Egrets.

4. DISSOCIABILITY OF THE FUNCTIONAL SYSTEMS

I believe it is justifiable to conclude that it is immaterial within the environment of a bird whether all of the described functions of the parent are performed by one and the same individual or not, since the instinctive behaviour patterns of the young bird affecting different functional systems are not disrupted when a different companion acts as the counter-function to each pattern. Of course, this must occur in a manner compatible with the instinctive behavioural system.

Young nidifugous birds will, for example, be content when the human being represents the companion in the functional system of guidance, whilst their drive to creep under and be warmed by a given object is satisfied by a petrol-stove as the counter-function. When the chicks awake and creep out from under the 'petroleum-hen', they pipe 'after' the guiding companion just as they cry 'after' the heat-donor when cold or tired.

In cases where the parent normally combines within itself the functions of guidance and feeding, a similar extensive independence of the two functional systems is found. I have repeatedly quite simply attached young jackdaws, which I could not lead myself for lack of time, to my flock of adult, free-flying jackdaws. The juveniles' following drive becomes directed towards the flock of adults without difficulty, and the birds never attempt to fly after me as long as they are not hungry. Neither will they allow me to lead them to places which they and the other jackdaws do not otherwise visit, whereas a young jackdaw attached to me by its following drive will of course blindly allow itself to be led anywhere. But the begging responses of such birds are still directed towards me, just like those of birds also primed to follow me. At the most, it is possible that the gaping drive is extinguished somewhat earlier in the former than in those birds which are attached to me by another bond because of their following responses. Such young jackdaws will very rarely exhibit begging from the adults satisfying their following drive.

Since the response to warning by the parents is completely innate and completely free of acquired effects, it is understandable that this functional system should show a particular degree of independence from others. Even the nature of the warning companion is of no account. A young bird for which the human being (or a particular human being) represents the parental companion in all other functional systems, will be aroused just as greatly by the warning of an adult conspecific as in cases where imitation of the parental warning calls lies within the human vocal range and where young birds located with their actual parents can be frightened by the imitation.

VI. The infant companion

We shall now define the converse concept of the 'infant companion', describing the image which the counter-functions of the young birds (corresponding to parental care instincts) form within the environment of the parents. This concept is more justifiable than that of the image produced by the functions of the parent bird in the young bird's environment and collectively referred to as the 'parental companion', although in some cases this image would have had to be dissected into guiding, feeding and warming companions according to its relevant functional systems. The second concept is more justifiable because the adult bird is usually much more capable of objectivizing stimulus compilation than the young bird, which often begins life at a much lower developmental level.

1. THE INNATE SCHEMA OF THE INFANT COMPANION

Young birds are in most cases instinctively recognized as conspecifics by their parents, which exhibit innate recognition of certain characters. It is in fact obvious that these characters cannot be acquired by imprinting, since the adult bird's own offspring are of course the first freshly-hatched conspecifics which it sees, and yet it must react to this first encounter with the entire repertoire of parental behaviour operating to preserve the species. The curious exception from this rule is the genus *Anser*, which has already been mentioned in a similar context and which represents the ultimate in instinctual deficiency. The Greylag goose appears to be literally the only bird species, among those subjected to close ethological study, in which the parents do not respond specifically to the sight of freshly-hatched young of the same species. Thus, in contrast to *all* other nidifugous birds examined in this respect, they will unhesitatingly lead young of other species and even of other genera, as long as the young birds themselves will adapt to the strange leader. For example, in an experiment which I conducted in 1934, chicks of *Cairina moschata* were led by the Greylag. Unfortunately, it is not known whether this involves genuine imprinting to conspecific young, i.e. whether Greylag geese which have already reared conspecific offspring will still be prepared to accept non-conspecifics in a later clutch. This would not be expected on the basis of known facts about other forms of imprinting. Corresponding behaviour has been reported for mother hens; it is stated that they had not accepted domestic chicks after once hatching out ducklings and even that they had led 'like ducks' into the water. I regard this as so improbable that I am inclined to interpret such reports as observational errors, even if Lloyd Morgan accepts such accounts. It would in any case be best if one were to *completely exclude domestic animals* from theoretical studies of instinct, for the reasons already mentioned. Erroneous generalization of the behaviour of the domestic hen to non-conspecific offspring has been a particular source of false interpretation.

In fact, non-domesticated gallinaceous birds stand out in having a specific, specialized response to the appearance and vocalizations of conspecific chicks. But even some races of the domestic chicken which are quite close to the wild type will by no means care for all of the young birds which they hatch out. One only has to read in Heinroth's *Vögel Mitteleuropas* how many of his young birds were hacked to death by the mother hen which had hatched them. Such foster-mothers will even hack at a scarcely-cracked egg if it does not produce the 'correct' *Bankiva*-specific hatching noises and vocalizations.

I observed how one of my Golden Pheasant hens (*Chrysolophus*) pecked at a pheasant chick (*Phasianus*), which it had hatched out with its own chicks. The hen exhibited quite peculiar behaviour, since she did not pursue the chick in her animosity, but only pecked at it when it was in front of her eyes. This occurred most often when the hen had settled down to warm the chicks, which would creep past beneath her beak one by one into the 'heated dug-out'. She would always fix her gaze *on the head* of the changeling and follow the movement of its skull with her head, just as birds follow insects which they do not dare to peck with a 'head-nystagmus', so that the image of the moving object is kept still on the retina and can be clearly perceived. The *Chrysolophus* hen would then regularly peck gently at the chick's head. This gave the impression that her drive to chase away the changeling was conflicting with her inhibition against seriously biting such a tiny chick.

One important aspect of this observation is the fact that the hen submitted the *top of the head* of this *Phasanius* chick to intensive examination. It in fact seems as if the head markings are the vital species-recognition signal incorporated in the plumage of the offspring of various nidifugous birds for the elicitation of species-specific instinctive patterns of parental care. The head markings of chicks are indeed usually more sharply demarcated and more colourful than the other body markings. In gallinaceous birds, the crown of the head is usually decorated with particularly clear-cut markings, and in the Water Rail there are luminous-coloured, naked areas on the font of the head. Heinroth had previously suspected that these markings represented species-typifying releasers. I have never observed the response to this releaser, or the effects produced by its absence, so clearly as with this *Chrysolophus* hen.

Morphological structures serving only for the elicitation of parental instinctive behaviour patterns are particularly common in various birds. The great variability of these releasers between closely-related species leads one to conclude that these typically brightly-coloured and conspicuously-constructed organs simultaneously perform the function of *species-recognition characters*. This applies particularly to the internal gape markings of young passerines, which were first fully appreciated by Heinroth and were comparatively described by him.

2. PERSONAL RECOGNITION OF THE INFANT COMPANION

It must be stated straight away that we know very little about the individual-personal recognition of offspring in birds. In very many cases.

where one would at first assume that parents would recognize a strange youngster among their offspring, more detailed investigation demonstrates that it is the contrasting behaviour of the strange young bird which draws the attention of the adults. Bees will kill a foreign queen which is placed in their hive, apparently largely because of the queen's conspicuous behaviour and not entirely because of her strange smell, as is generally assumed. If the new queen is made to starve for some considerable time before transfer to the foreign hive, so that she will respond to the first bee she encounters with greedy begging actions, then she will be fed at once and accepted without further ado. The less intelligent birds behave quite similarly towards conspecific offspring. If warblers or wagtails feed any young conspecific which begs for food in just the same way as they would feed their own offspring, we have no justification in assuming that the strange conspecific is distinguished from the true offspring. Acceptance of strange young birds is very often linked to the condition that they should be of the same age as the adult bird's own offspring.

Nevertheless, I assume that personal recognition of offspring is present in the majority of nidifugous birds. It is conspicuous that in these birds the offspring learn to recognize their parents personally far earlier than the latter recognize their offspring, in contrast to Night Herons and similar birds. I am convinced that personal recognition of the offspring is present in nidicolous birds which guide their young for some time after emergence, i.e. particularly in Corvids (with the exception of the jays).

In nidifugous birds, personal acquaintance with the individual offspring usually seems to take some time. It is, in any case, more difficult to smuggle a new member into a flock of older fledglings than into a younger group. Parent Greylag geese, which again behave in a generally non-instinctive manner, personally recognize even very young offspring and immediately notice a foreign gosling, even when it is of the same age as the offspring and is not characterized by incongruous behaviour. Heinroth has supplied the following observation, which is extremely characteristic for the rapid loss of certain groups of behaviour patterns in domestication and demonstrates how atypical the behaviour patterns of domestic animals can be. He writes:

> ... The following case is quite characteristic of the levelling of instinctive behaviour patterns or of the discernment capacity of domestic animals. After I had placed an orphan gosling with a goose family, in which the male was a pure-blooded Greylag gosling and the female was a domestic–Greylag hybrid, it emerged that the mother almost or completely failed to distinguish the foreign gosling, whilst the

male noticed it very rapidly and at first exhibited a great desire to attack the stranger. It took some time before the male had become accustomed to its presence, although the foster-child was, to the human eye, scarcely distinct from the other goslings.

3. THE FUNCTIONS OF THE INFANT COMPANION

In this instance, we shall restrict our considerations to those functions of the young bird operating as counter-functions to specific responses of the parent and constituting an interlocking functional system in combination with the latter. Viewed in this light, the functions of the infant companion exhibit the same simplicity and rudimentary nature that has already been seen in the functions of the parental companion.

(*a*) *The elicitation of feeding*

The feeding drive of adult birds is one which is quite powerful and virtually autonomous. In individually-caged, healthy birds it occurs very often as a vacuum activity. Robins, warblers and other birds often exhibit the following behaviour: Towards the end of the song phase, the birds can be seen hopping restlessly to-and-fro in the cage with food in the beak. At this time, one sees escape-searching movements usually seen in birds not yet accustomed to a cage. Heinroth describes similar behaviour in a corncrake. Since even wild captures, and birds which have not undergone imprinting of their directive drives towards the human being, perform such activities in the absence of any companion representing the young bird, it is actually surprising that the human being *never* acts as the infant companion in the environment of tame birds. Birds which I observed attempting to direct their feeding drive towards an acquainted human being always belonged to species in which the males feed their females, and the active birds were always males. Other behaviour of these males always demonstrated unequivocally that the food-receiving human being represented a female and not an infant. This is quite definitely correlated with the acute response of these birds to the relative strength of the companion: the powerful human being can never have the 'meaning' of an infant.

On the other hand, potentially-feeding cage-birds will feed non-conspecific offspring extremely readily, as long as the latter to some extent fit into the instinctive framework of the species. But this is probably not indicative that the feeding drive does not respond specifically to conspecific offspring. This behaviour is much closer to being a vacuum activity; the bird simply prefers to deposit the proffered food-morsel into a non-conspecific infant throat rather than to continue to hop around inside the cage with the food in its beak. The bird will of course take the food into the beak even when it is alone, as we have seen. Quite generally, one cannot immediately assume with an instinctive behaviour pattern performed 'with the wrong object' that the object is not innately determined, but acquired through imprinting. We are only justified in this assumption when the 'correct object' is ignored in preference for the substitute object when both are offered together. Otherwise we are concerned with satisfaction of the drive through a 'substitute object'. Such satisfaction of a drive is also found frequently among mammals, whereas imprinting to a species-atypical object has been previously observed only in birds.

A remarkable case of *satisfaction of a drive through a substitute object* is provided by the behaviour of various passerines towards the young *cuckoo*. It at first seems puzzling that the young cuckoo is able to elicit the feeding drive of *different* bird species in the way observed. The fact that the gape of the cuckoo exhibits no adaptations to that of the offspring of the host bird seems to be correlated with the active ejection of these competitors from the nest. Many other brood-parasites which do not do this (i.e. which grow up alongside their step-siblings) have undergone extensive adaptations to the external appearance of their competitors. In the widow birds, this adaptation extends to the tiniest details of head and (particularly) gape markings. If this were not the case, the adult birds would apparently respond better to conspecific young than to those of the parasite. It would be informative to conduct an experiment with the endemic cuckoo, in which the offspring of the host species would be returned to the nest when the ejection instinct of the cuckoo has been extinguished. Perhaps the adult birds would then no longer feed the cuckoo adequately. Thus, a brood parasite must either adapt the appearance of its offspring to that of the host's offspring so well that the species-specific feeding drives of the adult host birds also respond to the young parasite, or some 'mechanism' for the extermination of the true offspring must be developed. If the latter can be achieved, then the fostering of the young parasite becomes an indispensable substitute object for the 'abreaction' of the feeding drive. In its rôle as substitute object, the young bird does not need to possess the species-specific releasers of the host offspring, and this bears the great advantage that the parasite is then able to parasitize *different* species. The cow-birds (*Molothrus*) do parasitize different species, however, *without* ejecting juveniles of the host species from the nest. But in this parasite the young bird usually hatches somewhat before its foster-siblings and, according to Friedmann, makes up for the lack of specific 'suitability' of its releasers with the higher intensity of the stimuli transmitted. We have encountered a similar phenomenon in the eliciting stimuli of the gaping response (p. 150).

Just as the gape markings or head markings of the young birds can represent species-recognition characteristics, whose innately-determined recognition leads the parent birds to feed the 'correct offspring', such characters can be represented by species-specific *begging movements*, which (innately determined in their entirety) are just as characteristic as conspecific signals as the morphological characters already considered. Thus it is that practically every conceivable caper and bodily contortion occurs as a releaser among the various bird groups.

By itself, consideration of the variety and the characteristic quality of these species-specific characters leads to the supposition that they are not there 'for nothing' and that the parents must respond to them in a characteristic manner. This is in fact confirmed by the behaviour of hand-reared and tame birds which are imprinted on the human being in all responses involving an imprinted object, yet whose feeding instincts can only be elicited by young conspecifics. The function of food-provision is found to have a unique degree of significance in warblers and related types, where the gaping of the offspring evidently acts as a criterion for the parents in recognizing whether the young are alive. Warblers breeding in captivity are often induced by the easy availability of food to feed their offspring more often and with greater quantities than under natural conditions. This means that the young become satiated and cease to gape – something which apparently never occurs in the wild. The parents promptly respond to the abnormal lack of food-uptake by the entirely healthy offspring by regarding them 'as dead' and carrying them out of the nest to discard them.

Whereas passerines give the impression that the parents respond to their dependent offspring with a similar drive intensity to that exhibited by a male mating with a female, feeding by adult pigeons, herons and other birds has the appearance of almost reluctant surrender in the face of the turbulently begging juveniles. The Night Heron in particular gives the impression that the adult is never completely able to suppress its defensive response to the obtrusive behaviour of its offspring. When the adult arrives at the edge of the nest, it stands with its head raised and chin drawn in as if to withdraw its beak as far as possible away from the young, and will repeatedly turn away from the crowding throng of young until one at last manages to seize the parent's beak and induce it to regurgitate its stomach contents. Although the heron arrives at the nest with its stomach and pharynx full for the sole purpose of feeding, it nevertheless seems as if the adult must be virtually assaulted by its offspring before its regurgitation reflex is elicited. As I first observed this behaviour, I thought that the adult Night Heron had no innately-determined drive to regurgitate food into the juveniles' beaks and that the adult was instead merely prevented from regurgitation in front of the juveniles because of their turbulent crowding actions. This view proved to be erroneous when tested experimentally: One of my Night Herons had developed the habit of feeding one of its offspring on the roof of the aviary in the evening. The young heron habitually went to this spot even in the absence of its father, so one evening I exploited this fact in order to throw fish on to the cage-roof before the arrival of the adult and thus extensively satiate the young bird. There was a very unusual scene when the male heron arrived on schedule. The young bird did in fact beg, but it was not sufficiently motivated to seize the father's beak properly. The male stood in front of the youngster, adopting the described semi-defensive posture. But the juvenile was by no means as obtrusive as usual, and although the male repeatedly performed regurgitation motions (as if it had hiccups) it did not go on to carry out regurgitation in the way typical of parental regurgitation of stomach contents in front of very young Night Herons. The father retreated from the begging approaches of the young bird, performing repeated choking movements, but the youngster did not follow the adult actively enough to catch up with him. Extremely similar behaviour is sometimes exhibited by sexually-motivated female Muscovy Ducks (*Cairina*), a species in which the female is normally 'raped' by the male. If the male of this species should for some reason fail to follow with the normal intensity, one can often observe that he is unable to catch up with the smaller and more agile female, although the latter waits for the male in the copulation posture between the individual phases of pursuit.

The adult Night Heron is most probably instinctively adapted to the described behaviour of its offspring, and this behaviour is conditionally necessary for elicitation of the feeding response. At the time when food is regurgitated into the beaks of the young and not in front of them, the adult Night Heron can no longer adjust to the young – they *must* seize the adult's beak. This is especially worthy of note, since this is *not* the case in other heron species. I have only myself observed the feeding of the Little Egret (*Egretta garzetta*) and the American Egret (*Egretta candissima*), together with a film of feeding by the Common Heron (*Ardea linerea*). In all of these species, regurgitation into the offspring's beak is more extensively developed than in *Nycticorax*. The Little Egret and the ornate Egret at least exhibit a much earlier switch from external to internal regurgitation than Night Herons. They may even feed freshly-hatched young by internal regurgitation, as does the African Goliath Heron (*Ardea goliath*), according to Portielje. In these species, as in the Common Heron, the feeding adult bird moves quite purposively *towards one particular youngster*, just as a passerine does, and pushes its beak into that of the juvenile rather than having its beak seized by the latter. Young of the species mentioned also fail to show the pronounced attempts of young Night Herons to seize the adult's beak; they more typically exhibit begging with the open beak *pointing upwards*, which is very reminiscent of the gaping of passerines (particularly in the Little Egret). In contrast to young Night Herons, young of these species are adapted to the approach of the parent.

(b) Elicitation of nest-hygiene behaviour

Almost all song-birds and probably many others carry faeces of the young away from the nest. After feeding, one can observe the feeding adult bird remain for a moment on the edge of the nest and stare fixedly at the young birds. The adult is waiting for deposition of the faecal pellet. This response is particularly important in some tits, where the number of offspring is often so large that the young in the middle of the nest cannot easily reach the perimeter. In the young of these birds we in fact find a morphological elicitatory organ in the form of a specially-developed light-coloured wreath of feathers around the anus. This wreath of feathers is spread in a conspicuous manner before defaecation and directs the attention of the parent towards the offspring needing its assistance at any given time. The young of these species do not even attempt to discharge their faecal pellets over the edge of the nest, but hold them balanced on the vertically upwards-directed anus until the parent bird removes them.

(c) Elicitation of the guiding responses

We shall here define 'guidance' (Führen) as the sum of the instinctive behaviour patterns of adult birds which respond to the following responses of the young birds and themselves stimulate the counter-function of following by the young. Expression of the guiding instinct by the adult bird usually represents extremely specific and conspicuous behaviour, particularly in nidifugous birds. We can usually hear calls and observe movements which operate as releasers for the following instinct of the young birds. These movements and postures often have the simultaneous function of outwardly-directed threat, e.g. in the Mute Swan and the domestic hen. In fact, guidance and defence of the young are closely correlated in these animals.[66]

The influence exerted by stimuli emanating from the young on the elicitation and preservation of the guiding drive varies greatly from species to species.

There is a remarkable walking inhibition in guiding nidifugous birds: in order to prevent the young from lagging behind, the mother must never exceed certain speeds. The inhibition against walking too fast is easily lost (as is the inhibition against taking to the trees) both through mutation in domesticated forms and in cases of sickness in wild forms.

Brückner describes the behaviour of a mother hen which apparently exhibited very extensive loss of this inhibition. The bird always tended to lose its chicks very easily and had earned the name of 'galloping hen' on the farm because of the conspicuous difference between her mode of locomotion and that of a normal mother hen. I am not able to say whether the presence of the chicks is necessary for the preservation of this inhibition. In the early stages of the guiding drive phase, domestic hens usually proceed slowly even without chicks, whereas 'Hochbrutenten' immediately exhibit unhibited walking when they lose their chicks, although this mode of locomotion in fact gives the appearance of seeking and may therefore represent a specific response. I have no records of observations on the behaviour of the 'rapid walking inhibition' of pure-blooded wild birds after the loss of all the chicks.

The influence which is exerted by following on the guiding mother *Gallus* or *Cairina* (and *Chrysolophus* and *Anas*, incidentally) is in many ways very limited. The mother does not even look around to see whether all of her chicks are following. The factor which induces her to stop or turn back is the occasional 'lost-piping' of the chicks, referred to here simply as 'piping'. Piping is probably a characteristic type of vocalization for the majority of nidifugous birds. If one is aware of the significance of the call in the domestic hen or mallard, then one can also interpret the corresponding vocalization of the goose, crane, bustard or water-rail without further ado. But the adult birds of these species by no means understand the 'piping' of offspring of all the other species. My geese did in fact respond from the outset to the piping of *Cairina* chicks which they had hatched out, but this call differs little from that of a gosling. 'Piping' is absent in the chicks of the non-guiding gulls and terns.

The piping that can be heard when a chick is lost is no different to the human ear from the piping uttered when chicks are cold or hungry. At least, this is true of *Gallus*, *Chrysolophus*, *Anser*, *Cairina* and *Anas*, which are the only genera with which I am truly acquainted in this respect. But the mother does typically respond to piping in accordance with the situation, i.e. she will not, for example, settle into the warming position when lost chicks are piping in the distance. Response to lost-piping, with which we are concerned here, appears to be extremely reflex and involuntary in nature. In the mallard, the following occurs: The mother at first stops still in response to the piping of the lagging ducklings, extending her body and flattening her feathers, and then magnifies the guiding call, which she will normally be producing continually. If the stragglers do not follow in response and continue to pipe instead, the mother runs back to the lost ducklings, this time momentarily forgetting the ducklings close to her. Having reached the stragglers, she utters the greeting and 'conversation' call and is 'joyfully' joined in this by the reunited ducklings. The young are then relaxed and contented until the ducklings previously close to the mother (and now left behind) begin to pipe. Owing to the rapid return of the mother, these ducklings were unable to follow her, basically because of their behavioural inability to turn sharp corners or double back. The mother then repeats her previous behaviour, but the ducklings now close to her do make an attempt to follow her, in contrast to the first group. However, they only manage to follow her for a distance of a few yards, since the adult female in this situation loses her characteristic 'rapid walking inhibition', though they do manage to come somewhat closer to the other ducklings. The first group of ducklings has in the meantime exhibited 'piping on the spot' long enough to forget the previous walking direction, so that these ducklings are also behaviourally capable of following the mother for a few yards when she once more charges over to the piping second group of abandoned ducklings. In this way, the two groups are gradually brought closer together until they are at last able to unite, accompanying this with vigorous 'greeting-palaver'. In fact the chicks most probably respond with fear to any diminution of their number, as will be described later.

If we now consider the part played by this entire process within the mother's environment, it must be admitted that she is not capable of responding to dichotomy of the flock of ducklings. She always responds just to one part of the flock and completely neglects the other. The part to which she attends is always that which emits the stronger stimulus, i.e. it is always the 'piping' group. Each time the adult female turns back, she behaves as if she had lost *all* of her offspring and is returning to them as fast as possible. If the duck actually 'knew' what was happening, she would at least *lead* the second group (which, in contrast to the first group in the initial phase, is prepared to follow the mother) back to the first; but she never does this. At least, I have never seen this happen. She encounters the entire situation with the one response of proceeding as rapidly as possible to the source of distant piping. This poverty of response, this amazing simplicity of the instinctive structural framework is itself a wonderful thing. One is repeatedly surprised by the *limited number* of responses to stimuli involved in producing social behaviour, building an ingenious nest and in general preservation of the species.

I cannot say whether piping from hunger is somehow different from piping when tired. I must actually admit that I have never heard characteristic hunger-piping from chicks led by their mother. In an experiment, one must leave chicks beneath their 'warming companion' for some time before they begin to pipe from hunger. It is possible that such piping has a sharper and more penetrating ring than the typically tired and weak piping uttered by chicks wishing to creep under the mother. Perhaps the mother responds to these differences; but she may also behave according to her ongoing activities, warming in response to piping uttered whilst she is guiding and guiding in response to this call when she is in the act of warming.

(*d*) *The elicitation of warming*

In very many nidicolous birds, the mere presence of the offspring in the nest is the eliciting factor for various instinctive behaviour patterns, particularly those involved in the described function of warming. The counter-function of these nest-living juveniles to the warming instinct of the parent in fact consists only of their presence in the nest-cup. But the condition 'in the nest-cup' must be underlined, since many passerines respond to the stimuli emanating from the young only when the latter are located in the nest-cup. Very young warblers which had been carried to the edge of the nest by a young cuckoo were given absolutely no attention by the adult warming the cuckoo, although the beak of the adult projected over the edge of the nest only centimetres away from the undercooled offspring. By contrast, my storks exhibited adaptive behaviour with an egg lying on the edge of the nest in that they would insert fine nest-material under the egg and between the coarse twigs surrounding it, so that the egg was slowly raised and eventually rolled into the nest-cup. But this behaviour would appear to be unusual, since Siewert observed Black Storks which exhibited behaviour completely like that described for the warblers.

Nocturnal warming of the offspring during the guidance phase demands a particularly far-reaching modification of the adults' usual behaviour in nidifugous species which otherwise take to the trees to sleep. The females sleep with and warm the young chicks on the ground, whereas they take to the trees to sleep when alone. A stimulus emanating from chicks piping in response to cold is necessary for the subjugation of this quite powerful drive. I have twice been able to observe that a Muscovy Duck, whose ducklings had been taken away during the first night after hatching and which had therefore spent the night in a tree, would in fact accept the ducklings and lead them well on the next day and yet not be able to 'cope with' sleeping on the ground, so that the adult flew into a tree and left the chicks at their peril. Whenever there is something not quite right about drives concerned with care of the young, the absence of the inhibition against taking to the trees in the evening is frequently the first sign. In summer 1933, I arranged for a domestic hen to hatch out a flock of pheasant chicks and lead them. In the open, the pheasants would have doubtless lost their foster-mother immediately, since they exhibited hardly any response to her calls. In the limitations of the cage, however, they learned to recognize the hen as a source of warmth and the hen responded well to the low-temperature piping of the pheasant chicks by warming them. But in the evening, the hen regularly took to the trees and simply left the 'piping' pheasant chicks sitting on the ground.

(*e*) *The elicitation of defence responses*

Some birds exhibit an extremely strong response to the pain- or distress-calls of their offspring, whereas others actually defend their offspring by angrily attacking a predator in the vicinity but do not respond in any way to their young as a part of this behaviour. This latter category includes the majority of nidicolous birds.

Almost *all* nidifugous birds, at least all of the species whose behaviour in this respect is known, respond to the distress-call of their offspring by flying into a rage and attacking even a dormant predator with exemplary courage. This defence instinct is so strong in guiding nidifugous mothers that it is not always restricted within species and genus boundaries, in analogy with the feeding drive of some passerines. The birds then angrily defend young of other species, which they would not guide or otherwise look after. Only the defence drive can be elicited by chicks of different species (or even genera). In male gallinaceous birds, which otherwise take little care of their progeny, defence is often the only paternal response associated with the chicks. I was greatly surprised by a *Chrysolophus* cock which angrily flew at my face when I caught one of his half-grown chicks, although he had otherwise paid no attention to them and had trampled on them in the crudest fashion whilst courting the hen.

The response to defend the chicks tends to persist much longer in nidifugous birds than all other parental care responses. I observed with both *Cairina* and *Chrysolophus* that the adults responded to the distress call of mature juveniles, which they had long since stopped guiding, with an active attack on a human being. In late summer, when all of the chicks are normally grown-up, female *Cairina* and Golden Pheasant of both sexes defended *any* conspecific seized by a human being. On the 10th of August 1934, I caught with my bare hands no less than four otherwise extremely shy *Cairina* females in one go. As I had the first in one hand, the second flew at my face and I was able to seize her. A moment later, the third arrived, and as I lay flat on my stomach on the meadow, covering two of the birds (in order to prevent them from escaping) I just happened to have one hand free as the fourth duck came up to attack. Beforehand I had racked my brains to decide how I could catch these four birds, in order to sell them, without scaring off the free-flying mallard and the young Night Herons which had just fledged!

This serves as an illustration of the blind and unconditional nature of these defence responses!

In the elicitation of an active attack on a predator threatening the offspring, it seems to be very important that one of the young should have been seized by the predator. We will encounter a similar condition in the defence of companions by jackdaws. I was only able to elicit an active attack on my person *without* taking a chick into my hand in *Chrysolophus*, *Gennae*, *Gallus*, *Anas* and *Cairina* during the first days of life of the chicks. The mothers (both parents in the pheasant) later attempted to run into my way and to distract me from the chicks by fluttering and screeching, but they only attacked actively when I had a chick in my hand. Thus, as long as the chicks are still quite young and virtually unable to escape these birds respond to a predator at once as if it already had one in its claws, whereas they hold back until this actually does occur when the chicks are older and more agile, and instinctively 'rely' upon the agility of their offspring up to that point.

The behaviour of the guiding mother duck exhibits one peculiarity when one of her ducklings is seized by a predator. Her first action is directed not at the predator but at the duckling. The duckling is freed from the grip of the predator and thrown to the ground by a peculiar downwards sweep of the beak. I had some time previously observed that mother ducks always hit the duckling itself in attacks when one attempts to take away a duckling or push one under the mother. I had in fact involuntarily adapted to this behaviour to the extent that on such occasions I would hold my free hand protectively over the duckling. But I first realized that this seizure or downwards striking of a duckling held by a predator was no chance effect in summer 1933, when I saw a 'Hochbrutente' save a chick in this way from the beak of a Night Heron. I at first thought this was a chance occurrence, but I succeeded in eliciting exactly the same response by reproducing the situation and I later saw the same behaviour repeatedly. I am unable to decide whether the mother seizes the chick in its beak in rescuing it from the predator or whether it only pulls it downwards with its chin, since the movement is too fast. Since the mother typically beats at the predator with the wing elbows, the entire motor pattern has a similar appearance to that common in duck fights, in which the opponent is gripped with the beak, pressed downwards and simultaneously beaten with one wing elbow. But the coupling of beak- and wing-movements is not unconditional. The first duck, which attacked the Night Heron, in fact first pulled away the duckling with a beak-movement and then began to beat at the heron with its wing elbows. I have also observed a 'Hochbrutente' flying after a raven making off with a chick in its beak. It caught up with the raven about three feet up, seized it and brought it

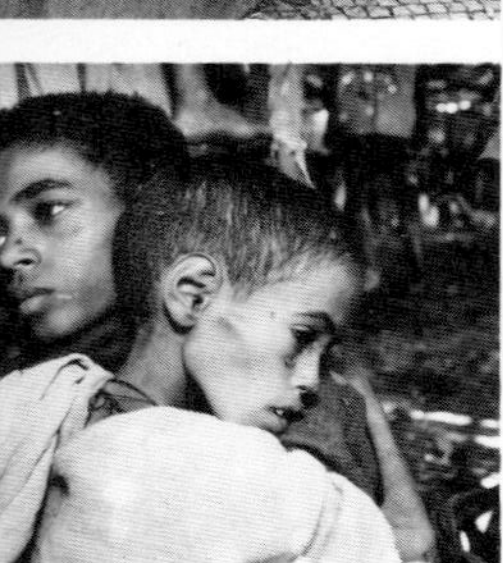

Other countries

AFGHANISTAN
ALBANIA
AUSTRALIA
Bondi Beach
Sydney
Woolagong
BELIZE
BULGARIA
BURKINA FASO
CANADA
Montreal
Toronto
● CHINA
Peking
Shanghai
CRETE
● Havanna
CYPRUS
EGYPT
● ERITREA
● ETHIOPIA
Tigray Province
FRANCE
Paris
GERMANY
Frankfurt
GREECE
● GUATEMALA
GUERNSEY
HOLLAND
Amsterdam
HONDURAS
HUNGARY
● INDIA
Bombay
Delhi
Calcutta
North
South
(Development)
● IRELAND
North and South
Belfast
Derry
Knock
● ISRAEL
Occupied Gaza Strip
Sinai
Jerusalem
ITALY
Florence
Milan
Sicily
Viareggio
JORDAN
West Bank
KUWAIT
● LEBANON
Beirut
Chouff Mountains
Sabra/Chatila
South Lebanon
LESOTHO
● MALI
● MEXICO
Acapulco
Mexico City
● NICARAGUA
PAKISTAN
PARAGUAY
POLAND
PORTUGAL
● EL SALVADOR
SAUDI ARABIA
SOUTH AFRICA
Soweto
Qwa Qwa Tribal
homeland
SOMALIA
SPAIN
Basque Country
Marbella
● SUDAN
SWITZERLAND
Geneva
Zurich
TANZANIA
THAILAND
● USA
Los Angeles
St Louis, Missouri
Philadelphia
New York
Washington
● USSR
Moscow
Odessa
ZAMBIA

● *Denotes extensive covera*

down. The duck then vigorously attacked the raven before it was able to escape and fly off, without its prey. The extent and biological significance of these defence responses is generally greatly underestimated, while the courage of the predator in attacking is to the same extent overestimated.

In nidifugous birds, defence responses are typically elicited by the situation 'predator near the nest' rather than that of 'offspring threatened by predator'. A young bird in another place elicits the defence drive just as little as any other parental care response.

This type of attachment of parental drives to the nest-cup is not found in Night Herons, which are not such specialized nidicolous forms as the passerines. In these birds, an infant is still an infant even if it is not located exactly at the spot 'where it belongs'. This is probably correlated with the early age at which the offspring begin to sit on twigs around the nest. In any case, one mother Night Heron (the very tame bird mentioned previously) flew down to the ground at once when I set down her twelve-days-old progeny there. She came up close to the young bird and attempted to warm it. Subsequently, she defended this young bird on the same spot, first from me and afterwards from a peacock, which (with the curiosity typical of its species) attempted to subject the young heron to closer investigation. However, this defence response *habituated* almost straight away. After chasing off the peacock twice, the mother abruptly ran away to stand and beg in front of her female keeper, who had remained motionless on the same spot throughout the experiment in order to observe. I was just able to prevent the peacock from raining blows from its spurs on the young heron – the mother would not have prevented this on this occasion. This effect, whereby a *less than adequate stimulus combination* elicits a response once and then fails to operate even at the second and third occurrence, is extremely frequently encountered in many different animal groups in experiments to elicit species-specific instinctive behaviour patterns with substitute stimuli. Lissmann has demonstrated this effect in exemplary manner in his refined dummy-experiments on the fighting response of the Fighting Fish (*Betta splendens*).[67]

The only case I know where a nidicolous bird exhibits elicitation of a complete defence response by stimuli emanating from its offspring is that of the jackdaw. Since this defence response is not restricted to the protection of offspring, however, it belongs in another chapter. Ravens, Night Herons, storks and most probably many other birds will defend the nest-site against the relevant predators without great dependence on whether the nest is full or empty. In the raven in particular, the drive to defend the nest awakens at a time when the nest as such is not yet present, when there is nothing but a site which has achieved the 'meaning' of a nest. I do not know whether this instinct undergoes a significant increase in intensity with the laying of the raven's first egg, but in the Night Heron this is so pronounced that with my birds I can reliably determine when the first egg is laid without chasing the birds from the nest.

It has already been mentioned that the offspring of nidicolous birds in many cases do not elicit parental behavioural responses (including defence patterns) when outside the nest.

Finally, we must include in the category of defence responses those instinctive behaviour patterns directed at diverting or distracting the predator from the offspring. I am only able to provide negative information about these responses, since I never succeeded in eliciting them from any of my captive or tame free-living breeding birds. The necessary stimuli seem to emanate from the nest-predator, and they evidently exert a powerful effect. With tame and semi-tame birds, a human being is not a sufficiently 'terrifying' predator to elicit these instinctive behaviour patterns and in fact only elicits defence responses in animals which in the wild would probably only succeed against very small predators. Since my experimental animals were also quite used to big dogs, so that not even the wolf (the largest 'adequate predator') would provide the stimuli relating to diversion of predators, I am only acquainted with this response under natural conditions. I should like to mention in parenthesis one of the refinements of these instinctive behaviour patterns, since it has to my knowledge not been previously described. Whereas mallard, partridge and other birds fake wing-damage in predator-distraction displays and are thus hindered in take-off, warblers and other small passerines mimic a generally *sick* bird – they ruffle their feathers, close up their eyes, stumble when hopping along and fly with the subdued wing-beat of an ailing bird. This is particularly interesting, since this refined development indicates that the pattern also bears some significance for the diverted predator.

(*f*) *The elicitation of rescue responses*

Whereas birds are thus equipped with a range of instinctive behaviour patterns adapted for protecting the young against predators, we find only a few responses which are adapted for rescuing the offspring in other situations.

For example, nidifugous mother birds are psychologically incapable of aiding their chicks when they have fallen into holes or fissures in the ground and are themselves unable to climb out. The solution of pulling out the chicks with the beak would present itself so readily for this situation, which apparently occurs quite often, that it is amazing that no species has developed the requisite instinctive behaviour pattern. I was thus particularly struck on observing the Blackcap (*Sylvia atricapilla*) performing behaviour which could not have been anything but an instinctive behaviour pattern for retrieving young from crevices. I had trapped a female of this species on the nest and placed her in a cage with her offspring. The female continued to feed her young well, and when they had fledged and began to fly clumsily around the cage it often occurred that one of them fell into the feeding vessel occupying one narrow side of the cage. The mother responded quite regularly to such an occurrence by rapidly leaping into the vessel, slipping under the fledgling by lowering her head and beak and pushing it up and out of the vessel on her back.

Heinroth once observed a mother midget duck turning over one of her offspring which had fallen on its back by pushing her beak beneath it. I observed the same with a *Cairina* mother in 1934.

This exhausts the recorded observations on the protective functions of parent birds regarding their offspring in such dangerous situations which are not produced by predators. One has the impression that a *situation* alone is not adequate to elicit such a strong response as the stimuli emanating from the predator, particularly since offspring in such emergency situations do not transmit such intensive distress signals as when seized by a predator.

(*g*) *The disappearance of the infant companion*

In many cases, the responses of the parental animal to the loss of an infant are quite characteristic. In decimation experiments even cats and other mammals react in a way which clearly demonstrates that they only miss their offspring quantitatively and not individually. In all birds so far investigated even this response to numerical reduction of the flock of young is absent. Even in nidifugous birds with many offspring (e.g. domestic hen and mallard) the counter-function of a single chick suffices to bind the parental instinctive behaviour patterns in such a way that there is no observable difference in the mother's behaviour from that exhibited with a normal-sized flock of chicks. This is all the more conspicuous in the mallard, since the ducklings themselves respond to a radical reduction of their number, as will be described later. Even if the chicks are not removed furtively without causing alarm, but are violently carried off amidst vigorous protective responses of the mother, the latter exhibits no alteration of her behaviour after decay of defensive arousal. The mother does not avoid the area where her offspring was carried off, nor does she exhibit increased fear of the 'predator'. She 'believes', so to speak, that her defence responses were entirely successful as long as she is left with at least some chicks which will release her parental instinctive behaviour patterns. However, I must leave it open whether these behaviour patterns observed in the domestic hen and the mallard apply to other birds such as Greylag geese. It is quite imaginable that the latter species would personally miss a lost juvenile, in roughly the same way as some birds miss a lost consort or social companion.[68]

In gallinaceous birds, particularly the domestic hen, the behaviour patterns of the guiding mother are not radically altered even when all of the chicks are taken away. She does not cease to utter the guiding call, to summon to food or to adopt the defensive posture, and she does not exhibit an adaptive search for her lost offspring. The mother in fact continues all activities associated with guiding offspring, though the vigour of these activities is markedly lessened. She will often stop still in a curiously abrupt manner, as if all her responses have jammed and require an impulse from the offspring in order to operate further. Very similar behaviour is exhibited by hens after the decay of the brooding drive when the clutch was infertile. Such hens also wander around 'guiding' for several days and show the same disruptions of 'spontaneity'.

Nidicolous birds in principle behave similarly to the described nidifugous types following loss of the infant companion. The disappearance of a single offspring from a large number is never noticed by the parents. Where the eggs and young are small, even removal of the entire contents of the nest does not always evoke an immediate response. After all the offspring of a jackdaw clutch had died in the summer of 1933, the parents carried away the corpses and yet continued to brood on the empty nest for several days. I once removed the infertile eggs from a pair of Night Herons after they had been brooded for the normal period. The birds returned to the nest following the resulting disturbance, greeted one another, and the female (which had been brooding previously) then settled into the nest-cup as if nothing had happened. The birds had left the nest a few hours later, however. I assume that the brooding bird had first noticed the disappearance of the clutch as the response of turning the eggs appeared (i.e. the next instinctive behaviour pattern directly related to the eggs which could not be performed in the empty nest-cup, in contrast to brooding). When the offspring have grown somewhat, the adult herons notice their absence immediately and there is even an excited greeting ceremony when the offspring are returned to the nest. Bernatzik has experienced this with the Little Egret and I have confirmed this for the Night Heron. Nevertheless, the actual fate of the young has no effect on the parents when their offspring are removed under their very eyes from the nest, despite their defence responses. In 1934 I did this with two young Night Herons and placed them in a small aviary, which was both easily visible to the parents and open to approach on all sides, located in the middle of the adults' everyday range. The parents did follow me as long as I held the distress-calling offspring in my hands, but they relaxed at once when this was no longer so. They immediately ate out of my hand again and straight afterwards flew back to the (now empty) nest to feed. In the course of time, they did realize that the young were in the cage, since they later sat on the roof with noticeable frequency. But as soon as they wanted to feed their young, they flew to the nest!

4. DISSOCIABILITY OF THE FUNCTIONAL SYSTEMS

The individual functional complexes in which the young bird is related to its parents as a counter-functioning object cannot be separated in an experiment to the same degree as other functional systems of instinctive behaviour patterns exhibiting instinct interlocking. Nevertheless, there are some responses which appear outside of the context of general rearing patterns.

It has already been mentioned that the defensive drive of many nidifugous birds appears in response to the young bird even when the latter is unable to elicit any other infant-oriented instinctive behaviour pattern. This operates in a purely reflex manner, for instance when a *Cairina* mother 'rescues' a mallard duckling with self-sacrificing courage from the experimenter's hands, only to bite it (and possibly kill it) a minute later when it attempts to mix with the mother's own chicks.

It is this automatic nature of these parental care responses which throws entirely new light on the unitary manner with which the juvenile companion is treated, even in different functional complexes. The low degree of dissociability of the functional complexes does not in this case depend upon better co-ordination of the stimuli emanating from the young animal (i.e. better 'objectivization' of the infant person). There is from the outset a predetermined mutual dependence of the instinctive behaviour patterns of parental care within the instinctive framework of the parent bird, which thus form, so to speak, *one* single behavioural chain. The unitary treatment of the offspring is thus determined within the instinctive framework of the adult bird and not in the rôle that the infant plays in its environment.

VII The sexual companion

Few animal behaviour patterns have been so subject to false, sentimentalist anthropomorphization as those seen in bird pairs. Poetry has particularly taken possession of species which do not prove to be particularly 'sentimental' on close examination, as for example the doves. Such false anthropomorphizations have been extremely damaging in one respect; because of their obvious psychological flaws, they have deterred serious scientists from paying attention to the great similarities which exist between men and animals in their sexual behaviour. By this, I do not mean the merely sensual effects of sexual drives, but the contrastingly finer, quite 'human' behaviour patterns of falling in love, courtship, jealousy and so on. Such previously widespread anthropomorphization is responsible for the fact that nowadays a purely factual description of animal love-life will do no more than stimulate slightly incredulous amusement, even in a listener with psychological training.[69] But it is actually not very surprising that this particular field incorporates the most extensive parallels between human and animal behaviour, since nobody would wish to deny that human sexual behaviour is particularly marked by the retention of purely instinctive behaviour patterns.

1. THE INNATE SCHEMA OF THE SEXUAL COMPANION

Innate releasing mechanisms associated with sexual behaviour patterns usually have nothing to do with imprinting. The responses concerned first awaken some time after the completion of all imprinting processes. Frequently, releasers of infantile instinctive behaviour patterns determine imprinting of the later sexual object. Generally, the conspecific inducing such imprinting (parent or sibling) does *not* possess any of the releasers relevant to sexual instinctive behaviour patterns, and if these are in fact present they are never presented to the young bird. Thus, there is no question of imprinting of the sexual companion of a young female through factors emanating from the father and brothers, or of the converse process of imprinting of young males induced by female members of the family. The releasing mechanisms determining imprinting have no immediate relationship to sexual responses – the imprinted characters are always characteristic for the entire species (i.e. for both males and females). Having said on p. 132 that a bird only acquires *supra-individual* characters of the companion through imprinting, we must add at this juncture that imprinting of the sexual companion only involves non-sexual species-specific characters.

Innate and imprinted factors initially combine to form a (so to speak) sexual schema of the 'conspecific', which determines the *species* towards which the sexual responses of the bird will later be directed. It is only at a much later stage, after achievement of sexual maturity some time after the cessation of receptivity to imprinting, that the bird begins to respond to those releasers which correspond to the innate releasing mechanisms of sexual instinctive behaviour patterns. Since the response to these releasers is innate and since these releasers themselves are the actual signals through which the bird innately 'recognizes' its sexual partner, it perhaps seems paradoxical not to include them in the innate schema of the sexual companion. However, in order to remain true to the definition already given, it would be best to discuss the function of these releasers in the section on the functions of the sexual companion. The independence of innate sexual releasing mechanisms from the imprinting process is often quite conspicuous in birds imprinted to the human frame, when the foster-parent chances to carry characters which correspond to such schemata. For example, the Heinroths observed that a partridge hen (*Perdix perdix*) just attaining sexual maturity fell in love with her female foster-parent because she responded to the latter's ruddy-coloured apron in the same way as to the red thoracic patch of a partridge cock. A male Barbary Dove (*Streptopelia risoria*), which had been similarly imprinted to the human frame, exhibited analogous behaviour towards me. I presumably represented a male 'conspecific' for him most of the time, since he usually exhibited nothing but fighting responses towards me. But when I presented him with my hand, held horizontally at a particular height in front of him, he would immediately switch to a copulation attempt, i.e. he reacted to sudden presentation of a level body surface in the same way as to the elicitatory pattern of a female inviting copulation.

Accordingly, the total schema of the actual sexual companion always consists of the innate and imprinted schema of a *conspecific*, to which must be added specific, sex-related signals which themselves play no part in imprinting. Hence the difficulty in applying our concept of the innate companion schema in this instance.

2. PERSONAL RECOGNITION OF THE SEXUAL COMPANION

To the casual observer, the fact that birds can recognize their mates among hundreds of conspecifics is extremely surprising, in fact almost unbelievable. The reply that all animals of the same species, all Negroes and all Chinese vary just as much among themselves as Europeans do is invalid. We Europeans are in fact much more distinct from one another than are the racially less diverse Negroes or Chinese, and these in their turn differ from one another far more than members of non-domesticated animal species. But even if the range of variation in animals is much narrower, all that is necessary is a corresponding refinement of the sensitivity of optical differentiation in order to provide as many possibilities for individual characterization as are present in the more divergent civilized human population. Even a human being is quite able, after practice, to perceive slight individual variations, as evident in the narrow range of variation in non-domesticated birds. In 1930, I could recognize each individual in a flock of fourteen jackdaws without glancing at the leg-rings and without making use of chance characters such as protruding feathers and the like. If we should ask *how* animals recognize one another, i.e. which characters are decisive in differentiation between individuals of the same species, the answer turns out to be quite astounding in one respect. For most birds, exactly as for us, the decisive features are the physiognomy of the head, the nature of the voice and the individual's mode of locomotion. Heinroth describes how a Mute Swan attacked his up-ending consort, because he could not see her head and took her to be a strange swan. It seems justifiable to assume that in very many birds the face is the region of the body bearing the visual characteristics of the individual. According to Heinroth, a mallard duck recognizes her drake by his voice. I can confirm this for a brooding female jackdaw which crept out of the nest-box in response to the summoning call of the male, without seeing the latter at all. This female paid no attention to the summoning calls of other jackdaws. For very many birds, the manner of locomotion is very probably an individual characteristic, as are the paths taken by a given bird and the perches which it uses. The latter must surely play a large part in individual recognition among herons, which have fixed path-habits and perches. I can myself distinguish individually all of the herons in my colony by their path-habits and perches, and my identification has always been confirmed by a check on their leg-rings.

3. THE FUNCTIONS OF THE SEXUAL COMPANION

The instinctive behaviour patterns of a bird pair serving reproductive functions exhibit what Alverdes calls *instinct interlocking*. The factor eliciting a chain of instinctive behaviour patterns in one partner is a behavioural act of the other bird, in many cases only the last element in a behavioural chain performed by the latter. This element, rather like the cue of an actor, provokes the partner to perform a new activity. Presumably, all mating responses of birds, and their reproductive responses in general, are cases of instinct interlocking. Only in animal species in which the female is simply 'raped' by the male does one find a lack of reciprocal behavioural patterns on the part of the female sex, so that the term instinct co-ordination does not apply. However, I think it is unlikely that such cases occur among birds.

(*a*) *Instinct co-ordination in pair-formation*

Demonstrative behaviour – Many responses which an animal (usually a male animal) exhibits when placed together with a conspecific are quite generally interpreted as courtship patterns. Similarly, conspicuous colours and shapes ('Prachtkleider') found only in the male sex are interpreted as a device exerting an effect on sexual selection by the female.

Both interpretations can be correct in individual cases, as Selous has shown with the Ruff (*Philomachus pugnax*), for example. The explanation of the origin of these instinctive behaviour patterns and morphological characters through sexual selection in its narrowest sense is to some degree a probable one, though it is of course not fully substantiated. On the other hand, it must be remembered that in many animals quite equivalent behaviour and quite similar morphological characters restricted to the male sex obviously elicit responses only from *other males*, as Noble and Bradley have shown for many lizards. In yet other animal species the behaviour concerned and the correlated organs are not restricted to the male sex at all, though such cases form a definite minority.

In all cases, behaviour and the associated morphological patterns ('Prachtkleider') are to be interpreted as *typical releasers*. Whether they operate on the same or the opposite sex, their effect is similar to the extent that they always provoke the performance of quite specific reciprocal behaviour patterns which are concerned with reproduction, in fact with pair-formation.

Doubtless, the most common (and probably the most archaic) effect of these releasers is represented by *simultaneous* elicitation of a positive response from the female and a 'negative' behavioural response from any other male. Heinroth aptly refers to all behaviour patterns which thus combine threat and courtship effects as *demonstrative behaviour* ('Imponiergehaben'). In the English literature on the subject, this is usually expressed briefly, and probably too generally, as 'display'. The concept of demonstrative behaviour incorporates the information that such behaviour is predominantly performed only when a receptive spectator is present.

Demonstrative behaviour in its wider sense is extremely widespread in the animal kingdom. It can be seen in presumptive form in the invertebrate world and is present in more or less the same form in all vertebrates, from teleost fish to man. Genuine demonstrative behaviour represents a constituent of the few undeniable instinctive behaviour patterns of man.

In its most generalized and obviously most primitive form, demonstrative behaviour does not consist of specific motor patterns performed only in this context, still less is it accompanied by special organs developed as supplementary features. Primitive demonstrative behaviour is in fact represented by the performance of all normal motor patterns of the species *with a conspicuous expenditure of energy, in excess of that actually necessary to fulfil the purpose of the behaviour*. The behaviour of a courting Greylag gander (*Anser anser*) incorporates this type of demonstrative behaviour in a particularly pure form. Each single movement of such a bird is performed with such an excess of muscular power that even an impartial observer would have the immediate impression of tense, affected behaviour. Both when walking and standing still, the bird sticks out its chest and holds itself upright. The stride becomes longer whilst the walking speed decreases. The normal side-to-side rotation of the body in walking is exaggerated so that it virtually looks as if the bird finds walking more laborious than normal. Whereas Greylag geese otherwise make little use of their wings and only fly off in face of actual danger, the 'demonstrative' gander makes use of every opportunity to demonstrate the strength of his wings. At such times, he will fly over quite short stretches which he would normally cover only by swimming or walking. In particular, he will fly headlong towards any actual or apparent opponent and then, after chasing off the intruder, fly back with just as much expenditure of energy to the female he is impressing. He subsequently lands by her with his wings raised and uttering a loud triumph-call. After landing, and following any other occasion for spreading the wings, closure of the wings is not immediate. Even in a goose which does not happen to be exhibiting demonstrative behaviour, we can see that the wings are held out stiffly for a moment before folding. This time-interval is greatly extended in demonstrative behaviour, and the ostentatious gander stands for a period of seconds with wings spread wide every time they are opened. In other Anatidae, exactly this behaviour has developed to produce a specialized display ceremony.

Demonstrative behaviour in man is extremely similar to that of the Greylag goose. Even extremely able men talented with powers of self-criticism can, under circumstances involving physical activity (for example when skiing or ice-skating), become considerably more vigorous and dashing when the number of spectators is increased by one attractive girl. In primitive or not quite grown-up men, this is often accompanied – as in the gander described – by attacking or at least molesting weaker pseudo-opponents. The most remarkable fact to be seen from this, however, is that many men perform the dissipation of energy characteristic of demonstrative behaviour with a *machine*, if they happen to be riding one. I have repeatedly observed that motor cyclists in such cases increase the noise-level and energy consumption of their machines by revving and simultaneously adjusting the ignition, without greatly increasing their speed. Sharp braking and much too rapid acceleration of motor vehicles can often be explained in the same way. The American writer Mark Twain, an extremely acute psychological observer, relates an example of collective demonstrative behaviour on the part of the crews of Mississippi steamers. At places where the passage of these ships could be observed from settlements on the shore, the speed always increased and a particularly large volume of smoke was produced by laying on particular kinds of wood.

I am of the opinion that all more highly-specialized forms of instinctive demonstrative behaviour have developed from origins corresponding to the behaviour described above for the Greylag gander. In the same way, I am sure that the supplementary morphological organs have been *secondarily* acquired. The questionable proof of 'phylogenetic' series seldom appears so convincing as in the comparison of specific components of demonstrative behaviour of related bird species. For example, the prolonged opening of the wings, as described above for the Greylag gander, has come to form an independent ceremony in other Anatidae. This component of demonstrative behaviour in *Anser* is only performed when the male has just flown or has just exhibited wing-flapping. (Opening of the wings in defence of the young belongs to a quite different complex of behaviour patterns!) The most that one could say is that a gander, when 'in love', is inclined to promote with particular frequency situations in which spreading of the wings occurs. By contrast, the male Egyptian Goose (*Alopochen*) spreads its wings in demonstrative behaviour without any further provocation. In the Orinoko Goose (*Neochen*), the response of flapping the wings with a demonstrative significance has been completely separated from the normal response of wing-flapping. During the demonstrative response, the male rears up to a position far past the vertical, spreads his wings and performs restricted, quivering wing-beats followed by a long pause with the wings spread. This is accompanied by utterance of the 'triumph-call' (Heinroth). In addition, *Neochen* has in fact developed a morphological releaser which underlines the operation of this behaviour pattern. The ventral side of the gander's body, which is prominently exposed in the posture described, is conspicuously coloured. The entire ceremony can immediately be recognized as 'the same' as the behaviour of a Greylag gander returning to its adored one and uttering the triumph-call, and the two displays are doubtless homologous. In *Neochen*, however, the display is far more specialized and far removed from the original biological function of the motor patterns performed. Similar relationships are highly probable in a great number of cases. Frequently, it is not possible to draw sharp boundaries between an instinctive behaviour pattern originally performed with a quite different function and a response operating only as a demonstrative behavioural element. In many cases, the relevant factors are not immediately obvious. For example, it is not easy to understand why a threatening cock must 'pretend' to peck at something on the ground and why a crane preens behind its wings in the same situation. In the European crane, this ceremony still has the appearance of normal preening, and anybody who is not aware of the significance of the ceremony would regard it as such. With the Manchurian crane, one would scarcely take the corresponding motor pattern as a preening action, although it is certainly homologous with the pseudo-preening of the European crane.[70]

Every thinkable kind of morphological organ occurs for the consolidation of demonstrative behaviour, and such organs, like the behaviour itself, are in many cases restricted to the male animal. These structures carry all of the typical characters of the releaser (p. 106) and are therefore to some degree similar to one another. Above all, it is impossible to recognize from the form of a releasing organ whether it is determined for the elicitation of a reciprocal behaviour pattern from a conspecific male or from a conspecific female. It frequently occurs, in fact, that the same motor pattern and the same organ are used both for intimidation of other males and for elicitation of a female behavioural response. When we are acquainted with the responses of a species, however, we can often separate the organs of an animal 'intended for the female' from those serving to intimidate other males. The elongated feathers of the neck-ruff of the Burmese Jungle-fowl and the domestic cock are predominantly employed against male conspecifics, whereas courtship of the female largely entails display of the tail-coverts. Hingston has expressed the opinion that all vivid colours and conspicuous shapes in the plumage of male birds, and in animals in general, operate exclusively as an intimidatory device. This is an extravagant generalization of a theory which applies only to certain animals, for example to certain lizard species.

It is, however, extremely probable that colours and shapes occurring in only *one* sex should generally be regarded as releasers.

If one is not to take recourse to an entirely transcendental, entelechial explanation, interpretation of male 'Prachtkleider' (display patterns) as *releasers* is the only theory which can explain the general improbability of the structures involved. The colour patterns restricted to the males are always more regular, more uniform and frequently simpler than the markings of the conspecific females. We have already discussed the combination of uniformity and simplicity in releasers on pp. 105–107. The colours of the 'Prachtkleid' as well as the marking pattern can be explained in the same way: reflection of a single, pure spectral colour from the wave-chaos of white light is in itself so improbable that a specific coloration of the feathers could alone become a releaser. The same probably applies to the pure and 'beautiful' notes of bird-calls and songs.

Any theory which explains the *details* of various morphological patterns as products of chance, conflicts with the laws of probability. In saying this, I am thinking mainly of the views expressed by Wallace in his publication *Colours of animals*. He maintains that the 'Prachtkleider' of the males are a 'consequence of the greater strength and vigour, together with the greater vitality of males'. This can be countered with the observation that in many birds the females bear the 'Prachtkleid' and not the males. In addition, in those animals where the 'Prachtkleid' of the males had more or less achieved its greatest degree of development, namely in spiders of the families Attidae and Lycosidae, the males are much smaller, weaker and less 'vital' than the females. The Peckhams have recognized the true releasing function of the male 'Prachtkleid' in this case.

My view that the typical appearance of 'Prachtkleider' is related to their releasing function is supported by the fact that in a veritably overwhelming number of individual cases the conspicuous, brightly-coloured areas of the male are employed as an organ in an elicitatory *behaviour pattern*. We do in fact frequently find releasing behaviour patterns which, so to speak, make use of already available material. By means of conspicuous body postures, spreading of wings and tail, ruffling of particular areas of the plumage and so on, a characteristic optical stimulus combination is produced, without particular differentiation of areas of the plumage to generate the stimuli involved. It has also been mentioned (p. 142) that this probably represents the archaic form of such behaviour. However, it would appear that the converse (i.e. markedly differentiated plumage areas which are not employed in specific ceremonies) is never found. In this case too, the impression is gained that the *ceremony is always more archaic than the correlated organ*, as explained on p. 142. For example, in the demonstrative behaviour (displays) of probably all species of dabbling ducks the feathers on the crown of the head, the wing elbows and the rump are rendered particularly conspicuous, since the ducks raise their outline as far as possible above the water by spreading the feathers on the dorsal aspect. This behaviour is also exhibited by males of species which have no specialized display feathers on these dorsal areas. Whenever such specialized feathers occur in a species, however, they are almost always located on one of the three sites named and sometimes occur on all three – as in the Mandarin drake (*Aix galericulata*).

Consequently, whenever elongated, iridescent or conspicuously coloured feathers, conspicuous areas of naked skin, inflatable organs and the like are found in a given bird species, it can be quite reliably assumed that the structures concerned play a part in a specific releasing behaviour pattern. This behaviour pattern need not necessarily be concerned in sexual processes, however – it can also perform a purely social function, as we have seen with the Night Heron (p. 141). But in cases where con spicuous elicitatory organs are only found in the male bird, they presumably serve either for the elicitation of a female sexual behavioural response or for the intimidation of other males; in many cases both functions are certainly involved. The functions of these releasers and the nature of the instinct interlocking systems which they set in operation vary so widely that they will be discussed in a special section.

The various types of pair-formation – The conjunction of two individuals for formation of a pair is guaranteed by an interlocking system of responses, which can vary radically from species to species. Nevertheless, the number of possibilities open for such a process appears to be limited, since we can find extremely similar behaviour patterns in different animal species which are taxonomically widely separate (e.g. birds and teleost fish). Of course, these similarities are merely convergent phenomena, as is illustrated by the fact that the pair-formation of the majority of birds is quite reminiscent of the behaviour of labyrinth fish, whilst pair-formation in a number of other birds is remarkably similar to the behaviour of another group of fish – the Cichlidae.

This being the case, there is no apparent justification for pronouncements on the phylogenetic relationships of the various forms of instinct interlocking. However, in order to give some degree of clarity to my account of these behaviour patterns, I shall attempt to select *three* types of such behaviour. In doing this, I shall steer clear of any phylogenetic assumptions, and I emphasize that I do not maintain that no other types of pair-formation can be postulated. The concept of the type incorporates the assumption that no individual concrete case will represent an ideal form of the type, and no more should be expected from the following described types of pair-formation. The purest forms are found among the lower vertebrates – the first type in some reptiles, the second type in labyrinth fish and the third in Cichlidae. These types are found in the birds with so little change in form that I have no hesitation in naming them after the animals mentioned and speaking of a lizard-type, a labyrinth fish-type and a Cichlid-type of pair-formation. This certainly does not mean that *all* representatives of the animal groups from which the names are taken exactly correspond to the relevant type in pair-formation. For example, I suspect that among the labyrinth fish the genus *Anabas* is distinctly different from the type of the Family. Similarly, the genus *Apistogramma* of the Cichlidae appears to have a labyrinth fish type of pair-formation. This said, we can pass on to a discussion of the three selected types of pair-formation:[71]

In many *lizards* the demonstrative behaviour and the 'Prachtkleid' of the males has no influence whatever on sex-determined instinctive behaviour patterns of the female. The latter in fact responds to the demonstrative behaviour and 'Prachtkleid' of the male in the same way as a weaker male – she flees. The male exhibits such demonstrative behaviour eliciting flight in females and weaker males towards *any* conspecific, but a male of approximately equal rank will respond with similar behaviour. In the former case, the fleeing animal is pursued and may be '*raped*', whereas it is only in the latter case that demonstrative behaviour reaches its zenith. If one of the displaying males does not flee in response to the threat of the other, a *fight* follows.

In this case, we are faced with the rare situation where a chain of behaviour beginning in a quite uniform manner can be continued in two different directions. It is as if the responses were moving along a track which forks into two at a set of points. The actual progress of the behavioural chain in one of the two possible directions is exclusively dependent upon the response given by the second animal to the male's demonstrative behaviour. Noble and Bradley have demonstrated experimentally that the demonstrative behaviour *always* carries on to the response of 'rape' unless presentation of demonstrative behaviour on the part of the other animal causes a shift to the track of fighting behaviour. In most species, if the second male is restrained from performing its demonstrative behaviour by fettering, narcosis or some other means it will be 'raped'. In some species with specialized development of a male 'Prachtkleid', however, the coloration of the restrained animal also had to be obscured (at least partially) in order to elicit this response. *In this case, the 'Prachtkleid' acts like a continuously-presented form of demonstrative behaviour.* The male of such reptile species responds in a purely male fashion in all cases, treating *any* conspecific which does not elicit his fighting behaviour with sex-determined *male* elicitatory patterns as a female. The demonstrative behaviour and the 'Prachtkleid' function exclusively as intimidatory agents and as releasers for reciprocal demonstrative behaviour from males of matching strength. This behaviour can therefore be biologically valuable in two different ways, firstly by preventing pointless fights through intimidation of the weaker opponent and secondly by preventing 'erroneous' copulatory behaviour through the elicitation of male responses.

In *labyrinth fish*, the situation is quite different. Whereas in the lizards only the male exhibits demonstrative behaviour, each of these fish will react towards any other of the same species with demonstrative behaviour. The only exceptions occur when one individual is much larger than the other, in which case the reciprocal response of the weaker individual is nipped in the bud. *This also occurs when the development of the sex-determined male demonstrative organs of one fish is far in advance of that of the second fish.* In fact, females will often not respond to a much weaker male with reciprocal display and will flee without further ado, as long as the fins of the male are sufficiently developed.[72] Even when the females are not sexually motivated, such that they do not respond to the demonstrative behaviour and 'Prachtkleid' of the male with female mating invitations, they will permit themselves to be intimidated by both factors, even when they are physically much stronger than the male displaying with its fins.

In labyrinth fish, demonstrative behaviour is similarly followed by fighting and mating responses. In this case, too, the actual type of response which emerges depends entirely upon the behaviour of the second animal. However, there is a vital difference from the behaviour of the lizards in that in labyrinth fish fighting responses are *always* elicited if the second animal does not divert the demonstrative behaviour of the first along the channel of mating responses with *female* elicitatory behaviour patterns. The direction of the continuation of the behavioural chain into fighting or amatory behaviour depends upon the sex and the physiological condition of the second animal. Even a female will be treated as a rival if she does not perform the elicitatory patterns of sexual receptivity. A further important difference between the behaviour of these fish and that of the reptiles described is to be seen from the following: females will also respond to any weaker (more properly, lower-ranking) conspecific with *male* demonstrative behaviour. Conversely, a sexually-motivated male can also be observed to exhibit female instinctive behaviour patterns in response to a stronger conspecific of the same sex. In many birds, this sexually 'ambivalent' behaviour is even more pronounced, and we shall need to deal with this in more detail later on.

Thus, demonstrative behaviour and the 'Prachtkleid' of labyrinth fish have a different function to those of lizards. They are similar in that they serve to intimidate rivals, but they do not prevent mating behaviour involving two males. However, they do perform the important function of suppressing male responses in a female exhibiting ambivalent responses. Further, they function as the eliciting agent for the initial female behavioural responses which switch off the fighting responses of the male and initiate his mating responses.

These latter responses must be considered in more detail at this point. Above all, it must be emphasized that they include prominent behaviour patterns which are adapted for sexual arousal of the female. These 'courtship' patterns have nothing to do with threat functions. It is far from my intention to deny this courtship function of these responses and to interpret all sex-determined male elicitatory behaviour patterns as demonstrative behaviour. The co-ordinated behaviour of pair-formation consists of a vast number of highly-specialized instinctive behaviour patterns exhibited by both sexes, and nothing would be more erroneous than to incorporate all male elicitatory behaviour under the heading of 'demonstrative behaviour'.

The first behavioural response which a sexually-receptive female labyrinth fish exhibits in answer to male demonstrative behaviour is in every way opposite in form. Whereas the displaying male always spreads his fins and places himself parallel to the other fish, a sexually receptive female always orients her longitudinal axis perpendicular to that of the male's and draws in her fins to lie as close as possible against the body. In addition, the female becomes pale in colour, thus providing another contrast to the bright, saturated colours of the displaying male. The continual attempt of the male to present his broad aspect to the partner whilst swimming along and the female's persistent endeavour to remain perpendicular to the male produce the circling behaviour of labyrinth fish pairs which is well known to any keeper of aquarium fish.

It is easy to see the radical difference between the mating behaviour of the lizards[73] and that of labyrinth fish. It has already been stated that *both* types of behaviour can be found among birds, and that we must therefore be extremely careful about generalization of principles which apply to a particular bird species. Noble and Bradley oppose the theory that male 'Prachtkleider' exert any influence on females and that their origin is possibly to be explained as a result of sexual selection. They state: 'This wonderful theory is complete nonsense when applied to the lizards, and if this is true for lizards, why not for birds as well?' This last question incorporates a basic misconception. Something which is completely wrong when applied to lizards may nevertheless be perfectly valid when applied to birds. Indeed, one is tempted to give in answer to this question: 'Because most birds do not have a functional penis, so that mating without the participation of the female is not possible.' This is an answer which can only be challenged on the basis of its causal structure.

One feature in which pair-formation in labyrinth fish corresponds with that of the lizards is that of the dominance relationship between male and female. In both, it is a precondition of pair-formation that the female should rank lower than the male.[74] We have seen on p. 195 that the 'Prachtkleid' of the male plays a part in the establishment of this relationship. To this extent, Hingston's hypothesis that all bright colours and conspicuous shapes in the entire animal kingdom function exclusively to intimidate other organisms incorporates an element of truth. Otherwise, this hypothesis is a precipitate generalization from facts which are in themselves correct.

Even the more careful interpretation of the same phenomena given by Schjelderup-Ebbe is not entirely correct. Although there is an (at least potential) struggle for dominance between the sexes in the majority of birds, there are enough species in which *no* dominance relationship exists between the members of a pair. In such species, the character of the relative rank of an individual is inapplicable in determination of its sex.

We have classified this type of pair-formation as the *Cichlid-type* after the *fish* which exhibit this behaviour in its most pronounced form. In the Cichlid species which I have investigated (*Aequidens pulcher* and *Hemichromis bimaculatus*), the *female retains her demonstrative behaviour towards the male throughout the entire cycle of reproductive processes.* If two mutually-unacquainted *Hemichromis bimaculatus* are set together, they will circle one another with the fins maximally spread and displaying their finest colours. In doing this, the two fish always remain parallel to one another. In other words, each turns its broad aspect to the other, just as two male (or, more exactly, two mutually threatening) labyrinth fish do. Now if the observer is familiar with the ritualized pairing behaviour of the labyrinth fish, he will often wait in vain for the fin-retraction and perpendicular posturing of the female. When this does not occur, the observer then erroneously expects fighting to take place at any moment, even if he is confronted with an actually compatible Cichlid pair. However, the behaviour which follows, *without* cessation of demonstrative behaviour on the part of the female, is composed of sex-specific elicitatory patterns. In *Hemichromis*, the assumption of markedly different colour-patterns apparently plays a part in this. *These* releasers inhibit the onset of dominance contests *without* rendering the female submissive and without overt fighting, in contrast to the male demonstrative behaviour of labyrinth fish.[75]

In these fish, collapse of the demonstrative behaviour of the female in response to that of the male signifies the failure of pair-formation: the male then responds unspecifically with pursuit, and possibly killing, of the female. For this reason, it is hardly ever possible to pair off Cichlids of greatly differing size. In particular, it is virtually impossible to induce a smaller female to hold her ground before the demonstrative behaviour of a much larger male without abandoning her reciprocal demonstrative behaviour and fleeing. The normal reciprocal 'display' which represents the introduction to Cichlid pair-formation is a typical case of the type of behaviour referred to by English-speaking animal psychologists as 'mutual display'.

We shall now investigate the extent to which the three described types of pair-formation are to be found among birds.

The purest form of behaviour corresponding to that of the *lizard-type* found in birds is provided by the Muscovy duck (*Cairina moschata*) and possibly in some related forms (*Plectropterus*, *Sarcidiornis*). Even in *Cairina*, however, the incipient development of female releasers can be demonstrated. In the first place, a female close to egg-laying and ripe for fertilization is in some way recognizable to the male, since one can observe that such ducks are pursued with particular vigour by males. The releasing signal may well be a particular, luminous red coloration of the facial tubercles of the receptive female (Heinroth). In addition, the females exhibit releasing *behaviour patterns* at the peak of receptivity. One can often observe, particularly when there are more ducks than drakes, that a female pursued by a male will abruptly freeze for a moment in the copulation posture before continuing flight from the drake. If the drake has not quite caught on, it can occur that the duck will actually remain crouching and wait for him. A similar phenomenon can be found even in the Class Reptilia. Peracca found with *Iguana tuberculata* that the females run away from pursuing males, but nevertheless exhibit a simultaneous female elicitatory pattern in the form of elevation of the tail and extrusion of the cloaca. In the Egyptian Goose (*Alopochen aegyptiaca*) the pursuit by the male and the flight of the female have become entirely 'ceremonial'. Heinroth states that the mating of this species has the appearance of 'prearranged mock rape'. In my opinion, many cases where we can observe so-called 'coyness' in the female during mating represent instinctive ceremonies based on the vestiges of originally genuine escape patterns. This would indicate that the lizard-type of mating also represents a 'primitive' form of behaviour in some birds.

Some ducks, particularly the mallard, exhibit – in addition to their actual highly-specialized type of pair-formation based on interlocking releasing ceremonies – an extremely peculiar behaviour pattern whose biological significance is difficult to comprehend. The males attempt to 'rape' *any* strange duck which they encounter during the entire breeding season. In contrast to the female *Cairina*, however, the duck really takes to flight and is in fact not normally overtaken by the pursuing drake. Since the female in this case is 'not guilty of' exhibiting the slightest responsive pattern, we are faced with the only instance of bird mating behaviour which corresponds to the pure lizard-type. For this reason, I shall consider this in some detail: In his relations with his consort, the drake shows the greatest consideration and will not attempt to copulate without an unequivocal invitation. In fact, he may fail to copulate even when invited. The drake thus behaves in the manner characteristic of most lizards only in response to *strange* females. However, this 'raping' response has yet other characteristics in common with that of lizard behaviour. In particular, the response is apparently elicited by any conspecific of either sex and is only inhibited by sex-determined male characters. The most predominant of these characters is evidently the male 'Prachtkleid'. In 1933, I had a pure-blooded male albino mallard, which I received through the kindness of Frommhold in Essen. It was extremely interesting to observe that this male was pursued by the other drakes and subjected to 'rape' attempts. Some years ago, I observed with a wild-coloured domestic drake that it pursued in the same manner male conspecifics which still bore the female-like juvenile plumage. Wild drakes never have the opportunity to exhibit such behaviour, since their breeding season has elapsed before young conspecifics are sufficiently grown. It must be said, however, that the male albino drake described was sickly and exhibited no sexual activity whatsoever. The male demonstrative behaviour can apparently replace the 'Prachtkleid', since this albino drake was entirely healthy and sexually active the next year and was never treated as a female by male conspecifics. I should greatly like to know whether castrated females, which possess the full 'Prachtkleid' but none of the male demonstrative behaviour, are 'raped' by the males. Probably, 'Prachtkleid' and demonstrative behaviour operate in exactly similar fashion to deter the 'rape' responses of other drakes and doubtless have no effect on the female pursued with intent to 'rape'. The female is not bothered with the presence or absence of impressive characters in the pursuer and simply attempts to escape in all cases.

All of this only applies to this particular form of duck mating, however. In the establishment of a lasting pair-bond, the 'Prachtkleid' quite obviously affects the females as well and is displayed to them in an unmistakable manner. Noble and Bradley's assumption that the 'Prachtkleid' of a male bird generally has the same function as that of the male of the reptile species which they investigated is thus only partially applicable even to these ducks and is even less valid in regard to other bird species.

We now come to those birds whose pair-formation corresponds to that of the *labyrinth fish-type*. This category includes the majority of bird species. The parallels with pair-formation in labyrinth fish extend into quite amazing details, as would be confirmed by anybody who has observed both male fighting fish and a Golden Pheasant during courtship of the female. Of course, this similarity is entirely due to convergence. In the fish, where external fertilization occurs, the co-operation of the female is just as important as in the birds, in which 'rape' of the female is not possible for anatomical reasons. It is certainly no chance phenomenon that the few birds which are known to exhibit the lizard-type of mating all belong to species in which the male has a functional reproductive appendage. Whereas the mating type of the labyrinth fish cannot have originated from a lizard-like 'rape' of the female, such a source would seem to be probable on comparative anatomical grounds for birds which exhibit extensively similar behaviour.

One property which is possibly latently present in all vertebrates emerges in those birds in which pair-formation occurs according to the labyrinth fish-type. In these birds, each individual possesses not only the complete complement of species-specific, sex-determined instinctive behaviour patterns of its own sex; it also possesses (though normally in latent form) the sexual instinctive behaviour patterns of the other sex. It may appear contradictory to speak of 'sex-determined' behaviour patterns at all in the face of such 'ambivalence', but male and female instinctive behavioural systems remain sharply separate within a single individual. The animal can respond to a specific partner only in one way – *either* as a female *or* as a male. The behaviour patterns of an 'instinctive outfit' are mutually bound as a very cohesive unit and the two units are not generally mixed. The membership of a behavioural component within the male or female behavioural units can therefore be just as exactly determined as if the two units were never exhibited by the same individual.

On the basis of the conditions under which female behaviour patterns can be experimentally elicited in a male (and *vice versa*), we have some knowledge of the factors which under natural conditions induce the emergence of the instinctive behavioural system corresponding to the gonads of the animal and suppress the opposite behaviour, causing it to remain latent.

Above all, it must be borne in mind that in bird species exhibiting the labyrinth fish-type of pair-formation *each* individual exhibits the tendency to perform the *male* behaviour patterns, and that it is the stimuli emanating from the sexual partner which suppress male behaviour patterns in the female and (so to speak) make room for female behaviour. An isolated female deprived of male company always inclines towards exhibition of male demonstrative behaviour (possibly in a fashion analogous to the development of the male 'Prachtkleid' in a castrated female). The decisive character of the partner which inhibits emergence of male instincts in the female is the *one-sided dominance relationship*. A female can only respond in a feminine way when she is presented with a socially dominant conspecific. A male can exhibit male responses when in complete isolation, but remarkably cannot do so when *only* accompanied by dominant conspecifics. A. A. Allen has observed with *Bonasa* that in a large number of males living together in one space the individuals lower in the rank-order not only failed to perform any male demonstrative behaviour, but even failed to enter into reproductive condition at all. If such low-ranking males exhibit any sexual instinctive behaviour patterns at all, then they are female in type. Allen refers to this entire complex of behaviour patterns quite simply as 'inferiorism', which expresses the decisive rôle of the dominance relationship.

It can scarcely be doubted that even in a natural society of a bird species whose pair-formation takes place according to the labyrinth fish-type, a female high in the social order will not respond in a female way and that a male standing on the lowest rung of the dominance hierarchy can exhibit no male behaviour.

In spring 1932, I had four ravens – a pair of adults, which were just preparing to breed, and two one-year-old females. I also regarded these two young females as a pair, however, since they exhibited vigorous courtship behaviour in which the older and stronger of the two sisters always took the male rôle. Because of a disruption in the sequence of instinctive behaviour patterns involved in pair-formation, the adult pair did not go on to breed, and the male eventually drove off the female for good with his persistent pursuits. When I subsequently allowed the two young ravens to join him, I was amazed to see the sibling which I had previously regarded as a male respond to the attempted approaches of the two-year-old male with all of the ritualized female behaviour patterns, such that the two formed a pair. I was particularly interested to see that this female did not at once cease to court her sister in male fashion and only began to treat her aggressively when she was firmly paired with the male.

Similar behaviour was exhibited by my adult jackdaw female Red-Yellow, the only surviving bird from my colony of tame jackdaws, which provided the observational material for my paper *Beiträge zur Ethologie sozialer Corviden*. I at first introduced four new young jackdaws to this surviving female, and in the next spring (1931) she formed a pair with one of the new birds and performed the rôle of the *male*. A normal nest was constructed, but no eggs were laid. In other words, the female responses of Red-Yellow remained completely dormant, for the jackdaw playing the female rôle was not yet mature enough to lay eggs. In 1932, the two birds constructed a peculiar nest reminiscent of Siamese twins, *two* nest-cups were built on a normal foundation. In *Coloeus*, as in the majority of passerines, the male normally constructs the rough foundation of the nest, whereas the preparation of the nest-cup belongs to the specifically female instinctive behaviour patterns. Both cups of this double nest were used for holding the eggs, though I could not determine their actual number because the nest lay in an extremely inaccessible deep cavity. The eggs were apparently unfertilized, since no young emerged. My suspicion that a female pair had been formed was confirmed by an unexpected event in autumn 1932. One of the jackdaws of my former colony, which I had long since believed to be dead, turned up again after an absence of more than two years. The returned jackdaw immediately behaved as if it had never been away and promptly demonstrated that it was a male by undergoing betrothal with Red-Yellow in the first few days after arrival. Red-Yellow immediately responded to the advances of the immaculate adult male and, in so doing, confirmed my assumption that the lack of an adequate male had been the cause of female pair-formation. In 1931, Red-Yellow had no male at all available, and in 1932 she had only an immature male born in 1931, in place of which she obviously preferred the female born in 1930 as a partner. The appearance of a competent male immediately evoked the response of all her female behaviour patterns. However, this surprisingly did not lead to a disruption of all relationships between Red-Yellow and her previous female consort. The male was initially very aggressive towards the latter, but gradually became accustomed to his wife's hanger-on. From this time onwards, the birds formed an inseparable trio. I do not know whether the male also copulated with the second female, but this is my suspicion. In spring 1933, the birds built a double nest similar to that of 1932 and filled it with such an enormous clutch that I am convinced that both females had laid, although (as in the previous year) I could not count the number of eggs exactly. The eggs were fertilized, but the small nestlings succumbed because of a disruption of nest-relief caused by the presence of the two females. The females always brooded simultaneously, each on one of the nest-cups, whereas the male (which had to relieve both females alone) was of course only able to cover one of the two nest-cups at a time. This had not affected the eggs, in fact, but it was most probably the cause of the death of the small nestlings.

This behaviour of female ravens and jackdaws plainly shows that the male instinctive behaviour patterns are also latently present in the female and can emerge in the absence of a consort companion of the opposite sex. In these birds, which exhibit little external sexual dimorphism and in which males and females are distinguished only by a gradual non-qualitative difference in the plumage areas employed in demonstrative behaviour, any strong individual with fine plumage will respond to a weaker and less impressively coloured bird by performing the rôle of a male. Nevertheless, formation of unisexual pairs does not appear to occur under natural conditions. In the normal, bisexual pairs, however, *the male is always dominant to the female*. Among the many jackdaw and raven pairs which I was able to observe, there were actually some in which strong, beautifully-developed males had selected disproportionately weak (in fact virtually sickly) females; but the converse never occurred.

A. A. Allen described similar behaviour in the American ruffed grouse *Bonasa umbellus*. In this case, too, strong females exhibited male demonstrative behaviour towards any weaker conspecific, regardless of the sex of the latter. Similarly, weaker males cannot maintain their male behaviour patterns in the face of greatly dominant rivals, and the female 'instinctive outfit' then emerges. Allen even succeeded in setting up an 'inverted pair' by introducing a weak, frequently defeated male to an extremely strong female which tended to exhibit male demonstrative behaviour, such that pair-formation with reversed rôles took place. Unfortunately, in the summary of the results of his extremely valuable study, Allen presents this observation in the form of inaccurate generalizations, in which he repeatedly writes 'the birds' instead of '*Bonasa umbellus*'. The sentence 'Birds are not sex-conscious' only applies to species in which pair-formation proceeds according to the 'labyrinth fish-type'. This does in fact include the majority of birds, but by no means all. In any case, introduction of the concept of 'consciousness' is unnecessary and misleading.

Wallace Craig had already investigated the phenomenon of sexual reversibility in 1908 and had conducted very impressive experiments with pigeons. Craig kept his experimental animals in isolation cages and studied the mutual influence of sexual behaviour of the birds by placing the cages together two at a time without actually allowing the birds to come together. It then emerged that actual physical domination by one of the opponents is not necessary for the production of what Allen later called 'Inferiorism'. When Craig gave a courting Barbary Dove (*Streptopelia risoria*) of average strength and normal temperament a particularly strong and vigorous old conspecific as a neighbour, the weaker bird ceased to exhibit courtship and even normal demonstrative behaviour. Soon after this collapse of male responses, the bird exhibited *female* elicitatory behaviour patterns. The fact that in this case demonstrative behaviour *alone*, without the slightest possibility of physical contact, was able to disarm the rival is extremely important. At first, one tends to regard it as improbable that simple display movements combined with bright colours and loud calls should exert a greater influence than sharp claws and a pointed beak. Craig's experiments demonstrate, however, that this is in fact the case. In rivalistic struggles between male animals, physical strength is never the sole deciding factor, and neither are weapons. Victory is predominantly dependent upon the determination of the fighting animal, or – to put it more exactly – upon the *intensity* of its fighting responses. In fact, the opponent can determine the intensity of this response from the demonstrative behaviour exhibited. In cases where there is a considerable difference in intensity between two mutually displaying animals, the animal responding at a lesser intensity develops an inferiorism, even if it does not have the chance to take part in actual fight.

The reversibility of sexual responses in birds whose pair-formation takes place in agreement with the labyrinth fish-type raises the question of how the formation of unisexual pairs is prevented. We must particularly examine this question with those species in which external signs of sexual dimorphism are absent. From what has been said, it would seem to be quite possible that even under natural conditions a stronger male might extinguish male behaviour in a weaker individual and evoke female behaviour. The fact that this, to all appearances, does not occur is probably closely associated with the process of individual choice of a partner. In other words – using a genuine analogy with human behaviour – the responses of falling in love are involved. Such responses play no part at all in pair-formation of the lizard-type. In the birds which pair according to the Cichlid-type, as will be discussed later, similar phenomena are in fact sometimes found, but in these it is already decided before the choice of the partner whether the animal will subsequently exhibit male or female displays. In a bird undergoing pair-formation according to the labyrinth fish-type, the subsequent performance of male or female behaviour depends upon the individual selection of the love-object. As has been mentioned, the choice of the partner represents the only hormonally-governed difference in the behaviour of the sexes: males only fall in love with subordinate animals and females only with dominant animals. If a partner of the opposite sex is lacking, these birds make do with a love-object of the same sex, and the animal immediately begins to exhibit the responses of the other sex. The personal bond to such a partner can be so strong that later arrival of a sexually-receptive animal of the opposite sex will not break it. This is, for example, true of pigeons, in which preference of the opposite sex seems quite generally to be less prevalent. They will in fact form male pairs and female pairs even when animals of both sexes are available. This predominance of personal relationship over sexual influences should be regarded as an effect of domestication, however. As we have seen, in jackdaws and ravens a love-relationship with a conspecific of the same sex is immediately broken if an adequate sexual partner appears on the scene, provided that the new arrival exhibits the correct dominance relationship.

Since pair-formation can only occur through a mutual understanding (more exactly, as a result of a mutual process of falling in love), under natural conditions the slight, genuinely sexually-determined difference in response patterns is sufficient to prevent the formation of unisexual pairs. In particular, the following must be emphasized: If we keep birds of a given species in a group composed exclusively of one sex, the responses of falling in love are oriented towards a substitute object, leading to the formation of unisexual mated pairs. If, however, a few males are present among a much greater number of females, all of the females become so decidedly in love with the few males that the dominant females will not respond to the female mating invitations of their subordinate sisters. Rigid attachment to a love-object once chosen, which has been described by Heinrich Heine in a classically simple fashion in *Ein Jüngling liebt' ein Mädchen,*[76] can be found in mallard, Greylag geese and jackdaws, even when superfluous individuals of the same sex are available. Uninitiated enquirers always respond with disbelief, however, when one maintains that a sexually-receptive bird refuses to undergo mating with an equally receptive conspecific simply because it is in love with another individual. I am inclined to the opinion that this 'Ritter-Toggenburg behaviour' is in many birds, particularly in social species, the only factor which prevents all superfluous animals of a given sex from pairing with one another.

This portrayal of the factors preventing the formation of unisexual pairs doubtless incorporates a sweeping simplification of the relationships which actually exist, but the description is nevertheless quite accurate for jackdaws and many gallinaceous birds.

We now come to those bird species in which there is *no dominance relationship* between the members of a pair and in which there is therefore no observable competition for dominance between the sexes – in contrast to the views of Schjelderup-Ebbe. This type of pair-formation has been referred to above as the *Cichlid-type*. We have already learnt that the preservation of demonstrative behaviour in the female is a characteristic of this type. In many cases, this in fact dictates the development of ceremonies in which the sexes stand opposite one another and behave almost identically. As an analogue to the parallel swimming of Cichlid pairs described previously, I should like to recall the rattling ceremony of the white stork. Similar behaviour is found in herons, cormorants, the petrel relatives and the grebes. This reciprocal demonstrative behaviour has been much discussed in the English literature, where it is called 'mutual display'.

In very many birds, a dominance relationship within pairs never appears because the partners never have to cope with 'differences of opinion'. This in itself is not sufficient for us to classify the species concerned as a 'Cichlid-type'. In very many cases of this kind, a dominance relationship is not only present at the *formation* of the pair but actually indispensable for pair-formation. This dominance relationship reappears at each disturbance of the harmony of the pair in that the female flees without a preceding fight.

The situation is quite different in the dominance-less pairs of the pure Cichlid-type, as in many herons. In the process of pair-formation of these animals, the two potential mates gradually approach one another whilst performing specific ceremonies, which inhibit the normal defensive response. However, threat and demonstrative behaviour of *both* animals repeatedly breaks through this pacific ceremony. Even later on the partners do not exhibit such complete mutual confidence as pigeon or Anatid pairs. A single abrupt movement of one of the birds, for example caused by stumbling or some other loss of balance, suffices to bring the other on the defensive, with the display feathers of the neck ruffled. A moment later, the partners are standing opposite one another, exhibiting the complete threat behaviour of the species and ready to fight. Frequently, a fight does actually ensue, and this (together with the manner in which it ends) deserves more detailed description. Verwey describes minor disputes, which he refers to as 'beak-fencing', in pairs of the heron. He observes that he originally believed that a ceremony was involved, but that he later concluded that serious fighting was actually involved. In the Night Heron, the situation is such that beak-fencing exactly corresponding to Verwey's description begins as serious fighting and then dissipates in the form of an appeasement ceremony. The animals rear up together and momentarily stab at one another with open beaks. Seconds later, beak-stabbing gives way in a smooth transition to beak-rattling of gradually increasing rapidity. The two birds nibble one another's beaks in 'placatory' fashion. Heinroth describes this nibbling beak-rattling as an expression of affection in his isolated hand-reared Night Heron. In actual fact, paired birds often exhibit this behaviour in the same context, *without* preceding beak-fencing. In accordance with what has been said, it appears very probable that affectionate beak-rattling has evolved from originally serious beak-fencing. This represents a typical case of evolution of a ceremony from a behaviour pattern which originally had an entirely different significance. I assume from Verwey's doubts and subsequent change of interpretation that in the heron the process of ceremony-formation from this response is incipiently present. At least, I believe that this author would agree with my interpretation on observing the correponding behaviour of the Night Heron.

I have described this behaviour of the herons in some detail because, to all appearances, in animals with dominance-less pair-formation *very many* affectionate ceremonies have been derived from threatening demonstrative behaviour. The ceremony always represents a truce without victory of one or other partner. It is characteristic of these pacification ceremonies that they develop from the posture and behaviour of threat before actual fighting occurs and before it has been decided which partner is the stronger. The behaviour of both partners of a pair during the process of becoming acquainted with one another is always reminiscent of the behaviour of two dogs during the same process: the tension, the threat posture, the expression of armed neutrality and finally the dissipatory pacification signal (in the case of the dog, tail-wagging). This is markedly different from the behaviour of labyrinth fish and that of most other birds, in which encounters are always accompanied by a fight, which always ends with at least a 'moral' victory for one of the animals.

A further vital difference from the labyrinth fish-type is that the individuals involved are not sexually ambivalent at the outset. If they were so, all individuals – males and females alike – would respond in male fashion, since neither of the two partners of a pair would have the opportunity to develop an 'inferiorism' (A. A. Allen). The male Night Heron enters the reproductive cycle without any relationship to a female, and the female never shows the slightest traces of male behaviour.

Apparently, the responses of individuals falling in love play little part in the Cichlid-type of pair-formation. In pairs, either of the partners *can be replaced* at any time by individuals of the same sex and physiological condition, even in those species where persistent pairs are formed. Schüz has demonstrated that this is so for the white stork, and according to Herrick the same applies to the Bald eagle *Haliaetus leucocephalus*. I suspect that pair-formation of the Cichlid-type is present in all birds in which there are on the one hand no sexual differences in the display organs, and on the other no relationships between the partners independent of the reproductive cycle.

The fact that three types of pair-formation have been presented by no means dictates that all forms of pair-formation in birds must fit into these categories[77]. This is simply a means of demonstrating the basic differences in pair-formation which can exist and the danger in making precipitate generalizations. There are more than enough transitional types. Within the Anatidae alone, a series of types from the lizard-type (*Cairina*) to the labyrinth fish-type (Egyptian Goose, Ducks) and on to the Cichlid-type (geese and swans) can be traced.

It would be of vital interest to know exact details of pair-formation in grebes, which exhibit extremely unusual sexual behaviour (Huxley, Selous). These animals exhibit reciprocal treading, with both birds often performing the rôle of the male. It would be important to know how the formation of bisexual pairs is ensured in these birds, which exhibit little sexual dimorphism and in which a dominance relationship probably plays little part. The beginnings of similar behaviour can be seen in some doves (*Streptopelia*), in which the male crouches low and is mounted by the female after actual copulation has taken place.

(*b*) *Synchronization of the reproductive cycles*

Not all so-called courtship behaviour patterns serve to attract a sexual partner to the performing bird. Many male birds in fact continuously court their females long after they have formed pairs which will definitely last. I need only remind the reader of the persistent courtship of the male Golden pheasant and the many continuous amatory ceremonies of pigeons, parrots, finches and other birds.

A. A. Allen has shown that in many, perhaps all undomesticated birds the male is not always in reproductive condition (as was formerly thought), but instead exhibits a quite short period of fertility fairly closely corresponding to the receptive phase of the reproductively-active female. Everything therefore depends upon temporal co-ordination of the receptive phase of the female with the reproductive motivation of the male. The 'synchronization of the reproductive cycles', as Allen calls it, actually plays a large part in the interlocking instinctive patterns of pair-formation. A wide range of behaviour patterns generally regarded as courtship activities are exclusively adapted to this one function.

In birds which only exhibit a brief encounter of the sexes for mating, such as the Black Grouse and the American ruffed grouse (*Bonasa umbellus*) investigated by Allen, the males doubtless find females of equivalent motivation as a direct result of their 'impersonal' courtship. However, Allen neglects the possibility of a subsequent synchronization of the physiological reproductive cycles of a pair of birds and believes that only birds which are 'synchronous' at the outset can form pairs. This is indeed true of the ruffed grouse and of many passerines, which only pair off for a single brood, but in birds which live in pairs persisting for several years it seems highly improbable that the chronometers of the two partners would require no subsequent regulation. Craig had noted for doves in 1908: 'Whenever one partner is ready to mate earlier than the other, the former is retarded by the influence of the latter and the latter is accelerated by the former.' Unfortunately, it is not always possible to determine the time of reproductive motivation of the male as exactly as Allen was able to do with the male *Bonasa*, by presenting a stuffed female mounted in the copulatory posture. An observation made by Heinroth shows how exactly synchronization by mutual influence can operate: Year after year, two female Chilean widgeon (*Mareca sibilatrix*), which acted as a married pair, laid twenty-two eggs in a common nest in eleven days.

In many species, exact synchronization of the partners is also necessary because the instinctive behaviour patterns of parental care follow one another in an exact temporal order. I call to mind the fact that in pigeons, for example, the male must also secrete the crop-milk used for feeding the offspring after exactly fourteen days of the incubation phase have passed, if the young are to survive. Even if male pigeons should be persistently in reproductive condition (which is extremely improbable), their internal physiological processes would have to be synchronized with those of the female for the purpose of timely crop-milk production.

Apart from this possibility of subsequent synchronization, which was not recognized by A. A. Allen, I should like to emphasize that Allen's experiments have not lessened the probability that the males of other bird species are capable of reproduction throughout the entire brooding phase. I regard an extrapolation of Allen's observations on a restricted number of species to the entire Class as entirely unsupportable. In particular, Verwey's observations on the heron seem to me to permit the conclusion that males of this species are in reproductive condition throughout the entire reproductive period, at least when they do not find a female. Verwey in fact emphasizes that the copulatory motivation of unpaired males exhibits a steady increase. The longer a single male must wait for a female, the more rapidly he proceeds to copulation after a female has arrived. In an extreme case, Verwey observed immediate copulation with a strange female, with complete omission of the 'betrothal' ceremony. This corresponds exactly with the behaviour of persistently reproductively-active mammals and of man, as Verwey himself pointed out. Increase in the motivation of a behavioural activity (in other words, reduction of the threshold of the eliciting stimulus) is a basic property of instinctive behaviour patterns which are not bound to specific cycles by physiological processes.[78]

(*c*) *The instinct co-ordinations of joint nest-building*

There are relatively few bird species in which only one of the sexes is involved in nest-building. In Tetraonids, the male pays no attention whatsoever to the female and therefore plays no part in nest-building; with weaver-birds, the male builds the nest almost entirely alone, while the female is only concerned with rudimentary lining of the nest-cup. According to Steinfatt, the latter is also true of the Penduline Tit.

In the great majority of bird species, however, both sexes are involved in nest-building. In very many birds, of which one normally reads that the males are not concerned in nest-building, the choice of the nest-site is determined by the male. It seems as if only this one, specific male activity remains from the entire building activity of the male. In mallard, the nest-search often appears in males some time before the females are really ready for nest-building or are actually prepared to lay eggs. The instinct concerned is frequently incipiently aroused in drakes even in the autumn, at the beginning of the betrothal period. Heinroth was also able to determine extremely conspicuous nest-searching in Eider drakes, although this is usually discounted in the literature. I counter the objection that Heinroth's observations were carried out on captive animals and are therefore not representative, with the observation that alterations of instinctive behaviour patterns caused by the conditions of captivity can only represent a *reduction*, a minus effect, and can never give rise to the production of novel behaviour. If Heinroth's eider drakes had *not* exhibited a nest-search, this would have been far from reliable proof that they would not do so in the wild; but I regard the converse behaviour as unconditional proof.[79] From observations on domestic cocks, it would seem that the male Burmese jungle fowl also determines the nest-site. I do not venture to decide whether this is the case with the Golden pheasant as well, since these animals are so extremely secretive at the beginning of the nest phase that I was scarcely ever able to observe the hen herself in the search for a nest-site.

Determination of the nest-site by the male seems to be far from indispensable to females of these species, however. At least, one is unable to observe hesitation or disturbance of site-selection in a female mallard, which has been unable to find a partner, in searching for a site at which to lay her unfertilized eggs. We are faced here with the phenomenon whereby one sex is able to perform an instinctive behaviour pattern which should actually be performed by the opposite sex and can thus compensate for the lack of a co-operating partner.

The determination of the nest-site occurs in an entirely different manner with the Night Heron, where the male in fact chooses the nest-site and begins to build the nest alone. The nest only achieves its proper form when the male has found a female; the latter sits continually in the nest-hollow and arranges the twigs carried up by the male. The fact that the female marks the centre of the nest by simply sitting still seems to be particularly important. In his initial, solitary nest-building activities, the male works in a completely unadaptive manner at different (though perhaps neighbouring) sites, so that the lack of a defined centre prevents circular arrangement of the nest-material. In the cases which I observed, the centre of the nest was determined by the arrival of the female, soon after which a rough circular design could be recognized in the heap of collected twigs. In fact, I believe that a solitary male *Nycticorax* would also be able to construct a round nest structure eventually, but it is certain that it is an important function of the female to determine the centre at an early stage of nest-construction.

The division of labour between the sexes in the collection of nest-material varies from species to species. In some passerine species, both sexes collect nest-material, each partner carrying the material it will incorporate into the nest (i.e. the male carrying the coarse material for the foundation and the female carrying fine lining material). Thus, the instinctive behaviour patterns of the two partners operate in an extremely independent manner. With hole-nesting birds, when the hole is so small that a special foundation is not necessary – in fact obstructive, since the nest-cup alone will occupy the available cavity – the male still carries in coarse nest-material, following the blind plan of his instincts, and this material must first be laboriously removed by the female in order to make room for the subsequent nest-cup structure. The male is therefore not affected by spatial relationships in performing his building activities, whereas the female is. I have observed such behaviour with jackdaws. Seton Thompson, in one of his short animal stories, tells of a case where a male sparrow continually carried pieces of wood into the nest-cavity, with the female regularly throwing these out as she collected feathers for the nest. Thompson gives an entirely anthropomorphic, and definitely erroneous, interpretation of this occurrence; but since he adheres to factual personal observations in his stories, I am convinced that the observation itself is reliable and that the same had occurred as with the jackdaw.

(*d*) *Elicitation of copulatory responses*

In most birds, copulation itself normally occurs only as an element in the chain incorporating the other individual behaviour patterns of reproduction, and it is only under quite specific circumstances that it can occur out of this context. This can occur, for example, when appreciable postponement of the response lowers the threshold for the eliciting stimulus, as was explained for the heron on p. 210. It is not known whether corresponding effects may appear in females. Bengt Berg reports that female Greylag geese which were paired with infertile male hybrids between Greylag goose and Canada goose nevertheless laid fertilized eggs, but the report does not indicate whether the author was aware of the significance of his information. A case of marital infidelity in the Greylag goose, confined to copulation itself, would be extremely interesting! Male birds of monogamous species in many cases respond promptly to copulatory invitations by strange females. The immediate invitation to copulation emanates almost always from the female. The major characteristics of this invitation are a low, crouching posture and immobility. In an experiment, either of these two characters will elicit the copulatory responses of a male. The Barbary Dove (*Streptopelia risoria*) mentioned on p. 187, which was imprinted on the human frame and always directed its sexual responses towards the keeper's hand, at first cooed and bowed continuously before the human hand without attempting to copulate, even when the hand was held still on the table. As soon as the back of the hand was held out flat at a specific height above the table, however, the dove began to exhibit copulatory responses. A. A. Allen has experimented with stuffed birds in order to determine the duration of male copulatory motivation under natural conditions. The stuffed birds were attached to trees in the territories of the males under investigation in such a way that they were bound to attract the attention of the resident birds. It emerged that the male birds, as long as they were in breeding condition, attempted to mount the stuffed conspecific without any preliminary behaviour, even if the stuffed birds were males and were not mounted in the copulatory posture. Heinroth describes the behaviour of a robin which similarly attempted to mate with a freshly-killed conspecific without any preliminaries. It must be remembered that the same male birds would have responded *differently* to a *living* female. They would have at first burst into song, chased the female or made other attempts to induce her to give an invitation to copulation. The immediate emergence of copulatory responses is to be explained on the assumption that the males responded to the immobility of the stuffed birds and the dead conspecific in the same way as they would normally respond to the immobility of a female prepared to copulate. In an animal which is in almost perpetual motion when awake, even complete immobility can function as an 'elicitatory act'.

In many birds, immobile crouching of the female is preceded by specific, obligate ceremonies. With the Rock Dove, for example, the female must first be fed by the male, after which the two birds preen themselves behind the elbow feathers in a peculiarly hurried fashion and then immediately proceed to copulation. All of these different introductions to mating, in particular those of the Anatidae, have been systematically considered by Heinroth.

(*e*) *The instinct interlocking mechanism of alternate brooding*

In many bird species, the two partners brood alternately and relief of one by the other is always associated with a series of innate behaviour patterns which belong in the category of *releasers* and are therefore of interest at this point.

In the Columbidae, nest-relief occurs in a simple manner, with the arriving bird squatting down beside the brooding bird and driving it from the nest, whilst (so to speak) simultaneously brooding. If one partner should leave the nest for some other reason, however, the mere sight of the departing partner operates on the other bird as a stimulus evoking return to the nest. This latter effect occurs similarly in many other birds which exhibit nest-relief, e.g. the Night Heron.

Nest-relief takes place at a quite specific time of the day in the Columbidae. In the majority of the well-known species, the female broods from late afternoon to the next morning and the male broods for the remaining period. The fact that the time alone can operate as an eliciting stimulus for nest-relief, without the sight of the partner playing an elicitatory rôle, was proven to me by a domestic pigeon, whose female partner was carried off by a cat before my eyes. Since I knew that the pair had young nestlings, I observed the male very closely after the unlucky event. He did not brood much longer than he would have done if he had been relieved by his partner in the normal way, but instead rose from the nest and went to search for food as if he had been relieved according to plan. In the evening, the male did not settle on the nest, but took up his usual sleeping position *alongside* the nest. Since a cold night followed, the nestlings were dead the next morning. The male nevertheless settled on the nest at about 10 a.m. and brooded on the corpses of the chilled nestlings until the afternoon. This behaviour continued for two days.

Whereas with the Columbidae the passage of a given time-interval can itself operate as a stimulus eliciting nest-relief, as shown, in other birds another 'mechanism' is found in the form of instinct interlocking. With my Night Herons, if the non-brooding partner chanced to wander into the vicinity of the nest (and the brooding partner), this sufficed to arouse the drive to relieve the brooding bird. Under natural conditions, the relieved brooding heron needs several hours before it has satisfied its food requirements and returns to the nest. This time-interval is determined by the environmental conditions and it is therefore not necessary for it to be innately predetermined. Because of feeding by human intervention, the clocks of my Night Herons were re-set. Whenever a brooding heron was fed, it immediately attempted to relieve its partner, even if the latter had also been fed just previously. The partner on the nest was frequently far from willing to get up from the nest, but it was then regularly induced to do so by the relieving partner in a quite specific manner. A heron settling on to an empty nest typically makes minor adjustments to the twigs on the facing edge of the nest – one could say that the heron *cannot* settle on the nest *until* it has done so. If the brooding partner will not make way for the relieving partner, the latter nevertheless exhibits the response described, bending over the brooding bird and beginning to perform building motions on the opposing edge of the nest. In doing this, the relieving bird presses down to some extent on the back of the squatting partner. Since almost all birds do their utmost to avoid dorsal contact, the squatting bird stands up 'irritatedly' and makes room for the other. I am convinced that my Night Herons relieve one another much more often than is the case with brooding birds under natural conditions. Similar behaviour, in which a temporal rhythm is determined only by external conditions and not by an internally-operated temporal mechanism, is found in the Cichlid fish which have already been considered as an analogy. In these fish, one of the parent animals maintains continuous watch over the nest-hollow and is relieved by the other from time to time. The sight of the approaching partner elicits in the watching fish a behaviour pattern which itself acts as a releaser – an excellent example of instinct interlocking. The fish which was previously guarding the eggs swims off with a peculiarly exaggerated swimming motion, always passing close to the approaching partner as if it were attempting to magnify the visual effect of its departure. The approaching fish then immediately takes over the guarding of the nest. In large aquaria, such nest-relief occurs at large intervals, but in very small tanks the fish alternate in nest-guarding almost incessantly, since the free partner cannot avoid coming close to the nest whilst swimming around in the small space available. Since the mere sight of the free partner elicits the described 'exaggerated' departure swimming, as an invitation to assumption of the guard-post, the fish never settle down, and this is decidedly detrimental to the rearing of the young since the responses habituate and lose their exactness. It is quite remarkable that the functional plan of the drives determining nest-relief should be disrupted by abnormal conditions in these fish in a manner so analogous to that observed with my Night Herons.

(*f*) *The disappearance of the sexual companion*

Some birds respond in an extremely conspicuous manner to the sudden disappearance of the partner. On one occasion when I gave a domestic drake (which was paired with a pure-blooded mallard duck) to be slaughtered, the widow exhibited conspicuous searching after the lost partner. She visited one after another all of the places in which the pair normally spent their time and continued in this behaviour for several hours. Thereafter, she flew off towards the Danube and was never seen again. (It should be emphasized that mother ducks which have lost their chicks never search for them in this rational manner.) A Carolina Wood-duck (*Lampronessa sponsa*) exhibited exactly the same behaviour in December 1933. Because of a sudden onset of cold weather, I had the entire flock of Muscovy ducks driven into a shed. By an unfortunate mishap, the female of the Wood-duck pair, which had been flying free for some months, was gathered up with these animals. I first noticed this after night had fallen and I preferred to release the female Carolina Wood-duck the next morning because of the great danger of a huge night-panic among the birds. However, when I intended to do so at the break of dawn the next morning, the male had already flown off. Two days later, he was caught in Heiligenstadt, twelve miles down the Danube, and reported to Rossitten, so that I was able to retrieve him. I had almost exactly the same experience with a mallard drake in 1934.

The behaviour of the Carolina Wood-duck pair on reconciliation was extremely interesting. As I released the drake from the transporting case, whereupon he flew to a meadow and began to call, the female came from an entirely different part of the garden and landed alongside him. The two birds then shook themselves and began to preen their feathers. Shaking of the body frequently occurs as a sign of inner relaxation; an animal lacking a greeting response has no possible means of responding to a situation of reunification, however significant it may be. With such reunited companions, one can only observe particularly intensive operation of reciprocal functional systems, which were interrupted in separation and whose behaviour patterns emerge at high intensity on reunification. The female Carolina Wood-duck, for example, incited the male partner to attack all the other ducks for almost the entire afternoon, and the male 'paid tribute to the occasion' to the extent that he actually did attack even the most formidable opponents – something which I had never observed previously.

4. DISSOCIABILITY OF THE FUNCTIONAL SYSTEMS

The few cases in which I have been able to perceive separation of the individual functional systems of the sexual companion are given below:

A pure-blooded female mallard born in the summer of 1932 courted in the winter 1932–3 a hybrid drake ('Hochbrutente' × mallard), which responded to her incitement behaviour with particular vigour. At the same time a pure-blooded male mallard courted the female in the shy, passive manner typically exhibited by males of those Anatid species in which the active part of courtship occurs in the form of female incitement behaviour. The hybrid drake, which responded very actively to the rivalry of the mallard, pursued the latter with particular rage, even taking to the air, despite the fact that the inherited effects of domestication normally acted against flying. It was just this reduced tendency to fly on the part of the half-domesticated drake which gave rise to extremely interesting behaviour on the part of the female mallard. In the spring, pure-blooded mallard exhibit distinctive restlessness, which is possibly correlated with the nest-search. In any case, the pairs fly over a large radius at this time, with the female always flying ahead and the male following. Even in the take-off, it is usually observed that the invitation to fly originates from the female. At the time when my female mallard began to exhibit spring flights, it emerged that the hybrid drake did not respond properly to her intention movements for inducing flight motivation. The male did not fly up with the female, or would turn back and land after flying a very short distance. Initially, the female immediately looked back for the male, began to circle and also landed soon after take-off. The pure-blooded mallard drake soon noticed that this was a field in which he was superior to his feared rival, and whenever the female took off the drake took wing somewhere in the vicinity and soon caught up with her. The female now stopped returning to her proper partner and began to fly off with the pure-blooded drake. From that time on, the female *flew* continuously with one drake and yet continued to stay in the company of the other at home. Although she was then sometimes to be seen with the pure-blooded drake at home as well, I never observed that she treated him as her partner there (i.e. incited him to attack other birds). She only exhibited incitement behaviour with the hybrid drake, and I frequently observed her copulating with him. The mallard drake was only a companion to her in the functional system of 'spring flight'.

Bengt Berg describes an almost identical case in one of his books. Although male Greylags were available, a pure-blooded Greylag goose (*Anser anser*) had met and paired with a male Canada goose (*Branta canadensis*) which was unable to fly. In this case, too, the greater fighting strength of the American bird doubtless played a part. However, a conspecific played the rôle of a flight companion for this Greylag goose. Reading between the lines, this male conspecific also determined the nest-site, in accordance with the male behaviour described on p. 210. During brooding, the flying Greylag male stood watch with the female in the species-specific manner, but after the offspring had hatched, the female returned to the father and guided the goslings together with him. As soon as the young could fly, however, the flying Greylag male once again took up his function. He treated the young birds as his own offspring, for example diverting the attacks of White-tailed eagles from the young and towards himself in the manner typical for the species. This function is apparently performed by the males of many goose species, in which the males actually swim and fly as the last of the train.

The question now arises as to whether this mallard and the Greylag 'knew' that different males always acted as companions in the air and on the ground, or – conversely – whether ducks and geese in which these two rôles are normally played by *one* male recognize this male 'as the same' in the air and at home. We cannot answer this question on the basis of our present knowledge, and we can only say that we have no right to assume that the companion in these different functional systems forms a unitary whole, even if one is inclined to assume that this is at least the case with the Greylag goose.

VIII The social companion

In many birds we find formation of flocks which are far more than simple aggregations of individuals. We find genuine organized societies, whose supra-individual function is based on specific social instinctive behaviour patterns and instinct co-ordinations of their members. This function can reach such complexity of organization in some colonially-breeding species and can appear so adaptive in terms of the general interests of the colony that one is driven to think of the social behaviour of colony-constructing insects. Such co-operation of individuals in a colony is based entirely upon instinctive behaviour patterns, just as in the case of the insects, and is nowhere based upon traditionally acquired behaviour patterns or upon the insight that co-operation in furthering the colony is advantageous to the individual. On closer analysis of the instinctive behaviour patterns which produce the co-ordinated co-operation of the members of such a highly-organized bird society, it emerges that apparently highly-complex behaviour of the whole colony is derived from *remarkably few*, *simple* responses.

1. THE INNATE SCHEMA OF THE SOCIAL COMPANION

The innate schema of the social companion is almost always *broad*, i.e. incorporates few innate signals, so that there is a far greater range open to object-imprinting. Up till now, no single conspicuously social bird-species whose social drives *cannot* be imprinted on the human frame in the young bird has actually been discovered. It is well known to amateur aviarists that hand-reared birds of social species remain tame and attached to human beings after the extinction of the infantile instinctive behaviour patterns, in contrast to solitary species. There is no pre-determined place for a 'friend' in the environment of a solitary bird, and for a nightingale or a robin the keeper is at most a useful food-automat, whereas he represents a fellow bullfinch or a fellow siskin for hand-reared bullfinches and siskins. For this reason, the attachment of such socially-tame birds has nothing whatsoever to do with the expectation of food.

The few characters which are incorporated in the innate schema of the social companion are most frequently found in the acoustic range, and they represent releasers or signals. Wherever display-calls more associated with the external environment are concerned (e.g. alarm-calls), the innate characters are dispensable. In other words, their absence does not have an observable disruptive effect on the filling of the innate schema with a species-atypical object (e.g. man). The absence of the summoning call in human beings, on the other hand, can have an effect on hand-reared birds. At least, such birds often exhibit a positive response to human beings who can give a good *imitation* of the species-specific summoning call, or there is a distinct effect on the imprinting process produced by the imitated summoning call.

It is possible that there are social species with a complex innate schema of the social companion. Unfortunately we know virtually nothing about the behaviour of the smaller (and presumably less intelligent) colonially-nesting birds.

2. PERSONAL RECOGNITION OF THE SOCIAL COMPANION

Although it has been said above that co-operation of individuals based on instinct co-ordinations in a highly-organized bird society is reminiscent of that of colony-constructing insects, we must at this point emphasize a major difference between the two. Just as in the higher mammals (including man), the major part of the social responses of birds is based on *personal recognition* between individuals. We already know from the studies of Katz, Schjelderup-Ebbe and others, who have performed experiments on domestic chickens, that a bird can individually recognize a fair number of conspecifics. I should like to add that this ability, and particularly *remembrance* of individuals, is *considerably further* developed in birds with a more highly-specialized complement of social responses than that evident in the domestic chicken. Jackdaws can, for example, immediately recognize a member returning to the colony after a number of months, and they definitely do not base this recognition on the Rossitten ring which the returning bird carries on its leg – in contrast to my own method of identification. The response to reconciliation is only pronounced when the birds are occupying brooding sites and are in reproductive condition. At this time, any stranger is driven off by *all* of the colony members with a joint attack. During the non-reproductive phase (particularly during migration), incorporation of new companions into the flock seems to occur with ease; at least my jackdaws repeatedly brought back strangers with them, and they were then treated as colony members without further ado.

We have already discussed the question of how and *by what features* birds recognize one another individually, on p. 187 (q.v.).

3. THE FUNCTIONS OF THE SOCIAL COMPANION

(*a*) *Motivational induction*

Whereas we are almost always confronted with instinct interlocking in the reciprocal responses of parents and offspring, in which one instinctive behaviour pattern of the first bird elicits a *different* instinctive behaviour pattern in the second; with interactions between the members of a society we are typically faced with a different type of instinct elicitation. Characteristically, a given instinctive behaviour pattern is elicited in one bird by the performance of the *same* behaviour pattern by a companion. In observing this behaviour, we must remember that this is *not* a case of *imitation.* No bird is capable of imitating an adaptive behaviour pattern. Even quite simple behaviour patterns, for example slipping through a hole presented in a cage-wire barrier experiment, are not imitated even by the most intelligent birds. At the most, such birds would find the hole more easily if at the decisive moment they happened to be following closely behind the leader, who is to provide the example of slipping through the hole. Under any other circumstances, however, they are unable to appreciate from the fact that another bird passed through the wire at a certain point that they could do it themselves. It is entirely erroneous to speak of imitation when one individual in a bird flock begins to perform a behaviour pattern and the other birds are thereby induced to perform the same activity. This type of apparent imitation is based on a phenomenon widely distributed among birds, whereby the sight of a conspecific exhibiting a particular motivation through specific display movements and calls evokes a similar motivational state in the watching bird. This is in fact why the display movements are present. If one is inclined to draw an analogy with human behaviour, then it could be said that the response concerned has an 'infectious effect', as does yawning in human beings.

The operation of this motivational induction effect must not be regarded as too extensive. A tired mallard returning home cannot be immediately induced to develop motivation to fly again by the flight motivation display of a conspecific, although a bird which has just bathed can indeed be induced to bathe again by a conspecific. Katz has shown in his paper 'Hunger and Appetite' that an isolated chicken which had just eaten to satiation, so that it had stopped eating of its own accord in the presence of residual food, could be at once induced to continue eating by addition of a hungry, greedily-eating conspecific. Conversely, my Little Egrets exhibited the following behaviour: In order to lure the birds into the cage at the beginning of the migratory season, I first allowed them to become very hungry and then presented excellent food in the cage, which was fitted with a number of lobster-pot style entrances for trapping the birds. After a number of the birds had slipped through the entrance, had eaten to satiation and had lazily settled on the perches with their feathers fluffed, the hunger arousal of the birds still outside waned to such a degree that they gave up their attempts to enter the cage and settled down to rest near the cage in the same way as their genuinely-satiated conspecifics. I was usually only able to catch the remaining birds when the Egrets which had already been trapped became hungry again and began to eat once more.

Such exceptional cases occur rarely in a free-living society, however. Birds of the same *non-domesticated* species in fact react extremely uniformly, in contradiction to the repeated statement that animals exhibit great variations in individual responses. This statement is derived either from misinterpreted observations of the effects of captivity or from unjustifiable generalization of the behaviour of domesticated species, which vary quite unpredictably in their performance of instinctive behaviour patterns. It is obvious that freshly-captured birds can react in extremely different ways to the disruptive effects of trapping according to their physiological condition at the time. Similarly, the different defects which their behaviour patterns acquire as a consequence of such disruption can persist for the entire period of subsequent cage-life and can give rise to enormous apparent individual differences. An animal keeper soon learns to recognize these phenomena, after a certain amount of experience, and to exclude pseudo-variations from his observations. The value and pervading accuracy of all Heinroth's observations are based upon a masterly acquisition of this ability.

In general, the members of a bird society respond to the impinging stimuli so uniformly because they are in any case normally in the same motivational state and have the same prospective behaviour. Even if they were not able to see one another, they would eat, bathe or sleep at about the same time. The display movements and acoustic signals which transmit the motivation of one individual to another therefore have the restricted function of transforming this *approximate* synchronization into *more accurate* synchronization. For this, it is also advantageous that behaviour patterns which actually do not need to be performed simultaneously (e.g. preening or bathing) should nevertheless be performed synchronously by the members of the society, because this ensures that the overall rhythm of the community remains in co-ordination. If the animals did not perform *everything* synchronously, there would always be the danger that one individual would remain behind, satiated and immobile, when the others are developing motivation for flight. For this reason, we always observe the members of a continuously-preserved bird society performing the same activities together – bathing, preening, searching for food or sleeping.

(*b*) *The elicitation of the attachment drive*

There are cases in which the approximate synchronization of the responses of the flock members, as determined by general motivational transmission, is not sufficient to ensure adaptive performance of responses for which it is of great biological importance that they should be performed *exactly* synchronously by all individuals of the flock. This applies particularly to synchronous take-off of a flock of birds.

The velocity of a bird in the air is so much greater than that on the ground that in social species 'precautions' have been incorporated to ensure that the birds do not lose one another, particularly when changing from walking or swimming to flying. Just as has been described for the following drive of young jackdaws and other birds (p. 159), the attachment drive of adult social birds is far more intensive in flight than in walking. Heinroth states that his cranes exhibited no attachment tendency whatsoever on the ground at certain times of the year, while they always remained close together in flight. All birds in which such variability of the attachment instinct has been observed in different modes of locomotion exhibit a remarkably differentiated response to the take-off of a companion, which elicits the attachment drive. Let us assume that a member of a jackdaw flock feeding on the ground were to take off and land again about ten yards away. In this case, none of the other jackdaws would be provoked to take off. However, if one jackdaw should fly up with the intention of flying some distance, the attachment drive of the flock companions is aroused long before the flying bird is ten yards away. The birds can distinguish in one another at take-off itself the intention to fly a short way off or to fly some distance, and they can even decipher to some extent the factor which induces a companion to take off. Whereas the abrupt escape take-off of a single flock companion can regularly carry off the whole flock, even without production of the alarm-call, this does not occur when take-off takes place for some less pressing reason. The fact that a very decisive take-off can carry off the flock companions even when the eliciting stimulus is positive in nature is demonstrated by the following observation: After the jackdaw (already mentioned a number of times previously) which was partially imprinted on hooded crows and partially on human beings had spent some time in the company of crows, it always responded quite vigorously to my summoning call, exhibiting a 'desire' for human company. This applied particularly when the crows had spent some time on the ground or in the trees. They were only interesting for the jackdaw as flight companions in the air and were an indifferent quantity when perching (p. 131). When I called the jackdaw on such occasions and the bird immediately flew off towards me, the entire flock of hooded crows regularly followed and only swerved away in fright when quite close to me. The response of following of one animal which 'knows' what it is doing provides much food for thought. Since these extremely intelligent Corvids collect a wide range of experience with increasing age and simultaneously become more purposive and decisive in their movements, I believe that the experienced, older leader plays a biologically extremely important rôle among his conspecifics.[80]

Considering the great importance of the responses of flying off with the flock, it is by no means surprising that in many cases specific releasers have been developed in association. Among such releasers, observable before take-off, we find both movements and conspicuous plumage markings, which are abruptly displayed at the moment of take-off.

Display movements which betray the flight motivation of the bird some time before take-off and induce the same motivation in flock companions are found particularly in birds which do not readily muster themselves for flight. In the sixth chapter, we have considered motivational induction within mallard pairs prior to take-off, and it need only be added that large flocks of these birds behave in the same manner. The behaviour patterns evoking flight motivation are without doubt derived from intention movements. As in the actual take-off from the ground, the body is held low and the head, together with the anterior part of the body, is bobbed briefly upwards. This is very similar to behaviour exhibited when the bird is really preparing to take off and then 'changes its mind' at the last minute, but there are certain differences. These movements induce similar flight motivation in any conspecific. By a long process of stimulus-summation, they lower the 'take-off threshold' of the other bird so far that the ultimate, very powerful stimulus provided by the actual take-off of the companion infallibly carries off the conspecific. Geese must also 'talk one another into' the necessary state of arousal before taking off, in a very similar manner to the mallard. They call with increasing intensity and simultaneously increase the frequency with which they perform a peculiar take-off display pattern – a brief lateral shaking motion with the beak. The head-movement of the mallard is immediately recognizable as a derivative of the movements of actual take-off, and has doubtless evolved from an initial movement towards take-off which was suspended at the last moment. This movement conveys the fact that the bird will take off, even to an amateur ornithologist who is not aware of its significance as a releasing ceremony; its relationship to the take-off response is far more conspicuous than its releasing function. With the genus *Anser*, on the other hand, nobody who is unfamiliar with the take-off ceremony would deduce a relationship between the ceremony and the movements of the take-off leap. Such a relationship is nevertheless present, in all probability, since there are other Anatid species which exhibit behaviour intermediate between that of *Anas* and that of *Anser*. In the Egyptian goose (*Alopochen*), the movement concerned is confined to the head and beak, as in the true geese, and its performance is similarly jerky. But the head is moved up-and-down, as in the ducks, and not laterally as in geese. The series formed by the three types of movement is extremely impressive. Somebody familiar with the behaviour of ducks would not immediately understand the movement of the goose, but anyone used to the take-off movement of *Alopochen* would recognize both the movement of *Anas* and that of *Anser* as falling in the same category. In this case too, it is highly probable that we are faced with the evolution of a releasing ceremony from a motor pattern which originally served a quite different function. I should like to suggest that the original impetus leading to the evolution of the ceremony was provided by quite simple motivational induction. Presumably, the companion originally responds to the intention movement of its conspecific with 'resonance', as described on p. 219ff, and this 'understanding' can lend a new biological significance to the intention movement and lead to its further development.

During take-off itself, many birds utter additional acoustic signals. Cockatoos emit a spine-chilling shriek, while domestic pigeons and rock doves clap their wings loudly together. In the latter the clapping noise indicates exactly how far the bird intends to fly and is not produced when only a short distance is to be flown. Whenever a long journey is ahead, however, the bird claps not only at the first few wing-beats but continues to clap loudly thereafter. One could speak of differences in intensity of the take-off responses and state that the magnetic effect of the bird's take-off increases in direct proportion to the intensity of the bird's own response.

Morphological releasers of take-off by companions are found in the form of conspicuous plumage markings, which are displayed at the moment of take-off. They are always so arranged that they are most easily seen from behind, but otherwise all possible arrangements and distributions of coloured markings are realized in different species. Space only permits me to select a few which occur with obvious regularity. For example, widely different groups of birds exhibit white feathers on the margin of the tail fan and just as often a white or conspicuously light-coloured coloration is present on the rump or lower side of the tail. A bird which appears to be uniformly grey in colour when perching, for example a Bean Goose, alters its appearance radically on take-off by exposure of the white-marked tail fan. The various wing markings of the ducks provide particularly good examples of morphological releasers of take-off in companions. The generally characteristic combination of simplicity and general improbability of such releasers is scarcely to be found elsewhere in such impressive form. Heinroth's presentation of different Anatid wings does actually give the impression of a plate taken from a book of maritime flags. In this connection, the following observation made by Heinroth is important: When there is a chance similarity in wing markings, one Anatid species flying past may elicit the take-off of another species, even if the two species are in no way related. The North-East European Ruddy Shelduck (*Casarca ferruginea*) has the same simple black-and-white wing markings as the South American Muscovy duck (*Cairina moschata*). Squatting Ruddy Shelduck respond to flying Muscovy duck in the same way as to flying conspecifics, whereas they normally exhibit no interest in the other species.

(c) Alarming of companions

Alarming, as a function of the social companion, can be analysed into two separate releasing functions even more easily than that of the parental companion. Once can distinguish elicitation of an *alert* and elicitation of *escape responses.* Nevertheless, these two behaviour patterns are so closely interrelated even in this case that we must consider warning as a unitary function.

The most basic, and at the same time most penetrating, form of alarming is the actual escape motion. The significance of this behaviour is immediately understood by the conspecific and an appropriate response appears. Whereas with the young of many nidifugous birds the alarm-call of the parent is itself sufficient to elicit an escape response, I know of no case in which fleeing of an adult bird was elicited only by the alarm-call of a social companion. The sight of the fleeing companion seems to be necessary for this. Ravens exhibit the same behaviour even in the response of the offspring to alarm transmitted by the parental companion.

Thus, the alarm-call uttered by the social companion, when it is not accompanied by an escape response, always elicits the *alerting response* alone. This contrasts with the effect of the parental alarm-call, which is often capable of eliciting flight in the young *without* the parents fleeing themselves. Of course, alerting can itself be elicited by the escape response of a conspecific as well, and also by the sight of an alerted companion. Overall, it can be said that the responses of differing intensity following one and the same type of stimulation each normally elicit only a response of lower intensity from the companion. If this relationship is not maintained, there is an inevitable avalanche-like magnification of the response in its transference from bird to bird. In species in which the response of the warned bird is already very close in intensity to that of the warning companion, such disruption can easily occur. We have already considered the panics which can arise among leaderless young jackdaws on pp. 165, 220. Very similar processes seem to form the basis for the disastrous panics which sometimes grip herds of various ungulates.

In contrast to jackdaws, herons respond even to the hasty escape of a conspecific only with alerting behaviour. They only flee when they have themselves perceived the frightening phenomenon which elicited the escape of the companion. A 'wild' Night Heron which bred in 1934 with one of the females of my Night Heron colony initially fled every time I approached and displayed all the signs of fright. In response, all of the tame conspecifics present flattened their feathers and adopted an alert posture. But none of them was ever provoked to flee. On the contrary, they were reassured so rapidly as soon as they saw me that they virtually gave the impression that they knew that the strange heron had 'only' fled from me.

This difference in the response of jackdaws and Night Herons to alarm has the consequence that with the first species *one* shy bird will make the entire flock shy, whereas no such effect emerges with the second species. When the birds are kept in a free-flying state, this difference becomes very obvious. Finally, it should be noted that *young* jackdaws are *not* able to make acclimatized old jackdaws shy; instead, they subordinate themselves to the 'tameness tradition' of the colony. This is remarkable, because it shows that under certain conditions the identity of the alarm-transmitting bird (one could almost say its 'authority') can be important in determining the response of the conspecific.

(*d*) *Elicitation of a communal attack*

Among all of the synchronizing instinctive display patterns, the only other responses which have a similar compulsive effect to that of the elicitation of take-off in conspecifics are the communal attack responses of various bird species. These possibly represent the instinctive behaviour patterns which are least affected by the prevailing physiological condition of the birds concerned. These 'altruistic' instinctive behaviour patterns of birds, which are reminiscent of the noblest behaviour of human beings, are in fact highly likely to be more reflex in nature than the majority of the other behavioural responses.

We shall regard as an attack any behaviour pattern in which the bird responds 'positively' to a predator, i.e. pursues the other animal, even if no physical attack occurs. Such behaviour is in fact exhibited by many bird species. Among small passerines, this represents the widest distributed form of social behaviour extending beyond the family. In addition, this behaviour is often characterized by *social co-operation between different species*, with the response extending over the species barrier.

Whenever a small passerine sights a cat (or, even better, a small owl) by day it approaches the predator, uttering the alarm-call. It is not quite correct to refer to this vocalization as an alarm-call, since it in fact has a significance quite different from that of the alarm-call of a gallinaceous bird or a goose. The vocalizations of small passerines corresponding to the calls of the latter bird species are usually termed 'fright-calls', in order to leave the term 'alarm-call' free for the response to be discussed here, though this (as has been said) is not quite appropriate. The bird is not *warning* conspecifics about the sighted predator – its actions are aimed (of course, on a purely innate basis) at a much greater effect. The biological function of the response is based on the fact that the predator is *driven off*, because there is a continually chirping ring of passerines following its every movement and thus preventing the predator from hunting. The co-operation of different species in this task, which defends the common interests of the participants, is guaranteed *in this exceptional case* by the fact that one species will frequently respond to the vocalization of another. Such comprehension of the vocalizations of other species is in fact *not* a general occurrence, contrary to repeated reports of such phenomena, so this exception strikes anybody acquainted with such things as particularly remarkable. The question also arises as to whether this response to non-conspecific attack-calls is inherited or acquired. The intensity of the behaviour patterns and their unconditional nature would lead one to assume that they are inherited, if it were not so improbable at a first glance that a bird should be innately equipped with a 'dictionary' of the bird-calls in its biotope. On the other hand, we often find a certain adaptation of alarm-calls towards co-operation between different species. There is often *conspicuous similarity of communal attack-calls extending far beyond the degree of relationship between individual species*. For many insectivorous birds (in the widest sense of the term) a 'tick' or 'teck' forms the basis of the attack-call, and for a large number of finch-like birds a rising, drawn-out 'zie' is characteristic. In some cases, there is a combination of both, for example in the 'fooit-tektek' of the redstart. Even the human being has no difficulty in comprehending the call of the shrike if already acquainted with that of the garden warbler or the nightingale. Characteristically, the house sparrow – which is quite generally a virtual outsider among indigenous birds – possesses an entirely different attack-call, the well-known 'tsereng-tsereng' (to use the unsatisfactory representation commonly given). According to my previous observations, this call is not comprehended by other small passerines.

Whereas disruption of the predator's hunting is confined to noise-production in the smaller birds, larger birds with similar communal attack drives can quite easily augment their behaviour to physical attack on a greatly superior predator, especially when they are in a large group. It is not inconsiderable that the bird perceives the number of its attack companions and that this contributes greatly in increasing the bird's own courage to attack. This is especially pronounced in Corvids, in which we find a very pronounced instinct to attack the predator *from behind*. Thus, whenever several ravens, crows or magpies are occupied with a single predator, the latter is pecked in the tail by one bird every time it leaps at another in the group. I once observed fourteen magpies performing this game with a large weasel. The response is of course entirely innate, and any healthy hand-reared magpie will exhibit this behaviour towards a dog or a cat. The interpretation that the black-and-white patterning of the magpie functions as warning coloration springs readily to mind. It is notable that the *schema of the object is completely innate*, i.e. the predator is instinctively recognized as such. This fact became particularly clear to me with a house sparrow which I had reared from a very early age. This sparrow immediately uttered the characteristic 'tsereng-tsereng' in response to a Scops Owl and tried to fly towards it, whereas it exhibited no response towards a cuckoo which was presented as a control. In this response, shape plays the predominant part in stimulation, for the responding birds will locate and pester even completely motionless owls or cats. It would be extremely worthwhile to determine the features by which predators are recognized, i.e. which characters predominate in the innate schema possessed by the responding bird.

Whereas the response to the predator, including the recognition of the object, is innately determined in the magpie, a hand-reared jackdaw exhibits no response whatsoever on sighting a cat or dog. On the other hand, the jackdaw will respond with a communal attack pattern if another jackdaw is seized and carried by any other organism. The innate schema 'companion seized by predator' eliciting this response is so amazingly simple, as regards incorporation of characters, that the response can also be given 'erroneously' in many inappropriate situations. It is only necessary that *any* black object should be carried by some living organism for this response to be elicited. The species and the individual carrying the object are immaterial. I once saw this response elicited by a jackdaw carrying a crow pinion feather. Similarly, the response can be elicited at any time by a dangling black cloth. Since I have given a very detailed treatment of the 'rattling response' of jackdaws in a previous paper, no more will be said here.

According to observations made by my friend G. Kramer, the situation 'companion in danger' is presented to a hooded crow when another crow is merely close to a human being. Free-living hooded crows exhibited their specific communal attack response even when Kramer bore the young crow with which he conducted his experiments simply squatting quite free on his arm. My jackdaws never responded to this situation.

It is characteristic of all of these communal attack responses that they can be elicited in an entirely similar manner by two different stimuli. In the first place, they are elicited by the original eliciting factor (i.e. the cat, the owl, the seized jackdaw, or a crow in the human hand) and secondly they are equally elicited by the attack-call of the social companion which first perceives the danger. Jackdaws also begin to rattle when they are demonstrably unable to perceive the situation which provoked the first jackdaw to start rattling. The absolute reliability with which such an intricate and 'self-denying' behaviour pattern can be elicited in an entirely reflex manner by a single call always strikes the observer as amazing. However, it is possible that similar responses of the higher mammals and man are elicited in a quite similar reflex-like manner. Even if the 'situation as stimulus' is extremely complex and specialized, the manner in which the normal civilized human being flies into a rage, for example on seeing a small child roughly mistreated, still exhibits the complete reflex nature of the rattling response of jackdaws, as does the speed and violence of the attack, which can startle the attacker himself. Uexküll recounts how he himself responded in this manner when an Italian coachman flailed a child with his whip. Uexküll's stick 'struck the coachman' before he himself realized what was happening!

An extremely peculiar, in fact unique, form of communal attack is found in the jackdaw. This attack response is actually directed against a *conspecific* if the latter should disrupt the 'peace' of the colony so far that a member of the colony should feel itself threatened *on its nest*. For such a situation, the nest owner possesses a specific, shrill call, which is taken up by all of the colony members as they gather around the initially calling bird. This gathering effect, rather than a physical attack on the miscreant, is largely responsible for separating the fighting pair and in relieving the harassed bird. Physical attack is usually exhibited only by the partner owning the nest. Characteristically, even the bird originally causing the disturbance takes part in calling, i.e. this bird is not aware of its original rôle in eliciting the response and joins in the motivation of its companions as a result of their vocalization. Despite the simplicity of this behaviour pattern, it effectively protects the jackdaw colony against tyranny on the part of an individual member, which would automatically arise if one were to attempt to force an equivalent number of birds *without* this response to breed together in a space as restricted as that evident in a jackdaw colony. Yeates observed with rooks that individual males attempt to rape strange brooding females and that they are prevented from doing so by a number of other males, which rush up from all sides.

(*e*) *The instinct interlocking mechanisms of the rank-order and of nest-protection*

It is a well-known and exactly documented fact that in the confines of an aviary or a chicken-yard, a quite specific rank-order of the individuals is developed, an order in which one bird is afraid of the other. At this juncture, I should only like to point out that this rank-order does not emerge in every society, although it can be observed in artificial aggregations of individuals, in 'associations' (Alverdes), which can by no means be interpreted as societies.

If a number of birds of any given non-social species are enclosed together in one cage, a rank-order forms among them just like that prevailing among free-living birds held together by their social drives. For this reason it may perhaps seem superfluous to attempt to divine biological adaptiveness in the rank-order of the latter species. However, if we should compare the relevant behaviour of an aviary association with that of a free-flying jackdaw society, we find *two important differences* which give food for thought.

In the first case, we notice that in a society the rank-order is much more persistent than in an artificially-created association. This is the very reason for the relatively low incidence of friction between jackdaws. The jackdaw exhibits the tendency to stabilize the results of the fights determining the rank-order over a long period of time and thus to estimate each companion once and for all as a subordinate or a superior. I regard this fight-suppressing process of incorporation into the existing rank-order as an adaptive form of 'national economy' behaviour, particularly since the individual bird is adequately protected against infringements on the part of dominant birds by the 'police response' described above, which is not restrained by any hierarchical boundaries.

Secondly, in an organized society we can continually observe tension between individuals *close* to one another in the rank-order, in particular between those which (as 'pretenders to the throne') are close to the position of the alpha individual. By contrast, *animals high in the rank-order are amicable towards low-ranking individuals.* In an artificial association, on the other hand, these two principles are generally reversed, with the dominant animals pecking at the weakest occupants of the cage with particular fury. This further difference between the society and an association can also be interpreted in terms of a biologically-adaptive form of behaviour in the former.

A peculiar type of rearrangement of a previously-existing rank-order is found with jackdaws, geese and other bird species in which the partners of a pair aid one another in fights. In such birds, every betrothal involves the extension of the rank of the higher-ranking individual to the other within the association or society. In the jackdaw, it is quite amazing how rapidly 'the word is passed' through the society following a rearrangement due to betrothal. In the autumn of 1931, I observed the following event, which was later shown to be a regular occurrence: A male jackdaw which had been absent from the colony for almost the entire summer returned, freshly moulted and obviously greatly physically strengthened as a result of his persistent wanderings. Following a brief bout of bitter fighting, this male dethroned the previous alpha-male. This itself is remarkable, since the returning male also had to face the attacks of the other male's partner, though her response was not as pronounced as it would have been if the fights had taken place in spring. The next day, to my surprise, I noticed when feeding the birds that small jackdaw, which I had already taken to be a female and which had formerly occupied a very low-ranking position in the rank-order, walked threateningly towards the former alpha-male and that the latter gave way without resistance. At first, I thought that the small jackdaw had perhaps attacked the dethroned male while he was still exhausted and intimidated from the fights with the returned male, or that the entire previous rank-order was disrupted by such a 'palace revolution'. However, the true explanation was quite different: the returned new leader had immediately undergone betrothal with this particular small female jackdaw! When one is acquainted with behaviour within rank-orders of other species, one is less surprised by the fact that all other members of the society had recognized within one day that a previous subordinate could no longer be attacked than by the fact that the *latter* should behave as if it were aware of its influential protection and quite suddenly cease to exhibit fear towards its previous despots. One only needs to think of the fact that swans which have previously been dominated and chased from the pond by a conspecific despot take several days after removal of the tyrant before they even notice his absence and dare to return to the water (Heinroth).

Finally, we must consider those birds which do not develop a real rank-order at all, as is the case with colonially-brooding *herons*. Just as in the case of the rank-order of the jackdaw society, where we are simply confronted with a behaviour pattern which also occurs in a less specialized form in non-social birds when they are forced to live together, the governing principle of the heron society reflects a phenomenon which is exhibited by almost all birds in a less conspicuously developed form. If a new bird is introduced to a cage-bird which is already accustomed to its cage and a fight develops, it is almost certain that the resident bird will win, even if its opponent is far stronger. The same applies, to a less marked degree, under natural conditions in encounters between birds of a species in which the pairs demarcate specific discrete territories; the territory owner generally prevails. Howard has carried out very extensive observations on the phenomenon of territorial demarcation.

Territorial disputes can be very easily observed with the common redstart. With these birds, one can often observe typical instances where a fight develops between two males on the border between their territories. The victor pursues the vanquished animal fleeing into its own territory and is then attacked there. It consequently flees back into its own territory pursued by its opponent, which is then defeated again, and so on. This happens until the birds gradually reach equilibrium, with the fight waning more or less on the original boundary. This behaviour is typical of territorial animals and can be found in a classical form in the stickleback, where to-and-fro pursuit across a fixed boundary can be observed continually.

Huxley has provided a very apt analogy to the territories of territorial birds with his suggestion of *elastic discs*, which flatten out in places where two meet together and develop a reciprocal pressure proportional to the degree of alteration of their form by a pressure operating from outside. The 'elastic behaviour' of such territories arises because the territory owner does not defend all of the included sites with the same fury, but instead places the greatest emphasis on a quite specific centre of activity. The intensity of territorial defence responses declines in direct proportion to the distance from this centre.

Territorial demarcation is lacking in a number of bird species, as for example in the house sparrow.[81] The latter species, owing to the remarkable compatibility of breeding pairs, represents a transitional stage on the way to colonial brooding species with a social composition similar to that found in the jackdaw. However, territorial demarcation is not lacking in all colonially-breeding species. In such species it is either completely absent, as in the types described, or it is developed to the limit, as in the herons. If territorial demarcation is regarded as a primitive behaviour pattern of birds, as I am convinced is the case, then here are two evolutionary routes leading to extremely close settlement of pairs and the consequent advantages of certain forms of mutual social assistance. Either the drive towards territorial demarcation is broken down (as in the sparrow) so that each pair tolerates the presence of conspecific pairs in its range, or the territory of each individual pair is gradually reduced in size so that the nests are eventually located close together, each lying within the rigidly-preserved boundaries of a tiny territory. This second route has been taken by the herons.

(*f*) *The disappearance of the social companion*

In the discussion of the sexual companion on p. 215, we have already covered a response similar to that with which we are concerned here. Probably, such behaviour is widely distributed in the Class Aves. I am only acquainted with a conspicuous response to the disappearance of a social companion in the jackdaw. The response itself is similar to that described for the disappearance of the sexual companion, but in the jackdaws – which are richer in display behaviour – the consequent search and uneasiness are more obvious than in the ducks.

The sight of capture of the disappearing companion, or of its fate in any form, has no effect whatsoever on the behaviour described. On one occasion, after two jackdaws of my colony had flown away with strange jackdaws before the eyes of all of the colony members, the remaining jackdaws exhibited exactly the same uneasiness as that exhibited after companions had been carried off unseen by cats.

It is important that this response is completely lacking when a member of the colony gradually sickens and dies. It disappears (so to speak) slowly in front of its companions, providing 'creeping stimulation' which does not elicit a response.

4. DISSOCIABILITY OF THE FUNCTIONAL SYSTEMS

Under experimental conditions, the individual functional systems of the social companion are utterly independent of one another. Since, in addition, the reciprocating objects of instinctive behaviour patterns oriented to social companions usually possess an innate schema (consisting of a restricted number of characters) which can be moulded to fit all kinds of other objects by imprinting, experimentation with hand-reared birds is actually supported by the possibility of permitting extremely different objects to function as reciprocating companions in the individual functional systems. However, it should not be forgotten that this extreme dissociability of the functional systems emerging under experimental conditions is itself an indication that imprinting plays a particularly *large* part in the formation of the companion schema, so that the innate schema is relatively sharply pushed into the background. When object-imprinting occurs in the manner normal for the species, this preponderance of the acquired schema insures a relatively great degree of integration of the companion (p. 133). When imprinting does not occur in this natural manner, the individual innate releasing mechanisms of the separate functional systems prove to be particularly independent of one another.

Frequently, no particular carefully-constructed experimental set-up is necessary to obtain a result of this kind. Since the different releasing schemata incorporate widely-differing numbers of characters, and since each of them incorporates *different* characters evident in the conspecific it is not at all surprising that in an isolated, hand-reared young bird they will respond to quite different living organisms in the environment. Thus it is a frequent experience, as with the jackdaw discussed on p. 131, that the accompanying flight drive of a bird which otherwise treats only human beings as social companions is directed towards conspecifics, or at least (as in the case of the jackdaw mentioned) towards related birds. For this reason, one often observes that a small number of birds hand-reared together will in fact become humanized in every respect, will only be interested in human beings and will not attempt to interact with one another, and yet in the air they will attach themselves to conspecifics just as normally-reared and free-living birds do. Such behaviour was exhibited by Heinroth's cranes and ravens, by many jackdaws which I kept, by an isolated hand-reared Quaker Parrot, and by many other birds.

Those functional systems whose elicitation is less 'conditional' (i.e. less dependent upon specific physiological conditions), for example the response of joint take-off, social attack and the response to the alarm-call or other danger signals emanating from the social companion, exhibit a particular degree of independence from other systems. The few simple characters which exert this unconditional compulsive effect on the bird are entirely innate and affect a bird imprinted on human beings just as greatly as a normal, free-living bird. This particularly applies to the communal attack responses. The instinctive nature of these behaviour patterns emerged with particular clarity in the jackdaw which has already been mentioned a number of times. This bird exhibited almost violent animosity towards the conspecifics with which it had to share the attic of our house. Since he was stronger and older than the others, I was frequently forced to protect them against him. Towards me, however, this jackdaw was the very essence of gentleness and affection and was quite reminiscent of a loyal dog, since I represented his social companion. Nevertheless, I only needed to elicit communal attack in this jackdaw, by seizing one of the other jackdaws, in order to provoke the old bird to fight me, the social companion, to the point of drawing blood in defence of the seized jackdaw, although otherwise the older bird spent all of its time attempting to fight and possibly kill the same conspecific.

IX The sibling companion

The fact that the behaviour of the young bird towards its siblings and the rôle which the latter play in its environment are discussed as a sequel to the chapter on the social companion, and not following the description of parental behaviour, is based on the large number of similarities exhibited between the mutual behaviour of siblings of species in which the offspring remain together for some time (forming a 'sympaedium') and that of the members of a society of adult birds.

1. THE INNATE SCHEMA OF THE SIBLING COMPANION

In bird species in which siblings remain together for a long time after emergence from the nest (i.e. particularly in nidifugous species), the innate schema appears to be remarkably specific, i.e. rich in incorporated characters, so that there is very little room for imprinting. A young Greylag goose will indeed readily accept a human being as a parental companion, but it will not accept a young Pekin duck as a sibling companion (Heinroth). A somewhat younger Greylag is at first attacked, which in itself probably represents a response to innate releasing mechanisms (the Pekin duck was not attacked!), but it is subsequently accepted as a sibling. Despite the simultaneously existing attachment drive oriented towards the parent, the attachment drive operating between sibling Anatid chicks is extremely pronounced. The sight of any flock of domestic goslings provides a very good illustration of this, with the chicks following the mother in a tightly-packed group. The parent bird does *not* form the nucleus of the flock – the offspring primarily congregate with one another and exhibit only secondary attachment to the parents. This clear-cut response to the sibling companion renders Heinroth's observation (above) extremely significant. In young ducks, the sibling companion schema also appears to incorporate a larger number of characters than is the case with their parental companion schema. Even young domestic ducks never attach themselves to young chickens or geese as sibling companions, whereas they will respond to either of these two species as parental companions. By contrast, my mallard reared from the egg readily accepted *Cairina* chicks of the same age as sibling companions and in no way distinguished between them and their true siblings. On p. 138, it was shown that young mallard, on the other hand, will not accept a *Cairina* stepmother as a parental companion. Nevertheless, this does not prove that in mallard, in contrast to Greylag geese and domestic duck, the innate sibling schema is more expansive than that of the maternal schema. This behaviour is much better explained on the assumption that a young *Cairina* is incomparably less distinct from a young mallard than is a *Cairina* mother from an adult mallard. In particular, the communicatory behaviour and calls of the chicks are virtually identical in the two species. With young Corvid fledglings, the sibling companion presumably plays no special rôle and possesses no innate schema distinct from that of the later social companion.

2. PERSONAL RECOGNITION OF THE SIBLING COMPANION

With nest-living young passerines it can be assumed with some certainty that personal recognition of the siblings is not present during the nest-phase. With the exception of the remarkable cases mentioned above, there exists in all nidicolous species a prevailing absolute non-hierarchical compatibility between the nestlings in any one nest.

This unconditional compatibility is extremely pronounced with passerine nestlings. With such nestlings, one virtually has the impression that the senses of the young birds are closed to the perception of all of the stimuli emanating from the siblings. They do not even respond defensively when a sibling treads upon them or roughly mistreats them in some other way.

The 'unconditional compatibility' between some nestlings described above frequently extends far beyond species boundaries. This provides a wonderful opportunity for determining experimentally which nestlings of other species can be placed in the nest of the species under investigation without evoking a differential response from the nest occupants. The absolute non-stimulatory nature of the natural siblings would permit determination of the characters on which the failure to respond is based.

I can state with certainty that the unconditional compatibility between nestlings has no exception among small passerines when unfamiliar conspecific nestlings or even those of related species are added to the nest. With herons, unconditional compatibility appears to be bound to individual factors. If growing nestlings from one nest are separated just for a short period and then reintroduced, they will fight with one another just as they do with strange conspecifics of the same age.

After emergence from the nest, very many young nidifugous birds exhibit individual recognition. Two young jackdaws, to which I gave preferential treatment within a flock of fourteen of the same age, by taking them for certain experiments in which they always remained together and made long flights following me in the sourrounding countryside, soon began to stay together even in the company of their conspecifics. Four freshly hatched Night Herons not only exhibited mutual compatibility but also performed defence responses of a communal (or rather 'sympaedial') nature towards an older clutch of Night Herons which had fledged some time ago. If the younger birds were attacked by one of the older conspecifics, they would at once run together even over large distances, press their backs together and threaten outwards with their beaks.

In some nidifugous species, mutual individual recognition appears to emerge among the siblings very early on. With 'Hochbrutenten' (a semi-domesticated type), the young exhibit mutual individual recognition *earlier* than the mother recognizes her young individually. I have experienced the situation where roughly eight-day-old chicks smuggled under a 'Hochbrutente' mother were readily accepted by the foster-parent and yet were attacked by the legitimate offspring *en masse*. The strange chicks were themselves far from restrained and a genuine battle developed in accordance with the ritualized fighting behaviour of the ducks, though the skirmish in fact remained harmless. At the beginning each chick of one clutch actually did face one chick of the other, and the opponents grasped one another by the ventral neck feathers and pushed against one another with all their might, just as fighting adult drakes do. Interestingly, these ducklings even exhibited the fighting response of beating with the wing-shoulder, which is typical of adult ducks, although ducklings of this age do not yet possess functional wings. The central pathway of the response is thus developed earlier than the organ that is activated by the response! The progress of this massed battle was also of interest. I soon saw two ducklings of the old brood fighting *against one another* and soon afterwards the same occurred with two of the introduced chicks. The degree of confusion gradually increased and this eventually led to a truce. The adult duck had watched this all without taking part – she possessed no response towards this phenomenon of a massed battle among the ducklings, which does not occur naturally, and this was therefore simply not part of her environmental picture.

In ducks, the structure of the sibling society is entirely different from that of gallinaceous birds. The cohesion between the ducklings of the Anatids, contrasting with the predominant central rôle of the mother in gallinaceous species, has the consequence that the sibling companion plays a much larger rôle even on the first day of life in Anatidae. I believe that the majority of gallinaceous birds first exhibit emergence of personal recognition among the individual siblings when a rank-order begins to develop within the sibling society.

At the moment, not much is known about the behaviour of other nidifugous species, although an investigation of the behaviour of young water-rail, in which the older offspring aid in feeding their younger siblings, would be extremely valuable. Unfortunately, my attempts to hatch moorhen chicks have so far failed due to unfortunate mishaps.

3. THE FUNCTIONS OF THE SIBLING COMPANION

(*a*) *Elicitation of the attachment drive*

When fledged young jackdaws are kept without parents, they exhibit an extremely strong instinct to attach themselves to one another, particularly in the air. They virtually adhere to one another and call plaintively when they lose their companions. However, if one observes their behaviour more closely – again particularly when they are flying – it is seen that their behaviour towards one another corresponds to the behaviour of any one individual towards the parental companion. In other words, the sight of the flying sibling elicits in the young jackdaw an attachment drive which is actually adapted for orientation to the parent. Each individual in such a flock of leaderless young jackdaws attempts in vain to find purposive leadership by a parental companion, and since (in anthropomorphic terms) each of *the others* 'believes' that this individual 'knows' where the journey is leading, a leaderless sibling jackdaw society of this kind can lose its way far more easily than an individual juvenile. In a family group of the species-typical kind, the jackdaw siblings are evidently so decidedly centred around the parents, that one cannot see very much of their mutual attachments. However, the two experimental jackdaws described above, which remained attached to one another even in a flock of conspecifics, show very clearly that there is also a close kind of attachment between siblings.

Apparently, it seldom occurs among nidicolous birds that the mutual attachment of the juveniles persists longer than the relationships of the offspring to the parents. The Little Egret appears to be an exception, for hand-reared birds of this species kept in complete freedom exhibit extremely close adherence to one another and unconditional sibling compatibility far into the autumn, whereas they are utterly hostile towards strange egrets. Since young Little Egrets presumably do not remain with the parents all that long, it is possible that under natural conditions they live in sibling societies independent of the parents during the latter period.

Whereas the sibling attachment instinct of all these nidicolous species obeys laws very similar to those governing the social attachment drive of generally social birds and its elicitation occurs in an extremely similar manner (pp. 159, 222), mutual attachment of nidifugous chicks exhibits some traits which must be given particular emphasis. It is obvious that adherence of nidifugous offspring to one another is biologically much more important than that of fledged nidicolous birds, especially since the leading adult bird (as we have seen on p. 183) does not respond to the absence of a single offspring, so that the latter must itself ensure that it remains together with its siblings. The mother only misses her chicks and 'searches' for them to a certain degree when the entire batch disappears. Even in chickens, where the chicks exhibit relatively little mutual attachment, we can nevertheless observe a very interesting case where one chick will respond to another and run after it. This occurs when one chick determines from the movements of another that it is responding to a movement made by the mother. In the same way that a social bird can observe in a companion the cause of take-off and the projected course of flight (p. 222), a chick can also see from a sibling whether it simply happens to move on under its own steam or whether it is afraid and hastily moving off after the mother in the distance. This extremely fine comprehension of the movements of each sibling permits chicks to stray relatively large distances from the mother, since it does not matter if some chicks lose sight of the mother as long as each can see a chick 'which can see one that can see the mother'. It would even appear that the sight of a sibling hurrying towards the mother elicits in gallinaceous birds a stronger degree of attachment than the sight of the mother herself marching away. This emerges particularly clearly with the extremely biologically-adaptive behaviour of flocks of chicks crossing open spaces, which are otherwise avoided as potentially dangerous. When the guiding hen crosses such an open space, she is at first usually followed only by the chicks which are close behind her – the others remain behind and only when the mother is on the other side of the open space will one run or fly towards her, followed one-by-one by the others at regular intervals. This behaviour can frequently be observed with country chickens which have not been too extensively subjected to artificial breeding. In the Golden pheasant, such behaviour is even prevalent within an enclosed aviary, especially since the Golden pheasant is a type which greatly prefers cover and is little inclined to cross open spaces.

Even in Anatidae, in which the chicks are much more strongly mutually attached than in gallinaceous birds, one duckling can determine from the movement of another the direction which the mother has taken. Since the chicks normally press together into a tightly-packed group during the first days of life and follow the mother (or the parents) as a unitary whole, this behaviour is first expressed somewhat later, when the offspring form the typical 'crocodile' train. With large flocks of chicks, the mother may already have disappeared around the next corner, while the chicks which are now unable to see her nevertheless follow her trail exactly. The ability to determine direction from the behaviour of siblings has the additional effect that the ducklings can in particular situations *imitate* the behaviour of a sibling, something which is otherwise unusual for birds. If a mallard sibling society is led to an obstacle by a human guide, on whom the ducklings are imprinted, and the obstacle is only surmountable at one spot, it often takes some time before one of the ducklings in the agitatedly piping group crowded behind the obstacle finds the passage. As soon as one individual had achieved this, however, the nearest duckling at once notices and hastily follows. All of the siblings then press forwards from all sides towards the gap in the obstacle (in the two cases observed by myself, a gap in a box-tree hedge and a worn patch in a high step). Such imitation of exit location does not occur in the most intelligent adult birds, nor (as far as I know) in the gallinaceous chicks, despite the otherwise great development of mutual responsiveness. This probably represents the major biological function of close mutual attachment among young Anatids.

(*b*) *Motivational induction*

As in the social group of adult birds, motivational induction from one individual to another plays a large part in the sibling society of the young birds and it is even evident in the first hours of life in some young nidifugous birds. It is a well-known fact that in their very first act of feeding, some chicks are provoked to feed by the feeding behaviour of their siblings. The fact that this does not represent imitation has already been mentioned. One can say, however, that feeding in small chicks has an infectious effect like that of human yawning.

Behaviour over and above this can be briefly covered. It is obvious that parallel activation of all life-processes through motivational induction in a sibling society, tended by the mother or both parents, plays a more important biological rôle than that in a society of adult birds. What would a mother hen do if the chicks were not to become tired simultaneously and did not require warming all together! In older sibling societies which have become independent from the parents, motivational induction occurs in exactly the same way as with persistently social forms.

(*c*) *Alarm behaviour*

I should like to mention mutual alarm behaviour for the simple reason that in some species very young chicks, despite their highly-specific response to the alarm-call of the mother, remarkably show virtually no response to the distress-calls of their siblings. This is at least the case with the domestic chicken; with the Golden pheasant I was unable to confirm this effect since I never reared chicks of this species away from the mother. With mallard, on the other hand, I was able to observe that at a very early age all of the ducklings would become very uneasy and shy whenever I grasped one of them and caused it to utter the distress-call.

(*d*) *Elicitation of co-operative defence*

I have only encountered mutual attachment of siblings in the face of a common enemy with herons, as has already been mentioned in the section on individual recognition among sibling companions. The attacking of unfamiliar conspecifics of the same age, which was described in the same section, does not fall under this heading since each individual young bird responds independently of its siblings, although all siblings respond in the same way. The response of young herons which does come into this category is predominantly represented by the *hasty* and quite purposive manner in which the attacked youngster runs to its siblings and takes cover from the pursuer *behind* them. The undisturbed siblings, previously subjected to no attack, then regularly oppose the attacker more courageously than the already vanquished bird could have done.

In spring 1931, I observed this form of co-operative defence performed by a sibling society of four freshly-fledged Night Herons, which were placed in an aviary in which two roughly one-month-older conspecifics had lived for some time. The resident herons immediately attacked one of the younger birds, which had dared to fly from the ground on to a perch. The younger heron consequently flew back to the ground from the perch, followed by one of the older birds, and ran back to its siblings, which formed a common front against the pursuer.

This response of the young Night Herons persisted for only a very short time. Soon after abandoning the nest, even siblings behave with extreme incompatibility towards one another. This is extremely interesting, since the sole cause is the fact that they no longer recognize one another. If such birds are forcibly confined together, so that they have no opportunity to forget one another, the compatibility between the siblings persists until the Winter. However, since the birds exhibit little mutual attachment to one another, they soon disperse under natural conditions and they respond to one another as to strangers in subsequent chance encounters. This does not apply to the Little Egret, in which fledged individuals exhibit an extremely strong instinctive attachment to their siblings and remain inseparable from them until late Autumn. In contrast to Night Herons, they adhere to one another particularly closely in flight. This attachment drive has the consequence that Little Egrets behave in exactly the same way under natural conditions as do Night Herons when induced to remain together by artificial close confinement. For this reason, the Little Egret also retains the described response of co-operative defence over a much longer period.

(*e*) *Disappearance of the sibling companion*

I am familiar with a pronounced response to the disappearance of a sibling companion in young Anatids, which (as has been frequently mentioned) exhibit the most intimate mutual relationships known for young birds.

With the flock of mallard ducklings which I reared and guided in summer 1963, I once attempted to take only the pure-blooded ducklings with me on my daily excursion and to leave the others, which were much less important for my observations, at home. This proved to be utterly impossible. Even after I had carried the mallard ducklings so far from home that they could no longer hear the hybrid ducklings, they would not run after me, but remained behind and uttered lost-piping. They did not calm down even after some time had elapsed, so that I was forced to fetch the remaining ducklings. Strangely enough, further attempts at leaving ducklings at home demonstrated that the absence of only one or two siblings is *not* noticed and that registration of the sibling companion is *not personal* but *quantitative*, in contrast to the responses which many birds exhibit following the disappearance of a social or sexual companion (pp. 215, 232), although the chicks are nevertheless able to recognize one another individually, as we have seen on p. 236.

(*f*) *Instinct interlocking in the rank-order*

In many sibling societies, no rank-order develops, namely in those in which the 'unconditional compatibility' discussed on pp. 235–236 excludes any conflict within the group. In the Anatidae, particularly in geese, this rank-less compatibility persists until the late autumn and then gradually gives way to a rank-order as sibling attachments are dissolved.

In gallinaceous birds, unconditional compatibility is disrupted at a time when the chicks are still attached to the mother. Everyone is familiar with the fights which break out among young domestic chickens at a particular age, and these are definitely not restricted to males, as is sometimes assumed by the layman. Golden pheasants remain compatible with one another for much longer, and it is of course possible that domestic chicks exhibit precocious maturity in comparison with the wild form. Contrary to some reports, which are based on the entirely erroneous assumption that the offspring of the wild type must take care of themselves much earlier than those of the domestic form, domesticated types are in fact prone to exhibit a reduced rather than extended period of ontogeny. One only needs to think of the domestic goose, which is reproductively mature in the first year in comparison with the three years needed by the wild type, and of the violent disruption of the chicken family by the hen following the early return to a gravid condition (as described by Brückner), in comparison with the persistence of the defence response of the mother Golden pheasant and *Cairina*.

When the rank-order has once been formed, the relevant behaviour of the group members in the ordered sibling society differs in no way from that of the members of a society of adult, social birds. It is, however, remarkable that with gallinaceous birds the mother in many cases apparently retains a non-hierarchical, unconditionally compatible relationship to the chick society at a time when the chicks have already developed a stable rank-order.

4. DISSOCIABILITY OF THE FUNCTIONAL SYSTEMS

In contrast to the functional systems of all other companion relationships in birds, dissociability has so far not been determined for those of the sibling companion in any species. The instinctive behaviour patterns adapted to siblings are also the only responses oriented towards conspecifics for which re-orientation towards the human guardian has never been observed in any bird species. This is doubtless partially due to the fact that the human being is in fact regarded as a parental companion by the bird.

X Summary and conclusions

This paper is not a description of a unitary investigation directed at a narrowly-defined research problem. It is rather an attempt to arrange a large number of observations, which have previously remained as unrelated entities, within an ordered system. It is in the nature of such an attempt that the result should predominantly have the character of a programme. The approach to the problems concerned inevitably means that the few answers which we already have at our disposal can only with difficulty be summarized in a few words. I shall attempt to surmount this difficulty by presenting the summary section by section and treating the few genuine results in the order in which they arise in this process. New questions will arise with some frequency, and I regard it as a merit of this study that it points the way to a broad field of experimental research which has as yet been largely untouched.

A. Serial review

CHAPTER I: THE CONCEPTS OF THE COMPANION AND THE RELEASER

From the many stimuli which emanate from one animal and impinge upon the sense-organs of a conspecific, we have attempted to select those which elicit in the latter social responses in the widest sense of the term. We have discovered that the stimuli and stimulus combinations to which the animal *instinctively* responds with specific responses belong to a category quite different from those whose elicitatory effect depends upon *acquired* properties.

The receptor correlate corresponding to an elicitatory stimulus combination (i.e. the basis of the specific tendency to respond to a specific key combination and consequently to spark off a certain chain of behaviour patterns) has been termed the *releasing schema*, in loose adherence to von Uexküll's terminology.

The instinctive, *innate* releasing schemata play a particular rôle in the behaviour of birds. If the releasing schema of a response is innately determined, it always corresponds to a relatively *simple* combination of individual stimuli, which as a unitary whole represent a key to a specific instinctive response. The innate releasing schema of an instinctive behaviour pattern is based on a restricted selection from the wealth of available stimuli, and the schema responds selectively to these stimuli in setting the behaviour pattern in operation.[83]

It is a biological necessity that these key combinations should possess a minimum of general *improbability*, which prevents elicitation of the response in a biologically-inappropriate situation (pp. 106, 192).

If the releasing schema of a behavioural sequence is *not innate*, this does not necessarily apply to the motor component of the behaviour pattern as well. There are behaviour patterns in which the motor component is instinctively determined, whereas the eliciting factor is not innate but must be acquired. Such behaviour patterns have been generally termed *instinct-conditioning intercalation patterns*. The individually-acquired releasing schemata setting such behaviour in motion correspond to extremely complex combinations of stimuli, in contrast to innate releasing schemata. In general, they respond to 'complex qualities', in which none of the immeasurably large number of incorporated characters may be altered without changing the complex quality and thus rendering the elicitation of the response unreliable. I regard this difference between innate and acquired releasing schemata as representing an important difference between instinctive and conditioned behaviour (pp. 103, 104).

The innate schema attains great importance in responses which have a conspecific as their object. With responses whose object is *some entity in the external environment*, the innate schema can only be adapted to stimuli which are an inherent property of this entity *from the outset*. Retaining the analogy with the key, the form of the key-bit is predetermined. The necessary general improbability of the schema has an upper limit, which is reached when the awaiting negative of the lock has reached the maximum degree of correspondence to the positive component represented by the key-bit. With responses directed towards *conspecifics*, on the other hand, both the development of the innate releasing schema and that of the relevant stimulus-key are within the evolutionary framework of one species. *Organs* and *instinctive behaviour patterns* which are exclusively concerned in the transmission of key stimuli (sign stimuli) achieve a high degree of specialization, always progressing in parallel with the evolution of corresponding, matching releasing schemata. Such organs and instinctive behaviour patterns are briefly termed *releasers*.

Releasers are very widely distributed within the Class Aves. It is scarcely an exaggeration to state that *all* particularly conspicuous colours and shapes in the plumage are related to some elicitatory function. The nature of the elicited response cannot be determined from the form of the releaser and it is entirely impermissible to relate all releasers to a given response, for example that of sexual selection by the female.

As far as I am aware, the interpretation of these structures and behaviour patterns as releasers is the only hypothesis which can explain their most prevalent and important property – that of a combination of simplicity and general improbability (pp. 106–107).

The high degree of specialization of the innate schema and the releaser leads in many birds a noticeably peculiar type of behaviour towards conspecifics. If two or more instinctive behaviour patterns are adapted to respond to the same object, there are two possibilities for ensuring that unitary and consistent response to the object occurs. The first possibility is a concrete, objective perception of the object by the subject, whilst the second rests in the connection of the different instinctive behaviour patterns adapted to the object through *the object itself*. With instinctive behaviour patterns directed towards a conspecific, there is an opportunity for maximal specialization of releasers and the corresponding innate schemata, so that under natural conditions consistent response towards the object is just as effectively ensured as would be achieved by subjective understanding of the objective identity of the object. Among the birds, a large number of species have followed this pathway, which appears inferior to us, and have thus *circumvented* the necessity for subjective understanding of the objective identity of the conspecific. The functional plan of their instincts transfers the unitary factor to the object transmitting the stimuli instead of incorporating it within the subject perceiving the stimuli – the object does not represent a unitary entity in the subject's environment. J. von Uexküll coined the term *companion* for a conspecific which is only responded to as an identical whole within a single functional system, and this expression has been employed in the preceding considerations.

CHAPTER II: IMPRINTING

I should like to emphasize, as the most important result of this investigation of instinctive behaviour patterns oriented towards conspecifics, the fact that *not all acquired behaviour can be equated with experience and that not all processes of acquisition can be equated with learning*. We have seen that in many cases the object appropriate to innately-determined instinctive behaviour patterns is not instinctively recognized as such, but that recognition of the object is acquired through a quite specific process, *which has nothing to do with learning*.[84]

With very many instinctive behaviour patterns oriented towards conspecifics, the motor component is itself innate, but the recognition of the object of the pattern is not. This is the case with many other behavioural chains, in which instinct-conditioning intercalation occurs in the course of ontogeny. However, there is a difference from the latter in the *manner of acquisition* of the elicitatory component.

An instinctive behaviour pattern adapted towards a conspecific, yet initially incorporated without an object, is fixated upon an object in the environment at a quite specific time, at a quite specific developmental stage of the young bird. This specification of the object can take place hand-in-hand with the emergence of the motor component of the instinctive behaviour pattern, but it can also precede the latter by a matter of months or even years. In the normal free-living existence of the species, the conditions are so organized that the choice of object of the instinctive behaviour patterns is reliably limited to a conspecific which represents the biologically-appropriate object. If the young bird is *not* surrounded by conspecifics at the psychological period for object-selection, the responses concerned are oriented towards some other environmental object, usually towards a living organism (as long as such is available) but otherwise towards some inanimate object (pp. 126, 135).

The process of object-acquisition is separated from any genuine learning process by two factors, and rendered parallel to another process of acquisition which is known from the mechanics of embryogeny and is referred to in that context as inductive determination. In the first place, this process is *irreversible*, whereas the concept of learning necessarily incorporates the condition that the acquired element can be both forgotten and revised. Secondly, the process is bound to a sharply-demarcated developmental condition of the individual, which often exists only for a few hours (pp. 126–131).

The process of acquisition of the object of instinctive behaviour patterns oriented towards conspecifics, which are initially incorporated without the object, has been termed *imprinting*. Imprinting involves a peculiar and extremely puzzling selection of the characters of the object: *only supra-individual* characters are fixated. If the imprinting processes of a young bird are experimentally channelled towards an animal of another species, after completion of imprinting the functional systems concerned are oriented towards the species to which the animal inducing imprinting belongs (p. 132). In this process, it remains a complete enigma as to how the bird is able to zoologically 'classify' the species to which it erroneously 'feels itself to be related'.[85]

Finally, it should be noted that cases are known from human psychopathology in which irreversible fixation of the object of specific instinctive behaviour patterns has been observed and which, on purely symptomatic grounds, exhibit a picture entirely similar to that found in birds in which object-imprinting has not taken place in the species-typical manner.

CHAPTER III: THE INNATE SCHEMA OF THE COMPANION

In no case are all of the characters of the conspecific acting as a companion to the bird within a particular functional system acquired exclusively by imprinting. There is always an *innate* framework present for the incorporation of the releasing schemata to be *acquired*, such that there is development of an innate total schema which can be broad or restricted according to the number and the nature of the components. Even in cases where this total schema incorporates only very few innate releasing schemata, i.e. is extremely *broad*, under the conditions normal for the young bird it has and performs the task of directing the imprinting process towards the appropriate object, which corresponds to the functional plan of the instinctive behaviour patterns. Under experimental conditions, the sign stimuli of the individual innate schemata can in some cases be determined by deliberately attempting to produce inappropriate imprinting of young birds to species-atypical objects. From the properties of these objects, one is often able to determine with some accuracy the stimuli which are indispensable for the complementation of the innate companion schema.

The interplay between the innate companion schema and object-imprinting varies greatly from species to species. There is every transition from species like the Greylag goose, which possess extremely broad companion schemata incorporating few characters, to birds in which virtually all of the releasing schemata are innately determined, so that no room is left for variability due to imprinting. It can be reliably stated that within the Class Aves the latter form of behaviour is likely to be the more primitive.

CHAPTERS IV TO VIII: PARENTAL, INFANT, SEXUAL, SOCIAL AND SIBLING COMPANIONS

In these sections, five different cases in which a particular conspecific plays a particularly important rôle in the bird's environment have been discussed. Following Uexküll, these companions attached to overall functional systems were classified as parental companion, infant companion and so on, although in some cases the dissolution of the environmental image of the conspecific had to be continued as far as the individual subsidiary functional systems, in order to be entirely consistent.

As far as their unitary structure permitted, the *innate schemata* of these companions represented in overall 'systems of functional systems' were investigated by means of discussion of a maximum number of cases in which they were occupiable by species-atypical drive objects.

Following a discussion of the innate schema, the *individual characterization* of the companion was considered in each of the five governing functional systems.

Subsequently, in each of the five cases the reciprocal functions of the various companions necessary for the performance of each of the individual component functions were analysed. In other words, the rôle performed by the companion (as an eliciting factor for responses which gave rise to a single functional system) was reconstructed.

Lastly, consideration was given to the extent to which the subordinate functional systems of each of the five companion systems are mutually interdependent and that to which they are dissociable in any given instance, particularly through imprinting to different substitute objects.

Since the observations on which these chapters are based are of value not only as illustrations of the interpretations presented here, but also for their own sake, space was not saved in their presentation. I am far more convinced of the intrinsic value of these observed facts than of the interpretation which I have given them. Anyone who is not interested in the individual observations can, however, omit four of the five special chapters on the companion systems without loss regarding understanding of the general part of this paper.

B. Conclusions relevant to the instinct problem

I regard it as a result of my study that the opinions concerning the nature of instinct incorporated as a working hypothesis in the second section (pp. 116–123) have been entirely supported by the emerging material. It must be stated that these hypotheses have not operated as guidelines for my investigations but were subsequently adduced as an abstraction from a previously-practised research technique, for which priority is definitely to be attributed to Heinroth.[86]

The assumption of a fundamental dichotomy between instinctive behaviour patterns on one hand and feats of learning and intelligence on the other has not encountered difficulty at any stage. On the contrary, it has aided us in understanding a number of otherwise incomprehensible behaviour patterns. There has been no cause for regret that the mutability of instinctive patterns through experience was flatly discounted and that the instinct was treated *as an organ*, whose individual range of variation can be neglected in the general biological description of a species. This interpretation does not conflict with the fact that some instinctive behaviour patterns can exhibit a high degree of regulative 'plasticity'. Many organs exhibit the same property. Despite the interpretation of the instinctive behaviour pattern in terms of a chain reflex, no specific adherence to either theories of specific pathways or to mechanistic dogma is intended.

The inseparable relationship between the development of an organ and the evolution of the instinct governing its use is virtually nowhere as conspicuous and undeniable as it is in the social ethology of birds. Whatever the factors which determine the biological adaptiveness in the development of the organ, they are doubtless the same as those which also govern the development of the relevant instincts. Within a sharply-demarcated group of birds we can see the formation of series in which the relationship of an instinctive behaviour pattern with the relevant organ is extremely clear-cut. In one specific exceptional case, we found that this relationship was so narrow that it seemed justifiable to incorporate instinctive behaviour patterns and organs involved in the same function within one overall concept. Organs and behaviour patterns have been termed 'releasers' without distinction where they aid in eliciting a social response in a conspecific (p. 106). We find with great frequency *'phylogenetic' series of releasers*, in which innate behaviour patterns without a corresponding underlining organ occur at one end, with highly-specialized organs developed for the emphasis of virtually-identical and doubtless homologous motor patterns at the other extreme. The various forms of demonstrative behaviour and demonstrative organs within the family Anatidae provide an example of such a series (p. 191). I should like to emphasize that nowhere in the entire field of comparative morphology would there seem to be series which are so utterly indicative of genetic relationships as those discussed here. This is probably to a large extent dependent on the fact that the releaser and the elicited behaviour are very little affected by environmental factors in the process of a (so to speak) internal 'arrangement' within a bird species, so that convergence can be excluded with great reliability from the outset. In this case, similarity *always* means homology. In consequence, we are often able to determine genetic relationships with a degree of accuracy seldom available to the comparative morphologist.

Nobody can deny that the phylogenetic mutability of an innate behaviour pattern possesses exactly the same characteristics as an organ and does not resemble that of a learning function. Its mutability is so similar to that of a particularly 'conservative' organ that the *instinctive behaviour pattern* actually carries great weight as a *taxonomic feature*, particularly when a releaser is involved. Nobody can prove that the individual mutability of an instinctive behaviour pattern concerns factors which are not involved in the individual mutability of organs. Unless the concept of learning is developed in an extremely wide sense, so that one could (for example) say that the hypertrophy of a frequently-employed muscle is a learning process, there is absolutely no justification in maintaining that an instinct can be modified by experience.

C. Discussion

In the course of this paper, particularly as a result of reports of observations, factual evidence which appears to be important in settling certain disputed questions has repeatedly emerged. Since this evidence is to some extent not directly related to the actual theoretical subject of this paper, I should like to consider it in a special section at this point.

(*a*) THE INSTINCT PROBLEM

1. It is by no means proven that an instinctive behaviour pattern can be modified by experience. All observations so far appear to support the opposite assumption. This conflicts with the views of Lloyd Morgan and many other authors who have analysed the instinct problem. A definition of instinct given by Driesch is in agreement with my own interpretation – it states that the instinctive behaviour pattern is a response 'which is complete from the outset'. This definition does, however, neglect the possibility of maturation processes discussed on pp. 118 and 119, which first take place when the behaviour pattern is already in operation.

2. Ziegler's definition based on theories of specific pathways is not able to explain the regulatory processes which have been demonstrated by Bethe's school. Bethe describes all regulatory processes as features of plasticity of the instinctive behaviour pattern. This statement is not in itself objectionable, but other authors have taken this to mean that the possibility of adaptive modification by experience is present (Alverdes). I reject this, and Bethe's experiments have rendered this possibility more unlikely rather than more likely. In any case, I do not regard it as very apt to use the term plasticity for a behaviour pattern when a specific unitary regulation appears immediately following a given disruptive operation (pp. 121–122).

3. Wallace Craig, in his well-known paper *Appetites as Constituents of Instincts*, made the important observation that the motivation for performing a specific instinctive behaviour pattern increases when elicitation is lacking for a period longer than normal. This leads not only to lowering of the threshold for the eliciting stimuli but also to behaviour which can be interpreted as a search for such stimuli. Craig summarized these phenomena as 'appetite'. In the second meaning of this word in

English, this term is certainly an apt one, but it is difficult to translate this into German, particularly since the phenomena summarized as appetite by Craig were not confined to 'positive' responses, as the author himself emphasized. The lowering of the threshold for the eliciting stimulus has particularly emerged with warning and escape responses (pp. 122, 164–166). The theoretically-attainable limiting value of the threshold lowering effect described by Craig is reached when the animal exhibits what I have described on p. 123 and also in my earlier paper as a '*vacuum response*', i.e. the response emerges without any preceding external stimulation. The vacuum response is not only important because it illustrates the operation of Craig's appetite factor, but also because it convincingly demonstrates the independence of the instinctive behaviour pattern from external supplementary stimuli (*behaviour supports* of Tolman) and shows the internal relationship of the behavioural chain.

4. William McDougall has demonstrated that particular instinctive behaviour patterns are dependent upon specific emotions as subjective correlates. ('Emotion' encompasses both the German concepts of 'Affekt' and 'Gefühl'.) After years of close contact with animals, one is forcibly struck with the impression that instinctive behaviour patterns are correlated with subjective phenomena which correspond to feelings and passions. No genuine animal observer can overlook the homologies which exist between man and animals and which virtually impel one to draw conclusions about the subjective processes within animals. No observer can refrain from such homologization. For this reason, all true animal observers are followers of McDougall, whether or not they are aware of it. Verwey writes: 'where reflexes and instincts are at all to be distinguished, the reflex operates in a mechanical manner whereas instinctive behaviour patterns are accompanied by subjective phenomena'. This is a somewhat bold definition of instinct, but I am in full agreement! Heinroth aptly responds jokingly to the entirely unjustified criticism that he treats animals as reflex machines with the statement: 'Animals are emotional people of extremely poor intelligence.' As can be seen, the opinion of the best of all animal observers and interpreters is in complete agreement with the teachings of McDougall. I have also professed to this opinion, but I do not regard it as sufficient to apply to animals the few instinct categories suggested by McDougall, which were in any case tailored for the human being. In addition, I do not see any opportunity for distinguishing, as McDougall does, between first-order and second-order instincts in animals, at least with birds. The great degree of independence which characterizes the component behaviour patterns of a functional system has the consequence that even those which are of apparently minor importance are just as important

as any others as autonomous mosaic stones for the function of the entire system. If we wish to say something about the accompanying passions and feelings, then to be consistent we must assume that there are as many different and independent kinds as there are autonomous innate behaviour patterns. In other words, we would have to assume that an animal has *far more* specific feelings and passions than are known in man, for which consistency requires that we should assume that the realm of human feelings must exhibit a similar process of simplification and abolition of differentiation to that known for human instinctive behaviour patterns. For this reason, emotional terminology derived from human language is insufficient from the outset for the description of the internal processes of animals, i.e. *the number of terms is too small.* Heinroth has therefore quite rightly created new emotional terminology in his studies, basing the terms on the terminology of the instinctive behaviour pattern and the German word 'Stimmung' ('mood' or 'motivation'). He talks of flight motivation, reproductive motivation and so on, and I have made use of the same terms in this paper (pp. 163, 209, 219, 221, 239). *Series* of responses which are arranged on a ladder of intensity of arousal and smoothly graded into one another in a chromatic scale must be regarded as belonging to the same motivational system. The duplex alarm and escape response of chickens described on p. 161 is an example of a case in which an animal possesses two intensity series, while human beings have only one corresponding scale. In any case, the term 'fear' is insufficient to categorize the duplex emotion of these birds; we must assume that fear of terrestrial predators and fear of predatory birds are present as two qualitatively different factors. These opinions incorporate a criticism of recent American papers such as H. Friedmann's paper *The Instinctive Emotional Life of Birds.* In addition, they imply interpretations exactly opposite to those expressed in H. Werner's views on primitive feelings and passions in his work on developmental psychology.

(*b*) BIOLOGY OF THE PAIR

1. Darwin expressed the opinion that certain conspicuous colours and shapes found in the animal kingdom had evolved as a result of sexual selection exerted on the animal equipped with such features by the other sex. Darwin largely based this interpretation on the 'Prachtkleider' (display costumes) of many male birds. Wallace entirely discounted the existence of sexual selection in this restricted sense and attempted to explain all sexual dimorphism as a sheer product of metabolic differences. The explanations of the nature and function of releasers presented here (p. 103) have demonstrated the untenability of Wallace's hypothesis.

Even if these explanations do depart in a large number of points from Darwin's views, they nevertheless demonstrate once again how Darwin's great predictive genius was far more accurate in interpreting the actually existing relationships than any of his opponents were able to do.

2. The findings of Noble and Bradley regarding the behaviour of some reptiles (p. 195) may *not* be generalized to birds in the manner indicated by the authors. The threatening and intimidatory effect of male 'Prachtkleider' is indeed present in very many birds, just as in the reptiles studied by Noble and Bradley, but this is never the *only* effect of these releasers.

3. A. A. Allen concludes in his paper *Sex Rhythm in the Ruffled Grouse* that birds are not conscious of their own sex. This only applies to quite specific species, though their number is quite large. The type of pair-formation exhibited by these birds has been termed the 'labyrinth fish-type' on p. 196.

4. Allen states in the same paper that males exhibit a period of physiological reproductive receptivity just as short as that of the females and that fertile mating can consequently only occur when the reproductive cycles of both partners are 'synchronous' from the outset. This doubtless applies to the species investigated by A. A. Allen (*Bonasa umbellus* and some small passerines), but these conclusions cannot, in my opinion, be simply generalized to cover other birds. I agree with Verwey that the male heron exhibits an extremely persistent period of reproductive motivation (p. 210) and I am in complete agreement with Wallace Craig's statement that subsequent synchronization of the reproductive cycles can also occur (p. 209).

(*c*) GENERAL SOCIOLOGY

1. The rank-order described by Schjelderup-Ebbe and other authors does apply to many bird societies, but by no means to all. There are colonially-brooding birds with extremely highly-specialized social responses, in whose societies no hierarchial gradients are formed. This category includes herons, cormorants, gannets and probably many other colonially-nesting marine birds.

2. It is to be recommended that *non-domesticated* bird types should be employed in investigations of sociological mechanisms, wherever this is possible. Even if there is nothing to stop us from taking the modified instinctive behaviour patterns of domestic animals as the object of an investigation, as Brückner states, I should nevertheless like to emphasize that we must avoid any source of error which is recognizable and avoidable in our researches. It should not be forgotten that

social behaviour patterns in particular are to a large part governed by instincts even in the highest animals.

D. The structure of the society

In conclusion, I should like to say a few words about the manner in which the interacting functions of individuals, the releaser of one animal and the released behaviour of another, produce the overall functional pattern of the supra-individual unit – the society.

It is part of the framework of modern techniques of unitary analysis that the whole is considered *before* its parts – the nature of the society is considered *before* that of the individual. H. Werner states in his book *Introduction to Developmental Psychology*:

> In every case it can be demonstrated that a total entity can be based on entirely differing foundations, that the so-called elements which combine to form the whole can change without altering the total character. For this reason, it is not a property of points that they can form a circle and it is not even a property of the unification of these points by synthesis. Any given component parts can form a figure of any kind, and in the same way the same figure can be formed by entirely different elements, such as crosses instead of points. Similarly, it is not at all a property of individual human 'points' that they should produce a total entity with a specific form. The synthesis of individuals never gives rise to a supra-individual unit.

At another point, he states: 'It is not the concept of creative synthesis but that of creative analysis which leads to progress.' Alverdes has criticized Uexküll's expression 'reflex republic' for the sea-urchin, based on the peculiar mutual interaction of spines, pedicellariae and ambulacral tube-feet, on the grounds that it 'conflicts with the assumption of totality'.

I should like to counter both of these views with the following observation: However much one may reject atomistic associative psychology or the theory of nervous centres, it may not be forgotten that the justification for unitary analysis is based on the existence of a physical totality, which is represented in the form of a governing, integrative apparatus – the unitary combination of the central nervous system. To seek such a unitary integration in places where it is not physically represented is a process which I regard as a departure into metaphysics, even though I reject all purely mechanistic interpretations.

Of course, genuine unitary integration of components can also lead to the formation of supra-individual units, but in such cases we must search for a concrete supra-individual integratory apparatus. Human society possesses such an apparatus in the form of the spoken language, which permits supra-individual aggregation of experience, supra-individual knowledge and in addition an extremely highly-developed co-ordination of the functions of individuals. In an animal society, however, the individuals *experience* very little of the existence and activities of their fellows. In this, they closely resemble the individual organs of the reflex-republic of the sea-urchin, which do not receive information about the activity of a neighbour organ through an integrative nervous system but only respond to that organ when the latter has a direct physical effect on them. This is just the case with the single individuals of an animal society, in which mutual interaction is definitely physical and noticeable. I hope that I have convinced the reader of this paper that this is so. In any case, I can think of no better term for the society of a bird species such as the jackdaw, which would describe its structure so completely and so fittingly as the term 'reflex-republic'.

In the sea-urchin, subordination of the parts within the unitary whole is so little advanced that despite all due acceptance and recognition of the value of systemic analysis, synthesis of the functions of the parts brings us closer to an understanding of the total function of the organism than the method of creative analysis of the whole entity suggested by Werner. The more unitary an organism is in organization and the further the process of differentiation and subordination of its parts is carried, the greater becomes the rôle which systemic analysis plays as a research technique and the less likely it becomes that the synthetic approach will bring us to our goal. Of course, the same applies to the supra-individual organism of the society. However, since in the majority of unitary animal colonies, e.g. that of termites, differentiation and subordination of individuals never progresses further than (or may only go part of the way towards) that of the individual 'reflex persons' of a sea-urchin, I am driven to contradict Werner's statements. I regard them as an insupportable generalization of opinions which may be entirely valid within the field of Gestalt psychology, but which are in no way transferable to the field of sociology.

'When a dog runs, it moves its legs,' states Uexküll, 'but when a sea-urchin moves, it is moved by its legs.' One can relate the unitary whole and the component in the same way with different types of society with different degrees of organization. When young people are growing up, they are extensively moulded by the society in which they develop. When young jackdaws grow up, they form a jackdaw society complete to the last detail with no prior image.

Werner states: 'The human being possesses, as a member of a supra-individual unit, properties which he acquires as a direct result of his subordination to the totality, and which can only be comprehended from the nature of his totality.' I must emphasize that two entirely distinct factors have been thrown together in this statement. The individual possesses properties which he achieves as a result of his membership within a specific, individual society in the form of certain behaviour patterns transmitted by tradition, such as a specific language in human beings. These properties would not accrue to the individual in the same way if it were a member of another society, another 'super-individual' of the same kind. In addition, the individual possesses properties which can indeed only be understood through an analysis of the society of the species concerned, but which the individual acquires as innately-determined (not traditionally transmitted) factors as a result of its membership of the *species* concerned and not because of its chance membership of a specific individual society of this species. The development of these properties is not influenced by the society.[87]

This, above all, shows us that we must separate innately-determined and traditionally-acquired characters as two fundamentally different categories. If we want to see clearly how different are the rôles played by these two categories in different animal species, we must observe the behaviour of isolated hand-reared individuals, in which the influence of tradition has been excluded from the outset. A human being treated in this manner would predictably be sharply differentiated from any normal member of society, and by studying such an 'element' of the human social group (however exactly) it would be impossible to construct even an approximate picture of human society. On the other hand, a jackdaw exhibits almost all of the properties and behaviour patterns which it would exhibit within the framework of a normal society, even when it is barred from any relationship with conspecifics from an early age. Many of these characters would of course not appear with regularity in the isolated individual, but this renders all the more remarkable the observation that some of them can be performed without object or purpose '*in vacuo*', which demonstrates that the behaviour patterns are functionally present and are simply dependent upon appropriate stimulation. Otherwise, the differences which are observed in the behaviour of the isolated hand-reared bird as compared with that of a control animal are to be traced to omissions in the imprinting process and not to a lack of acquired traditional transfer. Behaviour patterns carried over by tradition are of such minor importance that their omission is only noticed in quite specific cases (pp. 164–166).

In any case, the loss of traditional effects does not disturb us in our initial attempt to synthetically reconstruct the organization of the society from an exact analysis of the frequently incomplete innately-determined behavioural sequences of the isolated animal. This attempt is sometimes so successful that it repeatedly appears as a surprise to a biologist used to a more analytical approach. It always provided me with a pleasant surprise when I observed that the society of a bird species actually did exhibit the behaviour which I had reconstructed in some detail from the behaviour of the isolated, hand-reared young bird.

It is only when we assume that in human beings the innately-determined behaviour patterns play a rôle far inferior to that of acquired behaviour that Werner's analogy with a figure which can be formed in the same way with crosses or points may be considered to be justly applicable to the human society. I am far from prepared to make this assumption, but it can nevertheless be said that in human beings the influence exerted on the behaviour patterns of the individual by the society is far greater than with any animal species. The individuals of a social animal species can in every case only be combined to form *one* narrowly-defined form of supra-individual unit! If we should, like Werner, require an analogy, then we must state that these individuals are similar to the prepared stone blocks of an archway, which can only be combined to form one particular unit whose construction corresponds to the form of the elements. The sparse mortar of the individually-acquired behaviour patterns is not able to radically alter the form of the overall construction.

In comparing the human society, which is supposed to be largely based upon individually-acquired or even rational responses, with an animal society, we have so far assumed such differences in the behaviour of the component to the whole and of the whole to the component that it would appear to be necessary to study these two forms of supra-individual units with exactly opposing techniques. It is far from my intention to emphasize a contrast of this kind. I am far more convinced that sociologists in general, with the exception of McDougall, have greatly underestimated the part played by instinctive factors in the responses of human beings. It is an inherent property of instinctive behaviour patterns that they are coupled with governing emotions. However, the emotions coupled with the social instincts of human beings are regarded as something particularly high and noble. The last thing I wish to do is to deny that these instincts are really so, but this entirely justified high estimation of the emotions governing the social instincts of human beings has robbed many scientists of the psychological possibility of admitting that this quantity of high and noble behaviour is also present in animals and has also prevented them from recognizing instinctive behaviour in human beings. But this latter recognition is exactly what is needed for an understanding of our own social behaviour. Katz writes: 'In some ways there is a surprising correspondence in the social behaviour of animal and human groups, so that one is perhaps justified in hoping that animal psychology may eventually be employed for finding laws which govern the social behaviour of human groups.' This hope can only be fulfilled when we come to recognize that the instinct, governed by its own laws and fundamentally differing from other types of behaviour, is also to be found in human beings, and then go on to investigate this behaviour.

The establishment of the instinct concept (1937)

When two contemporary biologists attempt to discuss the problem of instinct, there is very often an astounding degree of reciprocal misunderstanding, based on the fact that each will connect a different concept with the word instinct. These differences in concept-formation, which are such an obstacle to mutual understanding, are probably largely due to the fundamental impossibility of providing a genuinely binding definition for a biological phenomenon and to the lack of realization of this fact. In addition, the erroneous belief that one can approach the problem of instinctive behaviour patterns with non-inductive methods and pronounce upon 'instinct' without defined experiments is the main reason for certain easily-refutable pronouncements which prominent theoreticians have made about instinct. This same reason underlies the formation of unwieldy concepts of instinct, particularly those which are far too *broad*. Such concepts are known from experience to provide obstacles to the progress of analytical research.

It is far from my intention to provide here a survey of any degree of completeness covering all of the concepts which have ever been connected with the word instinct. I shall rather attempt to demonstrate the fallaciousness (or at least the vulnerability) of some points of view and theories which have been put forward by prominent theoreticians concerned with instinct, and which find the most general distribution at this time. In doing this, I shall endeavour to show how closely such fallacies are often associated with inadequate and (especially) unspecific, broad interpretations of the instinct concept. From this critique, which I believe to be genuinely constructed on a basis of fact, a new and more closely-defined concept of the instinctive behaviour pattern should automatically emerge.

It is obvious that there must be facts which can be employed in the establishment of a more useful concept of instinct, since experience shows that all *practical students of animals* (zoo attendants, biologically educated amateurs or field observers) understand one another without difficulty when they come to speak about the instinct problem. They are evidently employing a concept of amazingly consistent structure, even if they might use different words for this concept.

Just a word about the term chosen: 'instinct' is no more than a word. Pronouncements can only be made about the *instinctive behaviour pattern* (Instinkthandlung), and our deliberations will be confined to this. In order to circumvent the diffuse nature of the word 'instinct', Heinroth employed the term 'species-specific drive activity' (*arteigene Triebhandlung*) instead of 'instinctive behaviour pattern', and his chosen term is doubtless more suitable. However, I am compelled to return to the term 'instinctive behaviour pattern' because the word 'drive' (Trieb) has recently been taken up in circles in which there is an obvious trend towards denying the existence of the very thing which we understand under that word. In order to avoid confusion with the (in my opinion) aberrant drive concepts of American behaviourists and psychoanalysts I am forced to drop the German term in favour of the Latin one.

At this point, I should like to give a brief summary of the opinions which will be criticized. Summarizing authors (so to speak) according to common errors could easily awaken an impression of deprecation, which I wish to avoid. For this reason, I state quite plainly that I owe much to these very authors, and that the last thing I wish to do is to underestimate their achievements, most of which belong in a different field to that criticized here.

One interpretation which is widely dispersed among biologists and even more among psychologists (in fact virtually representing a general assumption) is that instinctive behaviour is, both in the phylogenetic and in the ontogenetic sense, a *precursor* of the less fixed behaviour patterns which we refer to as 'acquired' and 'intelligent', or (following the recent American pattern) summarize under the inclusive concept of 'purposive behaviour'.

This interpretation can be largely traced back to Herbert Spencer and C. Lloyd Morgan. The latter, in his book *Instinct and Experience*, has developed in great detail his conception of the evolution of intelligent behaviour from purely instinctive behaviour patterns through a gradual increase in the influence of experience. The following, repeatedly-quoted sentence stems from Spencer: 'The increasing complication of instincts, which, as we have seen, involves a decrease in their purely automatic character, simultaneously brings about the beginnings of memory and intelligence.'

It is simply a logical extension of this interpretation for other authors such as Tolman, Russel, Alverdes (and to some degree Whitman and Craig as well) to deny the possibility of sharply demarcating the instinctive behaviour pattern from all other behaviour patterns and to interpret the instinctive behaviour pattern (indeed every component act of the longer instinctive behavioural chains) as 'purposive' behaviour. This interpretation is stated most distinctly in Alverdes' formula

$A = F(K, V)$, which is intended to indicate that every animal behaviour pattern is a function of a constant and a variable factor.[88]

McDougall's theory of instinct agrees with the Spencer/Lloyd Morgan school of thought in so far as the instinctive behaviour pattern is similarly interpreted as a form of purposive behaviour. This theory is otherwise marked by the fact that a limited number (in fact thirteen) of governing instincts is postulated, and these are regarded as employing the subordinate instincts, to some extent, as a means to an end. Proof of the purposive nature of instincts is seen in this means/ends relationship. This theory rapidly found followers in America. The concepts 'first-order drives' and 'second-order drives', which replaced McDougall's original terms in America as the word 'instinct' fell from fashion can be found in the writings of many recent English-speaking authors.

The Spencer/Lloyd Morgan theory and all interpretations based on this are sharply opposed to the interpretation of the instinctive behaviour pattern as a *chain-reflex*. H. E. Ziegler can be regarded as the main proponent of the latter theory, having established a histological definition of the instinctive behaviour pattern based upon theories of nervous pathways. The chain-reflex theory has found wide acceptance among zoologists with physiological training.[89]

The 'instinct theory' of behaviourists (in the narrower sense of the word), whose main proponent may be taken to be Watson, need only be fleetingly touched upon in this introduction. Complete ignorance of animal behaviour, which ails so many American laboratory research workers, is a necessary requirement to justify the attempt to explain all animal behaviour simply as a combination of conditioned reflexes. The presence of more highly-specialized patterns of motor co-ordination is denied flatly and with a certain passion by the behaviourists. Since this denial is based upon a simple lack of knowledge, a detailed refutation can be taken from the outset as unnecessary.[90]

I The Spencer/Lloyd Morgan theory

I shall confine my critique mainly to the two principles of the Spencer/Lloyd Morgan theory mentioned above, namely that of the susceptibility of instinctive behaviour patterns to the influence of experience and that of the assumed smooth transition from the most differentiated instinctive behaviour patterns to acquired and intelligent behaviour.

The first, and probably the most decisive, objection (as seen from the basis of our research principles) that I should like to make against the assumption of a modifying influence of experience on instinctive behaviour patterns is that the observational material underlying this interpretation is not reliable. Morgan cites learning to fly by young birds as a typical case of adaptive modification of an instinctive behaviour pattern through 'personal' experience. In doing this, he is neglecting the possibility that the changes and improvements which we observe taking place in the co-ordinations of flight can be based upon a *maturation process*. A developing instinctive behaviour pattern of a young animal can make its appearance *before* as well as *after* its final establishment, just as is the case with a developing organ. It is by no means necessary that the development of an organ and that of the instinctive motor co-ordination pattern determining its use should take place simultaneously. If the development of the behaviour pattern is more rapid than that of the organ, the facts are easily seen. For example, all ducklings have disproportionately small, and entirely non-functional wings. Nevertheless, within the first few days of life it is possible to provoke ducklings to perform a fighting response in which they exhibit exactly the same motor co-ordinations as adult conspecifics: the tightly-flexed wing-shoulder is used to beat at the opponent, the latter being gripped with the beak and held at exactly the right distance. The innate co-ordination of these movements is from the outset adapted to the bodily proportions of the adult bird, however, and the juvenile bird holds its opponent so far away that there is absolutely no chance of reaching the latter with its tiny wings!

If, on the other hand, the development of the organ is completed *earlier* than that of the appropriate instinctive behaviour pattern, it is not so easy to determine the relationships involved. In many bird species, the wings of the young are mechanically functional long before the co-ordinations of wing-movement are mature. When maturation of the co-ordination patterns is just beginning to catch up with the advanced development of the organs, the process has exactly the same external appearance as a learning process. Apart from the fact that the end-product is always the same, there is no external character to tell us that a maturation process is taking place along an exactly predetermined pathway. For this reason, experiments must be conducted. The American worker Carmichael maintained Amphibian embryos under persistent narcosis, which did not inhibit physical development but did completely suppress all motor patterns. When he allowed the larvae to 'awaken' at late stages of development, it emerged that their swimming movements were in no way different from those of normal control animals, which had 'practised' these movements for a number of days. My pupil Grohmann has performed corresponding experiments with young domestic pigeons, which were reared in very narrow, tube-shaped cases which did not even permit the birds to spread their wings. In addition, he obtained a curve for normally reared young pigeons in the following way: He selected various perches which the young pigeons had obviously preferred at various distances in their first excursions from the aviary. The distance of the perch reached was plotted on the ordinate and the age of the pigeon concerned was plotted along the abscissa. The normally emerging young birds exhibited a quite characteristic curve. In spite of muscular atrophy, which could not be avoided, the restricted birds exhibited curves with a *steeper* gradient of increase than the controls. The curves of the latter were matched in an extremely short time, usually in a matter of hours. In one extreme case, where the experimental animal was retained in the case for twenty-seven days after the normal date of emergence, the bird flew from the hands of the experimenter to the most distant of the recorded perches, thus giving a vertical line as its 'curve'.

These experiments conducted by Carmichael and Grohmann reliably exclude the possibility of learning in the developmental processes occurring in each case. If we should conversely wish to demonstrate the presence of a learning process to the exclusion of maturation, only one means would be available: the development of the emerging co-ordination patterns would have to take place in varying ways with variations in experience. So far, we know of no single case of such an observation in the entire animal kingdom, and certainly not for young birds learning to fly. The flight of a young bird developing in a room has never been observed to develop differently from that of a free-living bird, to the extent that specific co-ordinations have developed in adaptation to the spatial relationships concerned in a manner different from that occurring in the field. An example of such an adaptation would be for a young peregrine falcon to develop the necessary co-ordinations of hovering flight in a restricted space *better* than under natural conditions. This never occurs.

Another example of assumed adaptive modification of an instinctive behaviour pattern by 'personal' experience, which was shown to be invalid ages ago and nevertheless stubbornly reappears in the literature, is the statement that older, experienced birds build better nests than young conspecifics. This assumption is based upon misinterpreted observations of captive birds. With increasing age, captive birds exhibit a considerable improvement in general condition, particularly after the passage of one breeding season. Even the slightest degree of physical deterioration can lead to defects in the more delicate instinctive behaviour patterns, such as those of nest-construction. These defects give way to normal behaviour with the described improvement in general physical condition. This, and not individual experience, is responsible for the frequent appearance of later successful breeding following an initial failure of a brood. Proof for the correctness of this interpretation was provided by three bullfinch pairs which I kept as a young student. For the first year, two of these pairs lived in a large aviary whilst the other lived in a domestic cage kept by a friend. The first two pairs built extremely poor nests, which both fell and broke up before the offspring hatched, and the pair in the domestic cage did not build at all, although the partners proceeded to mate. The next year, all three pairs were accommodated in the outdoor aviary and all built exactly equivalent, perfect, species-typical nests. I had, in fact, no idea which birds were those which had already built once. I venture the statement that all known cases concerning better building by older birds are based upon the same phenomenon.

These two and other, equally vulnerable, examples of the modification of an instinctive behaviour pattern by experience are always presented in a manner suited to awakening the impression that the author concerned could provide innumerable further examples as necessary. However, when suspicion is raised by the stubborn reappearance of the same, only too well-known, examples in the literature and personal experiences are sifted for further (and this time reliable) observations, the search proves to be futile.

A certain knowledge of the variability of the instinctive behaviour pattern and of the laws governing this variability is necessary if one is to avoid the danger of describing particular effects as results of experience and products of resultant adaptation, when they are in fact produced by quite different factors. For this reason, I am compelled to discuss these effects briefly here.

In the first place, very many instinctive behaviour patterns, particularly the most simple among them (e.g. the co-ordination patterns of walking) incorporate a considerable capacity for regulative changes. Regulative plasticity need bear no relation to learning and experience, however. Many organs exhibit an entirely analogous phenomenon, once again particularly operative in the *least differentiated* among them. The experiments which Bethe carried out on the regulative capacity of walking movements of a wide range of animals showed that in all cases where regulative changes occurred, they were *immediately* evident after the given operation. In other words, they were not products of the influence of experience. Bethe repeatedly employs the term 'plasticity' for the regulative capacity which he determined. In itself, this term is inoffensive, but Morgan, Alverdes and others have taken this to mean the possibility of adaptive alteration of the instinctive behaviour pattern through experience, whereas the existence of this possibility is far from indicated by Bethe's experiments. One of his experiments in fact speaks unequivocally *against* this assumption. A dog, in which the two *nervi ischiadici* had been sutured in the crossed position, exhibited completely normal co-ordination of walking movements after the nerves had resumed function. Regarding sensitivity, however, there was no regulation, to the extent that the animal persistently responded with one hind leg to pain stimuli applied to the other. Combination of the appearance of regulation on the motor side with the absence of such on the sensory side is the clearest proof that experience plays no part in the emergence of motor regulation. If this had been the case, experience would have taught the dog the opposite.

A second phenomenon which is often erroneously associated with a regulative influence of experience is the following: Preceding events (experience in its widest sense, if you like) can determine *the intensity with which* a specific response is given to a stimulus of given strength or even determine which response is elicited by a specific stimulus.

Let us first of all take the variations in the intensity of performance of instinctive behaviour patterns. It is observed that with the instinctive behaviour pattern the converse (so to speak) of the all-or-none law applies. Almost all instinctive behaviour patterns of an animal species will emerge in the individual's behaviour as *slight indications* of the behavioural sequence concerned, even at quite low response intensities. These indications show the informed observer the direction in which the behaviour patterns of the animal will proceed when the requisite stimulus intensity is attained. Since they indicate the 'intentions' of the animal, such incipient behaviour patterns are often termed *intention movements*. Apart from the fact that in certain social animal species intention movements have attained secondary importance for the preservation of the species as 'communicatory' motivational induction agents, it must be stated that intention movements generally are far from operative in preservation of the species. Even in the special case mentioned, these movements perform no function in the direction of the adaptive efficacy of the completely-elicited response. One can find *all imaginable transitional stages* between the scarcely-indicated intention movement, only perceptible to the observer with knowledge of the behaviour pattern concerned, and the complete behaviour pattern fulfilling the adaptive function of the response. A Night Heron sitting in the branches of a tree in early spring indicates to the informed observer the emergence of the responses of the reproductive cycle of the coming year. The bird rather abruptly passes from a restful condition to an obviously excited state, bends forward, grips a nearby twig in its beak, performs once the motor co-ordination pattern of nest incorporation, and then abruptly falls back 'satisfied' into the previous restful state. If even more careful observations are made, it might be possible to observe the first traces of nest-building behaviour even earlier in the next year; for example, one might observe such behaviour in the form of temporary fixation of a twig coupled with an indication of the bowed posture frequently assumed in the nest later on. In the course of days and weeks, such traces develop in a smooth transition to give the complete sequence of nest-building behaviour patterns leading to construction of a nest.

The emergence of such scales of intensity is significant for the question of purposive consciousness of the animal. In the first place, the fact that the animal is just as satisfied with the incomplete behaviour patterns (which perform no adaptive function) as with those complete patterns which attain their biological goal is a clear indication that this goal is not the factor immediately determining the form of the animal's behaviour patterns and cannot be equated with a goal set for the animal subject. This is particularly obvious when the response intensity of the animal is somewhat higher but just insufficient for complete response, so that the behavioural act is broken off *just before* the biological goal is achieved. In captivity, such non-adaptive, incomplete instinctive behaviour patterns appear in some animals much more frequently than complete responses, something which is caused by the previously-mentioned defects associated with physical deterioration. Such incomplete and non-adaptive features are in fact the most frequent practical indicators of the instinctive character of a behaviour pattern in observations of animals. The uninitiated observer can scarcely be provided with a more convincing impression of the lack of purposive control, in an animal performing an instinctive behaviour pattern, than the observation of these incompletely-performed behaviour patterns. On observing the Night Heron cited above as an example, it immediately became obvious that the bird has not the slightest impulse towards the biological success of its behaviour (in this case towards producing a nest) but only towards the *release of the response concerned*. This impulse is in fact satisfied, at the prevailing low level of intensity, by shaking a twig once.

It is difficult to understand, in the face of these facts, why many authors still equate the behavioural goal set for the animal with the biological (i.e. adaptive) goal of the innate, instinctive behaviour pattern – almost equating the two. It is even more difficult for me to understand why an author such as Russel should say of the instinctive behaviour pattern in a book which first appeared in 1934: 'it is continued until either the goal is achieved or the animal is exhausted'. In fact, the exact opposite is true, and this has in fact been pointed out and correctly appreciated in the English literature. Eliot Howard made the incomplete behaviour pattern (resulting from inadequate intensity) an object of intensive study and has provided a large number of observational examples, all gleaned from natural situations, for the interpretation presented here.

All variations in intensity in the performance of instinctive behaviour patterns are of great importance for our question about the influence of experience, since, as has been indicated already, the intensity of a behaviour pattern can be determined by preceding events. Following repeated elicitation by a uniformly-maintained stimulus situation, the response intensity of a behaviour pattern can be either reduced, by habituation or adaptation to the stimulus, or increased by summation of stimuli.

Alteration of the intensity of instinctive responses through habituation and adaptation to the stimulus provides continuous series of response intensities. With gradual summation of the stimuli, the animal usually responds as to gradual increase in stimulus intensity. Phenomena similar to those produced by creeping stimulation emerge. In other words, final crossing of the threshold by the gradually-increasing stimulus effect is characterized by abrupt increase in the response intensity. For this reason, it is mainly complete and continuous intensity series produced by *adaptation* to the stimulus which permit us to demonstrate the association between different intensity levels of a given response. The observed form of a response as seen at two fairly distinct levels of intensity may show great variation. Only the presence of all possible transitions and the consequent impossibility of demarcating them from one another leads us to associate them.

A generally-recognized example of gradual decline in response intensity with stimulus habituation concerns the escape responses of wild animals as they become tame. The stimuli presented by a human being approaching the animal until a certain critical separation distance is reached are responded to with ever-decreasing intensity until the original, wild escape movements are replaced by mere alerting, or there is eventually no response at all.

The fact that a stimulus which is presented repeatedly remains objectively unaltered need not necessarily mean that the various behaviour patterns given in response are mere intensity levels of one and the same instinctive behaviour pattern. In the course of stimulus habituation, the animal behaves in exactly the same way as in response to reduction in intensity of stimulation. In this way, the same stimulus can elicit *different* responses, which are adapted to stimuli of different intensities. This can lead to a situation where, with gradual decrease in the intensity of a given response, abrupt transition to another response can be observed. For example, a pair of wild swans will flee to the nest when a human being approaches. As the swans gradually become tame, the intensity of this escape response decreases until this abruptly gives way to the response of nest-defence, which was previously blocked. We can then observe an abrupt change from low-intensity escape responses to high-intensity fighting responses. In this, the animals do not merely behave as if the perceived stimulus intensities were reduced, but literally react as if the human being emanating the stimuli were becoming smaller. In a situation which objectively remains the same, the swans first give the response which they would give to a human being or a wolf under natural conditions, and later perform the response which they would give to the approach of a weasel, a crow, or (at the most) a fox in their natural habitat.

In all of these cases, the performance of an instinctive behaviour pattern is in fact influenced by individual experience. Experience can determine the intensity with which the instinctive response is performed and can even specify *which* response is elicited by a particular stimulus. In certain cases, this type of effect may even have an adaptive character; but it must be emphasized once again at this point that actual adaptive modification of a behaviour pattern through learning, which forms the basis for the Spencer/Lloyd Morgan theory, has never been found. The result is never a *novel* behaviour pattern which has not been hereditarily fixed and predetermined in exactly the combination of movements found. The consecutively decreasing intensity levels of the escape response, occurring over a number of weeks in the taming process taken as an example, do not contain one single motor co-ordination pattern which is not definitely allotted to a specific intensity level of response and which could not *at any time* be elicited by a specific frightening stimulus of a particular strength, without any preceding experience. The responses corresponding to the individual intensity levels remain unchanged with truly photographic continuity, quite independent of the historical factors relating to their elicitation.

The same applies to the case where two different instinctive behaviour patterns are elicited by a stimulus which remains objectively unaltered. In this case, too, there is no appearance of motor co-ordination patterns which could not be elicited *at any time* in exactly the same form by an appropriate stimulus, as has been shown with the example of the pair of swans.

A further objection to the Spencer/Lloyd Morgan theory depends upon the broad structure of the associated concept of instinct. This concept in fact ignores a quite specific phenomenon which compels us to adopt an analytical procedure automatically leading to a different definition of the concept of the instinctive behaviour pattern. We owe our knowledge of this phenomenon to careful observation of the developing instinctive patterns of young animals, particularly of birds.

It is a peculiarity of very many behaviour patterns of higher animals that in a uniform behavioural sequence (i.e. one oriented towards a uniform, adaptive goal) *instinctive, innately-determined components and individually-acquired components can follow directly on from one another.* I have termed this phenomenon instinct-conditioning intercalation, whilst emphasizing that similar intercalation between instinctive behaviour patterns and insight-controlled behaviour can occur. At this point, where we are concerned with the influence of experience, it is primarily instinct-conditioning intercalation which is of interest. Such intercalation is represented by the incorporation of a conditioned element of behaviour (which must be acquired by the individual in the course of its ontogeny) at a specific, hereditarily-determined site in an otherwise innately-determined, instinctive chain of behavioural acts. In such cases, the innate behavioural chain exhibits a *gap* for which a 'capacity to acquire' is present instead of an innate, instinctive behaviour pattern. This capacity can be of a quite specific nature and be directly related to a quite specific variable factor in the environment, in fact more or less representing an adaptation to such variability. I call to mind the conditioning ability of bees, which can be termed an adaptation to the blooming times of various plants (as was shown by von Frisch).

The filling of gaps left in an innate behavioural chain for an appropriate acquired component understandably takes place under specific conditions presented in the natural habitat of the species, occurring in a manner which renders the intercalated component a biologically-adaptive and functional unit. Under conditions of captivity, disturbances and absence of the acquisition process frequently occur, even without intentional experimental intervention. It was in fact such effects which first drew our attention to the existence of two fundamentally different components in functionally unitary behaviour patterns.

An example of instinct-conditioning intercalation is provided by carriage and incorporation of nest-material by Corvids. In both ravens and jackdaws, the first behavioural component of the complex sequence of nest-building emerges in the form of a particular response: These birds begin to pick up objects of all kinds in the beak and to carry them over long distances in flight. This carriage of various objects is at first a completely independent response of Corvids, occurring without relationship to further building activities. As long as this response is the only factor in the bird's behaviour, there is in fact no preference for materials suitable for actual nest-building. Both ravens and jackdaws initially carried mainly fragmented pieces of roof-tile, which were the objects most frequently encountered near their aviary on the roof of our house. This, despite the fact that at the same site the birds had easy access to suitable twigs for nest-building. Preference for the latter first appeared when a further instinctive behaviour pattern of nest-building emerged – the peculiar lateral pushing and juggling movement, with which most birds attempt to secure the twigs at the nest-site. (When this occurs, site-conditioning takes place as a simultaneous effect, but this will be left out of the considerations here for purposes of simplicity.) In the motor co-ordination pattern of lateral pushing, only materials for which this preliminary building activity was determined in phylogeny (i.e. for twigs, stems, etc.) are suitable. The response is performed until the behaviour pattern either fizzles out, which is virtually always the case at the onset of nest-building, or until the object to be incorporated lodges somewhere and presents a certain resistance to the juggling and pushing movements, in which case it is released from the beak. This conclusion of the response is obviously consummatory for the bird, and since it only occurs following carriage of suitable nest-material the bird *learns* remarkably rapidly to select the biologically 'correct' materials in performing the carriage response itself.

On observing such behaviour, nobody who has ever seen a conditioning process deliberately procured by human intervention would be able to avoid the conclusion that the two processes are entirely similar. It is known from experience that conditioning carried out by a human trainer requires the operation of specific stimuli on the animal, stimuli which are termed 'rewarding' or 'punishing' by authors who endeavour to avoid subjective expressions. The behaviour of animals in acquiring a conditioned component in an intercalation sequence thus presents us with the question as to which factors operate as punishment or reward in such a case.

Wallace Craig, in his *Appetites and Aversions as Constituents of Instincts*, was the first to point out that an animal brings about, or 'attempts' to bring about, the performance of its instinctive behaviour patterns by means of what we term *purposive behaviour*. Following Tolman, this term is taken to cover all behaviour patterns which *exhibit adaptive variability whilst the goal remains the same*. This objective definition of purpose is extremely useful for the separation of conditioned and insight-determined behaviour from the instinctive behaviour pattern and provides us with a governing concept which incorporates all non-instinctive behaviour patterns. But it must at once be said that neither Craig nor Tolman perform a separation of this kind. Instead, the purposive behaviour through which the animal endeavours to enter the necessary stimulus situation for the elicitation of its instinctive behaviour pattern is interpreted as a *component* of the pattern concerned. I, on the other hand, separate these two types of behaviour as *fundamentally different* constituents.

Quite apart from this difference in conceptualization, it must be stated that the mere existence of instinct-conditioning intercalation provides excellent backing for the general correctness of Craig's formulation. One can really imagine no better proof for the endeavour to bring about the performance of instinctive behaviour than the fact that the 'appetite' relating to an instinctive behaviour pattern has the capacity to condition the animal to perform a specific, non-innate behaviour pattern, just as the appetite for a meat morsel has the capacity to condition a circus lion to perform such behaviour. The statement that an animal possesses an 'appetite' for the performance of an instinctive behaviour pattern, or the attainment of the stimulus situation necessary for the elicitation of such, applies quite precisely in many cases, but I nevertheless prefer to translate Craig's term 'appetitive behaviour' with 'Appetenzverhalten', since the word 'appetite' has a more narrowly-defined meaning in German. In the following text, this translated expression will be employed as synonymous with 'purposive behaviour'.

The necessity of separating appetitive behaviour from the instinctive behaviour pattern, as a category apart, is indicated by the very existence of intercalated behaviour patterns. We are provided with the possibility (and thus the duty) of dissecting functionally unitary behaviour patterns into components which, on the one hand, are purposive in nature and modifiable through experience, and those on the other hand, which are not modifiable and are inherited in exactly the same manner by all individuals of a species, just like morphological organs. We are given the possibility of distinguishing the behaviour of the shrike leading to recognition of thorns from a conditioning process. If we discover that in a specific 'instinctive behaviour pattern', previously regarded as a unitary whole, this process always intervenes at *the same site*, whilst leaving the remaining behaviour unaffected, there would be no justification whatsoever for extending the concept of the instinctive behaviour pattern so as to include the conditioned component which we have unequivocally identified. Further, conditioned behaviour is certainly not the only type of purposive behaviour which occurs intercalated in instinctive behaviour patterns.

As early as 1898, Charles Otis Whitman said: 'One may find mixtures and all kinds of mutual influence between habit and instinct, and these may be of great theoretical significance, but they cannot be exactly determined and therefore provide a dangerous basis for theories.' Every theory of instinct must of course be primarily concerned with the *pure instinctive behaviour pattern*. In my opinion, all authors who believe in insight of the animal into the purpose of its instinctive behaviour pattern and in the adaptive influence of experience on the pattern fall down just because they have chosen unanalysed intercalation patterns ('mixtures') as a dangerous basis for their theories. Thus it is that the instinctive behaviour pattern is credited with all the properties of intercalated acquired and insight-controlled behaviour components, whilst such properties are not just incongruous but quite opposed to the very nature of instinctive behaviour.[91]

It is our permanent duty to carry analysis as far as is humanly possible, and I find myself compelled to construct a working hypothesis on the factual evidence of intercalation patterns. I find it necessary to make a working hypothesis on the assumption that even the extremely complex behaviour patterns of higher animals and man, which are indeed 'constructed on an instinctive foundation' and yet incorporate intelligent components and predisposition for the effects of learning, must be interpreted as intercalation patterns. Even if these complex behaviour patterns are not open to the few methods of investigation available to us (and may possibly remain so), this is no reason for not rigidly carrying through the conceptual separation of the two elements. Only after such separation is it possible to pave the way for further analytical study. To reject the conceptual separation of the components, because animal and human behaviour patterns exist in which they cannot be clearly separated, would be roughly comparable to an attempt to reject the concepts of the germinal layers in embryology because there are organs of unitary function in which it is not possible to determine from the completely-developed condition which cells stem from a specific germinal layer. My suggested separation has been challenged with the accusation that it is 'atomistic' and incompatible with modern holistic (systemic) biological considerations. This accusation is just as unjustified as the assertion that it is detrimental to systemic considerations to distinguish cutis and epidermis in the skin. Just as there is little justification in maintaining that the fact that there is almost always participation of more than one germinal layer in the functional unity of an organ is a counter-argument against the germinal layer concepts, the fact that a functionally unitary behavioural sequence in a higher animal in the majority of cases incorporates instinct, conditioning *and* insight as intercalated elements should not be allowed to lead us astray. In further analytical studies of animal and human behaviour patterns, the most illuminating behaviour patterns are those with which we can demonstrate one of these three components in *pure* form, as suggested by Whitman. Investigators of instinct are thus bound to be initially interested in pure instinctive behaviour patterns and the simplest, most easily surveyable cases of intercalation.

We must recognize quite clearly the narrowness of the new interpretation[92] of the concept of instinct, which is forced upon us by the exclusion of purposive behavioural elements incorporated in intercalation patterns. Many behaviour patterns which play a part in animal behavioural sequences and occur intercalated with instinctive behaviour patterns are represented by the *orientating movements* which direct the animal towards or away from a specific goal in the environment. The movement which orients the animal in space can never be an innate, instinctive pattern, since the necessary co-ordination can of course not be determined in the special form required for any individual situation. The orienting movement is the most primitive and the most penetrating form of non-instinctive behaviour in the behavioural system. This represents the phylogenetic root of all appetitive behaviour. It is customary to characterize the orienting movement as a *taxis* in specific, particularly simple, cases, but we must acknowledge the fact that it is not possible to demarcate this behaviour rigidly from insight-controlled behaviour. When a frog responds to a fly with an orienting movement, by symmetrically orienting its eyes and then its body (using appropriate short locomotory movements of the feet) towards the fly, it is perfectly possible to describe the movement of the eyes, and perhaps the body-movements as well, in the terminology of the theory of taxes. However, this behaviour cannot be distinguished from behaviour controlled by the simplest form of insight. On observing the continuous phylogenetic series of corresponding behaviour patterns extending in a spectrum from protozoans to man, it must be concluded that we cannot distinguish between a taxis and a behaviour pattern controlled by the simplest element of insight. In the case of the frog, the insight element (in anthropomorphic terms) would be limited to the observation: 'There sits the fly'.

The orienting movement and the taxis in its narrowest sense must therefore be consistently interpreted as purposive behaviour patterns, simply because they typically exhibit variability whilst retaining the same goal, as required by Tolman. In this case, the 'purpose' set for the animal subject always[93] remains the attainment of the stimulus situation necessary for elicitation of an instinctive behaviour pattern. In the example of the frog, the necessary situation is attained with the symmetrical orientation to the fly.

In very many intercalation patterns, a vital rôle is played by the innate, instinctive readiness to respond to a quite specific combination of stimuli. Specific stimulus combinations often represent specific elicitatory *keys* (signals) for specific responses. These responses cannot be elicited even by stimulus combinations which are only slightly different from the norm. Thus, specific sign ('key') stimuli are associated with a receptor correlate, which only responds to a quite specific combination of stimulatory effects, rather like a combination lock, and then sets the instinctive behaviour pattern in motion. In another publication, I have termed receptor correlates of this kind 'innate releasing schemata'.

The innate releasing schemata achieve special significance in relation to instinctive behaviour patterns having a conspecific as their object. In this special case, the possibility exists for evolution and specialization of particular instinctive behaviour patterns and organs of an animal species in parallel with further differentiation of innate releasing schemata. I have briefly used the term 'releaser' for such elicitatory instinctive behaviour patterns and the supporting colours and structures (whose sole biological function is the elicitation of social instinctive behaviour patterns in the widest sense). In many animals, particularly birds, complex systems of releasers and innate schemata form the basis of the entire sociological organization and guarantee unitary and biologically-adaptive interaction with the sexual partner, with offspring – in short, with conspecifics in general.

Innate releasing schemata frequently play a large part in intercalation behaviour patterns. An intercalation pattern can just as easily begin with the triggering of an innate schema, followed by purposive behaviour, as it can be dependent upon acquired behavioural components for its elicitation. The frog previously taken as an example responds to an innate, instinctive releasing schema and immediately follows this with purposive behaviour in the form of oriented movements. Conversely, an animal can respond to an acquired eliciting factor with a pure, undirected instinctive behaviour pattern. For example, a duck may respond to the sight of a rifle (for which it of course possesses no innate schema, but for which it has developed a conditioned fear response) with the innate, undirected motor co-ordination pattern of escape diving. Thus, in an intercalation pattern both instinctive, innate behaviour and purposive components can be restricted to either the receptor or the effector limb of a response.

As can be seen, logical analysis of all animal appetitive behaviour leads to the division of the great majority of all functionally entire behaviour patterns into a chain of appetitive elements and the instinctive behaviour patterns whose elicitation is sought. But it must not be forgotten that these successive elements in any individual intercalation patterns are present in a finite, predetermined number. It is erroneous to believe that a behavioural sequence of this kind can be dissected into an infinite number of infinitesimally small purposive elements and a corresponding number of appetitive components. Tolman makes this assumption and states that the behavioural sequence is dependent upon supplementary guidance by orienting stimuli, referred to as 'behaviour supports', at *every* point. He neglects the demonstrable fact that within all intercalation patterns one can determine long and highly-differentiated behavioural sequences which are free from any variable orienting effect, are completely independent of 'behaviour supports' and show no evidence of appetitive components. In short, Tolman ignores the existence of what we recognize as the instinctive behaviour pattern.[94] In his opinion, the innate factors are limited to the guidelines of a pathway between successive intermediary goals, towards which the animal strives, in a manner which is left open to the individual. This opinion applies exactly to certain intercalation patterns of the highest mammals, in which reduction of the instinctive behaviour patterns involved has actually progressed so far that the latter (to use a simile) have a similar function to a series of tidbits set out in order to induce the animal to follow a specific path.[95] Since Tolman's personal experience is limited to higher mammals and he has only had the opportunity to observe intercalation patterns of the latter type as his 'instinctive behaviour patterns', his definition of the instinctive behaviour pattern as a 'chain appetite' is entirely understandable, but it of course does not at all define what we regard as the essence of instinctive patterns. It is without doubt characteristic for the instinctive behaviour pattern that it can become the goal of an appetitive behavioural act, but it is not the case that an intercalation pattern is, or must be, produced by repeated sequence of these two components of the behaviour pattern. The neglect of the possibility of intercalation automatically leads to the interpretation of all behaviour patterns which incorporate even the slightest instinctive component as 'instinctive behaviour patterns', without any attempt at analysis. This of course leads to the complete clouding of any distinction between instinctive behaviour patterns and purposive behaviour. The pathway to any further progress of analytical research is persistently diverted, since such conceptualization prevents us from extracting and describing the key properties and characters which are so overwhelmingly characteristic of the behavioural component which *we* regard as the instinctive behaviour pattern.

One author who systematically carries out this momentous equation of intercalation patterns (however complex) with instinctive behaviour patterns is Alverdes. He states explicitly:

> Some authors talk of instinctive behaviour patterns in man and animals as if a fundamentally different category were involved. Against this trend, it must be observed that any intelligent act incorporates a sizeable portion of instinctive, drive-bound behaviour, and that, conversely, no single instinctive behaviour pattern is performed completely automatically but always incorporate a variable, more or less situation-dependent, component in addition to the fixed, invariable components.

The one part of this representation by Alverdes is doubtless correct – every intelligent act involves instinctive behavioural components. As far as the participation of a variable, situation-dependent component in *every* behaviour pattern is concerned, however, Alverdes falls into the same (in my opinion utterly erroneous) train of thought as Tolman, who assumes that every behaviour pattern is influenced by supplementary orienting stimuli.

The complete independence of a *purely* instinctive behaviour pattern from orienting stimuli (Tolman's 'behavioural supports') is best demonstrated experimentally by means of a phenomenon which I term the 'vacuum activity'. If an instinctive behaviour pattern is not elicited for some time, the threshold level of the stimuli necessary for elicitation falls remarkably. (This is a phenomenon which we shall reconsider in great detail in the critique of the reflex theory of instinct.) The threshold reduction of the eliciting stimuli can reach a limiting value to the extent that the persistently withheld response eventually emerges *without* demonstrable stimulation. *It is impossible to imagine a more conspicuous and more remarkable characteristic of the instinctive behaviour pattern than the property of emergence* in vacuo *in the continued absence of eliciting stimuli*, quite independently of the supplementary stimuli which Tolman regards as necessary. It is peculiar that Tolman, in his argumentation for the purposivity of all animal behaviour and for the dependence of the latter on supplementary stimuli, should express the view: 'animal behaviour cannot "go off" *in vacuo*'. In his endeavour to extend the assumption of the existence on non-purposive animal behaviour patterns *ad absurdum*, Tolman in this sentence demands the very proof for the existence of such behaviour which we are able to bring in the form of demonstration of the vacuum activity.

The vacuum activity permits extremely clear conclusions regarding the parts of the behavioural sequence which are innately determined. This is extremely valuable when more highly-specialized, long, purely instinctive chains of behaviour patterns run off *in vacuo*. For instance, I once possessed a hand-reared starling, which performed the entire behavioural sequence of hunting flies from a look-out position *in vacuo*. This behaviour in fact included several components which I had previously interpreted as purposive movements and not as instinctive acts. The starling would fly to the head of a particular bronze statue and then search the 'sky' from this vantage point, looking for flying insects although none were present on the ceiling of the room. Abruptly, the bird's entire behaviour would indicate that it had spotted flying prey. Movements of the eyes and head were performed, as if the starling were following a flying insect with its gaze; the bird would then tense, take off, make a biting motion, return to its perch and perform the lateral beating and swinging movements which very many insectivorous birds perform with their beaks to kill the prey on the surface upon which they are resting. Repeated swallowing actions followed, after which the flattened feathers were loosened, and in many cases the shuddering reflex appeared, just as it would do after genuine satiation. The entire behaviour was such a deceptive replica of the normal, adaptive behavioural sequence and the behaviour just before take-off in particular was so convincing that I climbed onto a stool, not just once but several times, to see whether there were in fact any tiny flying insects which I had previously missed. But there were really no insects there! Optical pursuit of a moving goal which was in fact not there was compulsively reminiscent of the behaviour of some mentally-ill patients suffering from optical hallucinations, and this prompted me to ask what subjective phenomena operating within the bird are linked with the vacuum activity. The behaviour of this starling provided proof that the orienting location movement towards the fly is the only appetitive behavioural component which operates in this particular behavioural sequence.

Craig, like Tolman, applies the term 'instinctive behaviour pattern' to the entire sequence of behaviour, including the purposive search for the elicitatory stimulus situation. Since he regards this search for stimuli as an important *component* of the instinctive behaviour pattern, he interprets the latter as a purposive behavioural sequence, as does Tolman. In contrast to other authors taking this view, however, Craig presents us with the extremely important concept *that the release and performance of a behaviour pattern* (consummation of an instinctive action) *is the goal of the purposive behaviour*.[96] This itself largely prepares the way for the separation of appetitive behavioural acts from those motor co-ordination patterns which are (subjectively speaking) non-purposive and are performed autonomously, i.e. instinctive behaviour patterns by *our* definition. Even if Craig to some degree shares Tolman's view of the dissociability of all behavioural sequences into 'chain appetites', he nevertheless comes quite close to our concept of intercalation patterns in stating:

> When the behaviour pattern is very largely instinctively determined, it has the form of a chain-reflex. In the majority of supposedly innate chain-reflexes, however, the responses at the beginning of the chain or in the middle of the sequence are not innate, or incompletely so, and must instead be acquired by trial and error. *The terminal component of the chain, the consummatory action, is always innately determined* [my italics].

An example is provided by the feeding behaviour of the Peregrine Falcon, which is predominantly based upon innate motor co-ordination patterns. Appetitive behaviour is restricted to the search for the requisite stimulus situation, operating according to the principle of trial and error, and the exquisitely-specialized instinctive behaviour patterns of prey-catching typical of this bird species are elicited when the appropriate situation is reached. *The goal is already attained* when the bird reaches this situation; the subsequent motor co-ordination patterns are purely instinctive, apart from a number of orienting movements, and they can be observed often enough as vacuum activities. In man, in contrast to the falcon, the entire motor repertoire of the adaptive functional system of food-acquisition is subject to purposive behavioural control. Instinctive, and therefore 'rewarding', components forming the purpose of the entire behavioural sequence are the purely instinctive patterns of chewing, salivation, swallowing, etc. It should be noted that those stimulus situations which are best suited to elicitation of one of these functional responses are also the most 'appetizing'. Thus, even in human beings, the biological purpose of the behaviour pattern is by no means the goal of the behaviour pattern, at least in most cases. The goal is represented by the performance of instinctive responses.

My concept of instinct-conditioning intercalation has been criticized on the grounds that it is not sharply discernible from the Pavlovian concept of the conditioned reflex. Since these two concepts do exhibit similarities in content, it is necessary for me to justify the introduction of a new term. O. Koehler has described Pavlov's nomenclature as a 'dilution of the reflex concept' – a criticism which I readily share. In discussing the reflex theory of the instinctive behaviour pattern, I shall need to analyse the reasons for the necessity for a sharply-defined concept of the reflex in studies of instinct. The process of 'conditioning' is quite definitely not separable from a genuine learning process. Even if the following element of behaviour is of an extremely simple and undoubtedly reflex type, as was salivation in Pavlov's dogs, it is nevertheless basically insupportable to assume that the process of acquisition must be open to an equally mechanistic interpretation. In some cases, such an assumption doubtless seems justifiable, as for example with the surprising fact that the human pupil reflex can be 'conditioned' to a specific sound. It is quite certain, however, that it is a false generalization to assume that in many of the experiments conducted with dogs there is no participation of very many higher, more complex and more conscious processes. The expression 'conditioned reflex' leads one to overlook the importance and complexity of these processes. If we intend to make further use of the terms 'learning' and 'training', then we must consistently refer to the acquisition process involved in the conditioned reflex as a learning or training process or we must apply the term 'conditioning', instead of 'learning' or 'training', to the more highly-differentiated processes, as is in fact done by English-speaking behaviourists. I can see no obstacle to referring to 'reflex-conditioning intercalation' and I believe that this distinction could be advantageous to both the reflex concept and the concept of conditioning. Pavlov's own neglect of the duplex nature of the processes involved is presumably to be explained by the intentional avoidance of the psychological approach which was so typical of that author. It is, however, impossible to understand why Pavlov overlooked the virtually suicidal extension of the reflex concept which arose from neglect of the two separate components involved.[97]

Before leaving the question of intercalation patterns, I must give consideration to a peculiar acquisition process which operates to fill out predetermined 'gaps' in specific instinctive behaviour patterns, which incorporate a specific capacity to acquire the missing component. In birds, many of the instinctive behaviour patterns *oriented towards conspecifics* incorporate this feature. It has been known for some time that social birds reared in isolation from conspecifics will fixate their social innate behaviour patterns (in the widest sense) upon some object in their surroundings. This is usually the human foster-parent or some other living organism, but it can be an inanimate object if the former are not available. Subsequently, such birds do not respond in any way to their true conspecifics.

The process of fixation of the object of an instinctive behaviour pattern adapted for interaction with conspecifics differs in a number of highly-important points from those true learning processes which we encounter in the filling of gaps in instinct-conditioning interpolation patterns. These peculiarities have drawn me to introduce a specific term for this fixation process; in an earlier article, I referred to the process as 'imprinting'.

In the first place, this acquisition process lacks all of the basic features of conditioning. The animal does not behave according to the principle of trial and error, as it does when acquiring an instinct-conditioning intercalation component, and there is no guidance towards the correct behaviour by reward and punishment. Instead, exposure to specific stimuli during an extremely limited period operates to determine the entire later behaviour of the animal, without (and this is a vital feature!) such behaviour necessarily undergoing practice at the time of operation of the stimuli. This latter feature definitely excludes the possibility of learning. In cases where there is a long interval between the stimulus-determined object-fixation and the performance of the instinctive behaviour pattern this is particularly evident. My previous observations indicate that in the jackdaw (*Coloeus mondula spermologus*) the object of sexual instinctive behaviour patterns is determined during the nest-phase of the young bird. Young jackdaws which are adopted by human beings at the time of fledging will typically re-orient the behaviour patterns normally directed towards the parents towards the human foster-parent, but this will no longer happen with their sexual behaviour patterns. The latter are only re-oriented to man when the jackdaw is adopted at a much earlier age. A Muscovy drake (*Cairina moschata*), which had been brooded with four siblings by a pair of Greylag geese and led for about seven weeks, subsequently showed itself to be attached to its siblings, and therefore to conspecifics, in its social responses. However, as its copulatory responses emerged in the following year, they proved to be fixated upon the species of the foster-parents, which had been ignored for more than ten months.

A second peculiarity of the acquisition process concerned is that it is correlated with a quite specific developmental stage of the young animal, as has been found with the jackdaw and the Muscovy duck. More exact statements about the duration of the sensitive period can be made for object-acquisition in the following response of some young nidifugous birds. In young mallard (*Anas platyrhynchos*), pheasant (*Phasianus*) and partridge (*Perdix*), the sensitive period of object-fixation only lasts a few hours, beginning shortly after the chicks have dried.[98]

The third special peculiarity of object-imprinting of instinctive behaviour patterns directed towards conspecifics is its irreversibility. After passage of the physiological period of acquisition, the relationship of the animal to the object of its behaviour is the same as if it were innately determined. As far as we know at present, this relationship cannot be *forgotten*, whereas the capacity to forget is an intrinsic feature of true learning, as has been particularly emphasized by Bühler. If the capacity to forget were absent, the possibility of alteration would of course be excluded. In the light of the relatively tender age of our knowledge, it is strictly speaking impermissible to regard the irreversibility of the imprinting process as proven. The basis for this assumption is provided by a limited number of observations, largely collected by chance; but these, in fact, point without exception in the same direction.[99]

The emphasis upon these three peculiarities of the imprinting process itself indicates the parallels which I am attempting to pinpoint: The influence of living conspecific material, the restriction to a narrowly-delimited ontogenetic phase and the irreversibility of the entire process are three features which distinctly separate imprinting from learning and indicate a far from insignificant parallel to processes of acquisiton known from the mechanics of embryogeny. One is greatly tempted to apply the terminology of developmental mechanics and to speak of *determination* of the object of the instinctive behaviour pattern by *induction*.

However far one may wish to follow these analogies, they do show that in the ontogeny of the instinctive behaviour pattern factors operate which are very similar to those operating in the ontogenetic development of *organs* and are in any case more similar to these than the factors operating in the development of intelligent behaviour. *In this case, too, the instinctive behaviour pattern has the same properties as an organ, and it is this one fact which is important in drawing all of these parallels.*

With this, I have said virtually all that we know about individual variability of instinctive behaviour patterns and about the relationships of this variability to experience and insight. I believe that I am justified in saying that *all* of the observational facts that we have so far accumulated speak against adaptive modification of the instinctive behaviour pattern by experience and through insight on the part of the individual animal.

We now arrive at the second major tenet of the Spencer/Lloyd Morgan theory – the assumption that further differentiation of the instinctive behaviour pattern has led to acquired and insight-controlled behaviour in smooth progression *in phylogeny*. I should now like to attempt to throw some light on the few facts which are possibly suited to providing answers to the following questions: (1) How does the instinctive behaviour pattern as such behave in phylogeny? (2) What are the phylogenetic relationships linking instinctive to acquired and insight-controlled behaviour patterns?

In the attempt to reconstruct the phylogenetic origin of an instinctive behaviour pattern, we are dependent upon sources of knowledge other than those involved in the study of the phylogeny of an organ. Palaeontology tells us nothing and ontogenetic recapitulation of ancestral forms is scarcely ever indicated. Nevertheless, there are a few such cases. For example, with pipits, larks, Corvids and some other Passeres, which walk on the ground by placing one foot in front of the other instead of hopping, we presume that the motor pattern involved is a secondary acquisition and that it does not represent a primitive form of behaviour in comparison with the two-legged hopping motion of the vast majority of passerines. Since in the Class Aves walking must necessarily be regarded as the more primitive mode of locomotion, it is an important source of support for the former interpretation that freshly-fledged pipits, larks and Corvids at first hop in the two-legged fashion typical of other passerines, prior to maturation of the motor co-ordination pattern of walking.

In individual cases, the behaviour of hybrids provides conclusions about the process of phylogeny of instinctive behaviour patterns. We know that hybrids are frequently not intermediate between the two parent species, but that both the instinctive behaviour and some morphological characters exhibit a throw-back to phylogenetically older forms. For example, Heinroth was able to show that a pair of hybrids between a shelduck and an Egyptian Goose exhibited in the mating ceremonies a type of behaviour corresponding to the widely-represented Anatid form (i.e. presumably the phylogenetically older type), although both of the parental species exhibit entirely different, highly complex and mutually divergent pre-mating ceremonies.

In general, however, we are dependent upon the relationships of instinctive behaviour patterns within the zoological system in investigating their phylogeny. We are thus faced with a field of study whose enormous size roughly corresponds to that of comparative anatomy. This field is virtually untouched. As far as I know, there are four papers – two by Heinroth, one by Whitman and one by Kramer – which have been exclusively concerned with the task of systematically examining the behaviour of instinctive behaviour patterns within a selected group of related species. Verwey, in addition, has investigated the relationships of particular instinctive behaviour patterns in the various herons in his well-known paper on the common heron.[100] Despite the pitiably small extent of this literature in comparison with the enormous size of the field involved, it has produced one uniform conclusion which is of great importance to us: It has been clearly demonstrated that every instinctive behaviour pattern which could be traced through a section of the zoological system, large or small, could be employed as a taxonomic character just as reliably as the external form of a given skeletal element or some other organ. In the study of groups whose internal systematic relationships are well known on other grounds, it actually emerges that in many cases a particular instinctive behaviour pattern may prove to be a *particularly conservative* character, in that it is evident in the same form in a *wider* range of species than a given morphological organ. In many of the larger taxonomic groups, there is no single organ, in fact not even a specific combination of developmental patterns of several organs, which can be found without exception in all members of the group concerned, whereas it often occurs that a given instinctive behaviour pattern is literally found in every species of the group. The following diagnosis of the relatively uniform Order of the Columbae is taken from a modern zoology textbook: Carinate, nidicolous birds, with a weak beak (expanded in the region of the nostrils), tapering wings of medium length and short legs with a squatting or cleft foot. Not a single one of these characters is consistent in the Order. The Crowned Pigeon (*Goura*) is not nidicolous; *Didunculus*, has a beak of aberrant form; short, round (in fact, typically gallinaceous) wings are found in *Goura*; and there is a number of terrestrial species which are characterized by quite long legs. Thus, not even a combination of organ characteristics provides a diagnosis of the group without exceptions.

If, on the other hand, we were to classify the Columbae on the basis that the male broods the eggs from the early morning to late afternoon, whilst the female sits on the eggs for the rest of the day, there would be no member of the Order which exhibits an exception to this behaviour. Further, I know of no other Order of birds which might give rise to confusion because of similar nest-relief behaviour. It is far from my intention to suggest a systematic scheme based solely on instinctive behaviour patterns as taxonomic characters; I merely wish to show that the instinctive behaviour pattern must be recognized as *one* type of taxonomic character.

This applies particularly to the extremely distinctive group of instinctive behaviour patterns whose function is the *elicitation of social responses* from conspecifics. If these releasing behaviour patterns are subjected to special comparative study within a systematic category of some size, it emerges that they represent group characteristics which are more constant and less prone to change than other instinctive behaviour patterns. Apparently, this is due to the fact that the elicitatory pattern (releaser) and the elicited behavioural response together represent within a species, an 'arrangement' which is particularly independent from environmental factors. The contrasting development of tail-wagging in Canid carnivores as an appeasing signal and of a similar movement in Felids as a threat gesture is a pure effect of 'convention' established between the releaser and the innate schema of the animal species concerned. As far as the function is concerned, the 'arrangements' could just as well have been the reverse – the actual function does not determine the specific form of the arrangement. Since the specific form of the releaser is determined purely historically, like a cipher or morse alphabet, similarity between two elicitatory behaviour patterns virtually always indicates *homology*. It is extremely improbable that similarity in the releasing ceremonies of two different groups of animals would ever arise by convergence.[101] This interpretation is supported by the fact that in the comparative study of sizeable groups we often find such complete and evidently cohesive evolutionary series that genetic relationships are indicated as convincingly as in any of the evolutionary series which I have encountered in comparative anatomy. The possibility of definitely excluding convergent evolution in many cases permits the comparative student of instinct to make pronouncements about genetic relationships with a degree of certainty seldom justified in morphological studies of phylogeny.

In the discussion of variations in intensity of instinctive responses (p. 265), it has already been indicated that incipient behaviour patterns – actually nothing more than specific, incompletely-performed behaviour patterns – can attain a secondary significance in that in social species they can transmit motivational effects from one individual to another. In such cases, the primary feature is without doubt the development of instinctive 'understanding', of 'resonance' with the intention movements of the conspecific. In this way, an originally functionless intention movement can acquire a novel function. Apparently, on this basis higher differentiation of the intention movement can lead to development of an elicitatory behaviour pattern. In parallel with this process of development of the stimulus-transmitting factors, the resonance-motivation is differentiated to give a more narrowly-defined releasing schema – a receptor correlate corresponding to the specialized releaser pattern in an amazing wealth of detail.

In all probability, the responses which ensure simultaneous take-off of acquainted individuals in some social Anatid species have passed through a developmental process of this kind. The releasers which pave the way for take-off in the mallard can be at once recognized as the operation of a take-off response which is curtailed at the last second. Even a bird-watcher who is not aware of the special releaser function of these patterns gathers that the bird will take off sooner or later. The head and thorax are thrust briefly upwards from the lowered body, just as in the actual take-off leap from the ground – though there is an interesting difference. In the Greylag goose, on the other hand, the movement is far removed from the original form of the intention movement and would not betray the take-off motivation of the bird to anybody not familiar with its significance. This movement consists of a peculiar lateral shaking of the beak, which looks as if the goose were attempting to remove water from its beak. However, the relationship with the intention movement of the mallard is rendered extremely probable by the behaviour of other Anatids which exhibit various intermediate forms of the movement. For example, the take-off movement of the Egyptian Goose (*Alopochen*) is restricted to the head, as is that of *Anser*; it is not lateral, however, but 'still' up-and-down in form, as in the ducks. The series formed is very impressive; anyone acquainted with the behaviour of *Alopochen* would without doubt immediately understand the movements of both *Anas* and *Anser*.[102]

In individual cases, the relationships between elicitatory instinctive behaviour patterns in the zoological system permit us to draw conclusions about their age, at least with respect to certain morphological structures in the species concerned. For example, a number of closely-related ducks of the genus *Anas*, including our endemic mallard, exhibit a specific social courtship ceremony with photographic identity of form. These species are by no means similar in coloration, but in the male sex of each species relationships can be determined between the conspicuous structures and colours and this courtship ceremony. All of the markings are located at points which are particularly emphasized in the movements common to all species. Since species *without* such colourful markings possess the same courtship motor patterns, however, I believe I can justifiably conclude that the instinctively determined *movements* of the ceremony are *older* than the structures and colours which support their elicitatory function in many species. In one case, we can even make an estimate of the possible absolute age of a given ceremony – the greeting ceremony which is exhibited in exactly the same form in all night herons of the genus *Nycticorax* and in the otherwise extremely aberrant South American heron *Cochlearius*. The structures supporting this ceremony (that is, elongated and peculiarly differentiated head feathers) are utterly different in *Nycticorax* and *Cochlearius*; but in both they are so formed that they operate in full along with the same motor co-ordination pattern. Since we can make certain assumptions about the age of the separation of the genus *Cochlearius* from the heron stock, we can in this case estimate the minimum geological age of a particular instinctive behaviour pattern.

I believe that, with the few examples quoted (unfortunately representing quite a considerable fraction of the entire knowledge actually available), I have shown that comparative studies of instinct must probably be initially conducted along the same lines as comparative anatomy; that is, as a descriptive science. Thus, we would first have to *collect* and *describe* instinctive behaviour patterns of a wide range of animals. The act of collection itself introduces the necessity for experimentation, without which we cannot determine whether a given behaviour pattern is innate and instinctive or not. Field observation usually tells us little about this, and this leads us to the necessity for keeping animals, indeed for keeping them under particularly good menagerie conditions, since the slightest physical detriment of the animal leads to extensive deficiencies in the range of instinctive behaviour patterns. Mere collection of knowledge about the instinctive behaviour patterns is therefore extremely demanding, and above all very expensive. Apart from this, the greatest difficulties are presented in giving an unambiguous description from which the behaviour pattern can be recognized with genuine reliability. The first requirement which emerges is that for a useful and uniform nomenclature. The expressions which have automatically arisen in discussions between those acquainted with animals are often revealingly reminiscent of those used in early morphological studies. Instinctive behaviour patterns, just like organs, are frequently referred to by the name of the original discoverer. For example, one discusses whether Verwey's 'snapping movement' in the heron is homologous with the similar response which I described for the Night Heron, and so on. In addition, one never describes a response with the formulation: 'This or that species generally behaves in this way', but always with the statement: 'This species *has* this or that response.' Even after the establishment of a practical nomenclature, mutual understanding of observations, particularly in comparisons of observations, is greatly hindered. It is a sad sight to see how completely serious investigators must stoop to vocal and balletic mimicry of animal behaviour in order to understand one another at all. There is of course only one way out of these difficulties – the photographic medium, where possible the cine-film.[103] I am in the act of planning a modest excursion in this direction, taking the described social courtship behaviour of ducks of the genus *Anas* as the object. I intend to study the extremely similar ceremonies of relatively distantly-related species (and of hybrids between them) and to preserve the observations in film in order to provide support for my oft-attacked statements about the homology of instinctive behaviour patterns. In fact, I am of the opinion that intermediate behaviour exhibited by hybrids speaks for genuine homology between the two instinctive behaviour patterns of the parental species.[104]

If we survey the whole range of available facts regarding the relationships of instinctive behaviour patterns in the zoological system, we must conclude that they demonstrate a parallel between the evolution of instinctive behaviour patterns and the evolution of organs, just as was the case with the facts that we were able to assemble about the ontogeny of instinctive behaviour patterns. This is in full agreement with Whitman, who had already stated in 1889 that in phylogeny instinctive behaviour patterns are developed according to the same laws and *within the same time-spans* as organs. We do not know what factors govern the phylogenetic development of organs and instinctive behaviour patterns. But we can draw one conclusion: There is no justification whatsoever for including individual experience among these factors.

A quite different question, separate from the mutual relationships of instinctive behaviour patterns in phylogeny, concerns the relationship between these patterns and acquired or insight-controlled behaviour. The interpretation suggested by Spencer, that the highly-complex and greatly-differentiated instinctive behaviour patterns lead directly on to variable behaviour patterns, has been energetically countered above.

On asking after the basis of the interpretation that the instinctive behaviour pattern is the phylogenetic precursor of the acquired or insight-controlled behaviour pattern, we only find the one fact that in the higher vertebrates species with higher intelligence have doubtless evolved from others which were instead equipped with highly-differentiated instinctive behaviour patterns. This phenomenon is, however, completely restricted to the vertebrates. Even a quite superficial survey of the zoological system forces the conclusion that there are no relationships, which can be expressed in a sentence of such simplicity, between higher specialization of instinctive behaviour patterns and the evolution of the capacity to exhibit learning and intelligent control of behaviour. At the most, it could be said that an inverse proportionality can be demonstrated in the development of the two types of behaviour, but this itself only applies to extreme cases, such as the colony-building insects on the one hand and the Anthropoids on the other. One reliable statement which can be made about species particularly highly differentiated in one direction or another is the following: Extensive development and specialization of the instinctive behaviour patterns *inhibits* further development of the variable behaviour patterns, and, conversely, the development of the latter must be preceded by extensive reduction of the instinctive behaviour patterns. In higher vertebrates, the development of powers of intelligence has certainly taken place in parallel with a corresponding reduction of the instinctive behaviour patterns, and the functional replacement of instinctively-determined behaviour patterns by plastic purposive behaviour easily leads to the assumption that the latter has evolved directly from the former. If we consider the insects instead of the vertebrates, however, and use the same one-sided approach as Spencer, then we come to exactly the opposite conclusions, since with these animals species with highly-specialized systems of instinctive behaviour patterns have certainly evolved from others in which the variability of behaviour was greater. The conditioning ability exhibited by the domestic roach is not less than that of a bee and in fact probably exceeds the latter in some respects. If we were to repeat Spencer's analysis here, we would reach the conclusion that instinctive behaviour patterns have evolved from acquired and insight-controlled behaviour. This interpretation was, in fact, put forward by supporters of Lamarck: The reader is reminded of the 'habit theory' of instinct suggested by Romanes.

If, instead of comparing extreme forms such as insects and Anthropoids, we compare closely-related animals with almost identical instinctive behaviour patterns, it emerges that the capacity for learning and insight can be amazingly varied despite this similarity in instinctive behaviour. This variability in higher capabilities with entirely similar instinctive co-ordinations can be illustrated with the concealment response of two related Corvids. Ravens and jackdaws, which were placed in the same genus in earlier nomenclatures, have exactly the same instinctive motor co-ordination pattern for concealing food remnants. Nevertheless, the two species exhibit the following differences in the *employment* of these entirely similar motor acts: When a jackdaw has in its crop a food morsel for concealment, it will exhibit an appetite for a situation in which it can effect concealment, that is for location of a small cavity of some kind. In general, this appetitive behaviour is restricted to a mere orientation movement, which always takes place in the direction of the deepest and darkest of the available cavities and crevices. The jackdaw is not equipped to learn from experience that the function of the concealment response is not fulfilled if another jackdaw is allowed to watch the process of concealment. Nor does the jackdaw realize that certain sites which can only be reached on the wing are inaccessible to human friends and that objects hidden in such places are safe from confiscation. The raven, on the other hand, can at an early age grasp the fact that the concealment response only leads to later recovery of the food morsel when nobody observes the concealment. Similarly, repeated removal of the hidden objects by the human foster-parent is sufficient to induce the raven to hide its morsels only in elevated places inaccessible to human beings. The motor co-ordination patterns themselves, however, are no less fixed in the raven than in the jackdaw. In an experiment, it can easily be shown that even in the raven the concealment response is an adequate goal in itself, since this response will be performed to excess in a non-adaptive and non-functional manner under the requisite conditions of captivity, just as by the jackdaw. It can also be demonstrated that the raven has no insight into the nature of 'concealment', to the extent that the concealed object should be rendered invisible.

The difference in behaviour between the jackdaw and the raven is thus restricted to the components of the behavioural sequence which we have termed appetitive behaviour, in this case a specific orientation movement which is incorporated in an instinctive behavioural sequence. We find two functionally extremely different intercalation patterns in which the instinctive, innate components are *absolutely similar* in the two birds. It is the incorporated non-instinctive behavioural components which have undergone an extensive change. These, derived from a simple orienting movement with the appearance of a mere taxis, have developed as acquired or even insight-controlled behavioural components (assuming that we are justified in assuming that the behaviour of the jackdaw is to be regarded as primitive with respect to that of the raven). Whereas we must postulate an extremely slow tempo for the phylogenetic variation of the instinctive behaviour pattern, according to all that we know of the relationships in the zoological system – exactly as for the variability of a given morphological organ of a particularly conservative nature – the capacity for intelligent behaviour appears quite *abruptly* in the system. This abruptness of the evolution of intelligent capacities, to which man owes his extreme distinction from his closest zoological relatives, is repeatedly found, though to a lesser extent, in the animal kingdom. One could name an enormous number of examples in which zoologically closely-related species exhibit a surprising degree of variability in their capacities for acquired and insight-controlled behaviour, just as has been described with the jackdaw and the raven.

Since, as we have seen, some authors interpret the concept of instinctive behaviour patterns so widely that appetitive behaviour is simply included as a part of instinctive behaviour, it is only a logical extension of their views when they maintain that higher, intelligent behaviour evolves from what they regard as the instinctive behaviour pattern. All the same, one is surprised at the omission of the observation that it is only the appetitive behavioural components which are comparable with acquired and insight-controlled behaviour patterns and which must be regarded as their forerunners. At least, the omission of this observation is surprising in the works of Craig, who expressly separates the instinctive behaviour pattern into the appetitive behavioural component and the consummatory act.

This observation, which would presumably raise no objection from Professor Craig himself, will now be developed with all possible emphasis. I believe that every piece of behaviour in animals must be fundamentally divisible into appetitive behaviour and the performance of an instinctive act, as long as the behaviour represents an entire functional unit. Attainment of the biologically-adaptive goal remains a common attribute of the two types of behaviour. Either one or the other can undergo further specialization, and within a given behaviour pattern specialization of one component frequently occurs at the expense of the other, leading to complete disappearance of the latter in some cases. Animals with extremely specialized instinctive behaviour patterns, for example bees, are born into the releasing stimulus situation in such a way that with many of the instinctive behaviour patterns involved we see nothing of appetitive behaviour necessary for attainment of the situation appropriate to elicitation of the responses concerned. In the opposite extreme case, the instinctive behaviour pattern whose performance represents the subjective goal of the behavioural sequence can be so constricted towards one end of the sequence that the entire motor activity developed to perform an adaptive function is left open to purposive control. The greater the intelligent capacities of a given animal species, the greater the extent to which the goal can be left open to purposive behaviour, until eventually the typically instinctive conclusive act of the behaviour pattern is restricted to an emotional or motivational response to a situation. In a weaver-bird, advanced intelligent capacities are just sufficient to bring about the stimulus situation in which the highly-specialized instinctive acts of nest-building are elicited. In this animal, the appetitive behaviour involved in nest-building is predominantly restricted to the attainment of this stimulus situation, in which the presence of suitable nest-building materials, a forked branch and so on plays its part. (That is, leaving aside a few orienting movements which occur later.) A human being in an approximately parallel situation performs the entire activity of home foundation through purposive behaviour. The instinctively-determined conclusion to his sequence of behaviour is the emotionally-based situation of being at home and secure. Bühler assumes that emotional influence upon such consummatory situations increases with decreasing representation of instinctively-determined motor activity, and that the magnification of the task left open to appetitive behaviour is compensated by an intensification of the underlying motivation.

I consider it as an *important* characteristic of the instinctive behaviour pattern that it can achieve results which are beyond the intelligent capacities of the animal species concerned. For this reason alone, it would seem to be impossible for an animal to improve its own instinctive behaviour patterns through learning or insight. In practice, we are unable to decide absolutely whether an instinctive behaviour pattern is fundamentally unmodifiable by learning or insight. We can only observe that such adaptive modification does not occur in any animal because the solutions to the problems set in the environment of the animal achieved by instinctive behaviour patterns are *always* far in excess of the intelligent capacities of the species. Apparently, the ability to solve a problem by learning or insight is never present *alongside* an instinctive motor co-ordination pattern which solves the same problem. The basis of this fact lies, in all probability, in the phenomenon that *whenever* the ability for conditioned or insight-controlled solution of a problem arises in the phylogeny of an animal species, the solution must always be *much more favourable* for the survival of the species (as a result of adaptive plasticity) than the solution to the same problem provided by fixed instinctive behaviour patterns. This probably represents the main reason for the reduction of instinctive behaviour patterns in species of advanced intelligence.

The presence of an instinctive behaviour pattern would similarly seem to inhibit the development of a conditioned or insight-controlled response with the same function. This is at least the case with human beings. One only needs to consider the behaviour of eminent people (otherwise equipped with a great ability for self-criticism) when exhibiting the doubtless instinctive response of mate-selection through 'falling in love' for the correctness of this statement to become obvious. The example of the jackdaw and the raven quoted earlier does actually demonstrate that further development of intelligent powers is possible within certain limits without reducing the innate, instinctive components of a behaviour pattern. However, with further developments of this kind the instinctive patterns must doubtless give way.

I conceive of this process of reduction as predominantly based upon the emergence of *new gaps* with incorporated appetitive behaviour *within* pre-existing, purely instinctive patterns and the instinctively-determined sections of intercalation patterns.[106] A similar view was expressed by Whitman, who had the following to say about the easily demonstrable reduction of instinctive behaviour patterns in domestic animals:

> In non-domesticated species, a high degree of invariability must be maintained with respect to the instincts, whereas they are *reduced* [my italics] to varying degrees of variability in domesticated species, such that the latter exhibit a correspondingly wider freedom of behaviour, though this naturally entails a simultaneous increase in the probability of occurrence of irregularities and so-called 'errors'. In my opinion, these 'instinct errors', far from representing the first signs of psychological degeneration, in fact provide the first indications of increasing plasticity in innate motor co-ordination patterns.

At another point, Whitman stated that intelligence exhibits the tendency to 'break-up' instinctive action, and at yet another juncture he made the following statement: 'Plasticity of the instinct is not in itself intelligence, but it represents the open door through which the great teacher experience can make her entrance to bring about all the wonders of intelligent behaviour.' If we replace the expression 'plasticity' with our more narrowly-defined concept of an 'ability to acquire' incorporated within a sequence of instinctive motor co-ordination patterns, we can then give full agreement to Whitman's interpretation, as set out in the above statements.

We must not overlook the fact that the degeneration of the instinctive part of any animal (and presumably human) behaviour pattern will stop at a specific point. The reader is reminded of the statement made by Craig (quoted on p. 277), that the end of a behavioural sequence is always instinctive. In a great number of cases, even in man, we see preservation of motor activities of an instinctive nature representing the goal of the correlated appetitive behaviour leading to attainment of the appropriate releasing stimulus situations. I call to mind the fact that the most 'appetizing' meals are obviously those which transmit particularly intensive elicitatory stimuli for salivation, chewing or swallowing. Some dishes are regarded as particular delicacies although they only elicit one of these responses, since their effect is very pronounced. Take, for example, the easily-swallowed oyster or certain, almost tasteless pastries which strongly elicit chewing because of their especially 'crispy' consistency. In other cases, as mentioned on p. 289, instinctive motor patterns can disappear entirely, the instinctive goal of the behaviour can be reduced to an emotionally-governed (and therefore attractive) stimulus situation in which no further behaviour patterns are elicited.[107]

In the solution of a specific problem set in the environment of an animal, appetitive behaviour and performance of instinctive patterns are *mutually substitutive* to the extent that greater participation of one type of behaviour will always automatically lead to relief – and thus to reduction – of the other in the performance of a given function. In this way, particularly pronounced differentiation of one or the other type provides a specific evolutionary possibility and *direction* for the animal's behaviour pattern. At extreme levels, specialization in one of these directions is almost definitely irreversible and will exclude a later development in the other direction. At the same time, behaviour patterns already differentiated in the other direction presumably undergo *reduction*.

Thus, our conclusion regarding the phylogenetic relationship between instinctive behaviour patterns and purposive behaviour conflicts with the interpretation taken by the Spencer/Lloyd Morgan school, just as was the case with our conclusions about the effects of individual experience on the ontogenetic development of instinctive behaviour patterns.

II McDougall's theory of instinct

McDougall's theory of instinct agrees with that of Spencer and Lloyd Morgan to the extent that it includes the assumption that there are all possible continuous transitions between the instinctive behaviour pattern on one hand and learnt or insight-controlled behaviour on the other. McDougall presents a particularly broad interpretation of the concept of instinct. Every behaviour pattern which incorporates the slightest instinctive feature is at once classified as an instinctive pattern. It is thus understandable that the instinctive behaviour pattern is essentially interpreted as purposive behaviour.

The foremost feature of McDougall's theory, however, is the assumption of governing 'instincts', which employ subordinate 'motor mechanisms' as *a means to an end*. In America, where the use of the term 'instinct' has recently fallen from fashion, the terms 'first-order drives' and 'second-order drives' are used in the same, or very similar, sense as the terms 'instinct' and 'motor mechanism' as used by McDougall. McDougall and a number of more recent authors regard the relationship between the two, which is assumed to be based on the employment of the innate motor co-ordination pattern as a means to an end in the service of a governing instinct directed towards a particular goal, as proof of genuine purposivity in the first-order instinct.

McDougall groups the instinctive behaviour patterns of man and animals, according to purely functional considerations, into thirteen (!) governing instincts. He takes little interest in considering the phylogeny of the instinctive behaviour patterns and their relationships within the zoological system or in considering the fact that functionally analogous instinctive behaviour patterns can arise independently of one another in different animal stocks. The phenomena of homology, which are so important for our approach, and for comparative zoological considerations in general, carry no weight with McDougall, and he therefore regards function not only as a classificatory principle but as the very essence of instinct. For this reason, it is never stated that the instinctive behaviour patterns of animals and man can be divided into a particular number of functional groups; the existence of thirteen instincts is laid down in a fairly dogmatic fashion.

Let us first take a closer look at the concept of governing and subordinate instincts. For example, McDougall assumes that there is a 'parental instinct' which employs as a means to an end all of the individual, innate instinctive motor co-ordination patterns which are concerned in the care of the young in a given animal species. It has already been explained, in the discussion of intercalation patterns of instinctive and purposive behaviour, how the individual instinctive innate component functions and the appetitive behaviour oriented towards them represent one functional unit in intricate correlation with one another under natural conditions. But we have also seen above how easily this unity can be completely disrupted by the failure of an apparently quite unimportant component of the behavioural chain. A further example of this is the following: I was able to demonstrate experimentally with female ducks leading offspring that the various parental responses which are exhibited towards the offspring are completely independent in their elicitation and are only combined to give a programmed functional unit through the unification of the appropriate releasing characters in the conspecific ducklings. The unity is at once disrupted if these releasing characters are presented through *separate* objects. A Muscovy duck (*Cairina moschata*) will defend a mallard duckling just like a conspecific duckling and yet treat it with utter enmity after 'courageously saving' it from the hands of the experimenter. This behaviour can be explained on the grounds that the distress-call of the duckling, which elicits the defence response in a reflex-like manner, is virtually the same in mallard and Muscovy ducklings, whereas the species-typical head and back markings, upon which other parental responses depend, are markedly different in the offspring of the two species. In my opinion, the fact that the functional unity of a component functional system supposedly governed by one 'parental instinct' can be disrupted by the absence of a minor morphological character is proof of the autonomy and equivalence of the individual behaviour patterns involved. *We would obviously be justified in postulating the existence of a unifying, directing instinct governing all component responses only if we were able to observe a regulative factor operating over and above the experimentally-demonstrable regulative capacity of the individual responses.* I have never observed the operation of such a first-order instinct to compensate for disturbances in the co-operation of the component responses and to restore unity through co-ordination of the individual responses. I believe that we are not justified in assuming that such an effect exists. Our interpretation, that a large number of autonomous individual responses are only correlated to give a functional unit by means of the phylogenetically-developed structural and functional plan of the species, seems far-fetched to one who is only acquainted with instinctive behaviour patterns in their normal adaptive performance and is not aware of the easily procured experimental breakdowns which so strongly support the assumptions set out here.

McDougall speaks of 'an instinct' when a system of species-specific instinctive behaviour patterns seems to be combined to a unit through common function. We can without doubt carry out such a classification according to purely functional principles, referring, for example, to all instinctive behaviour patterns involved in care of the young as 'parental instincts'. However, there is no place for applying the *singular* of this term. All of the instinctive behaviour patterns defined as such here would belong under the heading of 'motor mechanisms' in McDougall's scheme.

To interpret any one of the instinctive behaviour patterns involved in a functionally unitary behavioural sequence of some length as subordinate to or executive upon another pattern is only possible from a quite specific viewpoint. This viewpoint, however, is (as far as I know) at no point taken up by McDougall – it is rather to be attributed to Wallace Craig. In order to illustrate this with an example, let us assume that a blackbird, following a long period of rest, exhibits an appetite for the appropriate stimulus situation for elicitation of the co-ordination patterns of food-seeking. The blackbird will consequently exhibit purposive behaviour directed at the attainment of a situation in which the pattern of boring for worms (common to many birds of the thrush group) will be successfully performed. This purposive behaviour will itself incorporate a wide range of instinctive motor co-ordination patterns, such as those of walking, hopping, flying and so on. As a whole, this behaviour – including the ultimate goal response of boring for earthworms – provides a typical case of an intercalation sequence. We are driven to conclude that there is an entire range of instinctive behaviour patterns, which normally occur almost exclusively as components of intercalation sequences, whose purposive goal is represented by the performance of *another* instinctive behaviour pattern.[108] They therefore do not normally represent the goal of appetitive behaviour directed solely towards them. Such behaviour patterns are usually 'simple' motor co-ordinations, particularly the various types of locomotion, the patterns of visual control, grasping, pecking and so on. These motor co-ordination patterns actually operate like *tools*, like *organs*, which can be employed by the animal for various purposes. But it must be particularly emphasized that these patterns are not employed as the means by a 'governing' instinct; they are *tools of purposive behaviour*, even if the goal of the appetitive behaviour is the performance of a given instinctive behaviour pattern. It is extremely characteristic of these 'tool responses' that they can be employed, like organs, in the service of *different* forms of appetitive behaviour, without a consequent change in their own form. Just as the beak of a bird shows no 'adaptive variability' in its form in the wide range of its modes of use in food-seeking, fighting, nest-building and so on, these behaviour patterns never exhibit a change in the innate co-ordination of motor activity. When a fly-catcher has to feed its offspring, it performs the motor patterns of its usual fly-hunt 'in order to' permit performance of the parental feeding response instead of the otherwise normal eating response. But the motor co-ordination pattern employed is the same in both cases. The uniformity of the motor pattern and its resistance to effects of learning is illustrated particularly well by the following example: A female canary, which I had not set up for breeding, performed the motor patterns of securing the nest foundation with the green fodder which was provided. In performing these patterns, the canary placed one foot on the stem involved in weaving and manipulated the free, projecting end with the beak until it was wound around the branch and thus secured. Since edible green fodder was involved, this bird soon learnt to hold the stems still with one foot and to bite off pieces. Thus, the bird learnt to make use of the motor pattern for another appetitive category (eating), although it was actually biologically adapted for the purpose of nest-building and in this context does not at all give the impression of being a 'tool response'. The motor patterns which the canary performed as a result were extremely similar to those of a tit, a raven or any other passerines exhibiting stabilization of a food morsel with one foot as a species-typical instinctive behaviour pattern. Interestingly, this female canary was only 'able to' employ this acquired stabilization of the food with one foot whilst it was in physiological condition for nest-building. As summer approached, this ability was lost, although the opportunity for performance of the behaviour was in fact deliberately presented in the form of large salad leaves. At this time the bird just did not 'possess' the response of treading upon an object and its intelligence was insufficient to permit reconstruction of this 'tool response', although it had been sufficient to permit employment of the motor pattern for purposes other than its biological goal while it was available. There is a considerable difference between the acquisition of a new application for an inherited tool and the spontaneous development of a novel tool.

These observations of the female canary distinctly show that even instinctive behaviour patterns which under conditions natural for the species would definitely not be interpreted as 'tool responses' or 'second-order instincts' can occasionally acquire a novel application through learning and be subordinate to another appetitive category. Conversely, instinctive behaviour patterns which are normally almost exclusively performed in the service of an appetite for performance of another pattern can under experimental conditions become a goal in themselves, thus representing the goal of a particular form of appetitive behaviour directed towards their elicitation. In order to be able to state that a given instinctive behaviour pattern is subordinate to another and is only performed for the purpose of performance of the latter, one must always carry out analysis of individual cases. The statement that a given response is generally speaking a subordinate instinctive behaviour pattern is erroneous from the outset, as I shall shortly demonstrate. It therefore seems extremely misleading to apply any term other than 'actual' instinctive behaviour pattern to these 'tool responses'. Tolman, for example, refers to what we have here termed a subordinate instinctive behaviour pattern as an 'innate skill' and distinguishes this from an instinct. A sharp distinction of this kind cannot be made, because responses which in their natural form serve as tools of other appetitive behaviour can also become independent goals, and because instinctive behaviour patterns which are normally completely autonomous can conversely be subordinated and employed as tools for a completely novel category of purposive behaviour (as was shown with the example of the female canary). The blackbird discussed in the earlier example will hop and fly under natural conditions only in order to move to a particular place, whereas a caged bird will hop and fly restlessly up and down in its cage. But it should by no means be thought that the abnormal conditions of captivity are necessary for the motor co-ordination patterns of locomotion to become a goal in themselves. Even after very slight relief from service under the yoke of other appetitive behaviour categories, it is obvious that these patterns themselves can come to form the goal of purposive behaviour. A high-spirited dog (or a healthy raven) cannot be driven to perform so much locomotor activity in the service of purposive behaviour that it will completely omit the independent performance of the motor patterns of running (or flying) in its spare time. If the purposive compulsion to perform such motor patterns is completely eliminated under experimental conditions, most animals will be seen to perform these patterns as vacuum activities almost as persistently and as frequently as under the pressure of the goal which must normally be reached. Under experimental conditions, typical tool responses will exhibit the highest degree of independence from the conclusive act which normally terminates the behavioural sequence. In the greylag goose, the motor patterns of grass-pulling and up-ending, whose performance occupies a large part of the daily pattern of activity, are completely independent of the biological goal (in both cases, the acquisition of food). If a greylag goose which has been grazing all day in the open is brought into a room in the evening, it will begin to perform the grazing patterns on all possible (and impossible) objects after only a few minutes have elapsed. One must see the intensity of the vacuum activity oneself in order to gain an impression of the truly elemental force with which even the simplest motor patterns of this type are driven to emergence. It is just as impressive when greylag geese in a flock perform the motor patterns of up-ending on an entirely barren lake just as persistently as if the geese were supplying their entire food requirements, although the only real source of food is to be found on the shore in the form of a food-bowl. One always hears from uninitiated observers astonished speculation as to what the birds might find in the clear water. Thus up-ending, just like grass-pulling movements, can occur as an utterly autonomous instinctive behaviour pattern. On the other hand, over and above the necessity to perform the response for its own sake, the same motor co-ordination patterns can be employed as a tool for attaining a stimulus situation appropriate to performance of another instinctive behaviour pattern. Thus, a greylag goose can just as well up-end *in order* to arrive at the stimulus situation of eating. To keep to the same example, a human observer seeing the goose first exhibit up-ending and then eat the dredged material is only too inclined to unhesitatingly make the latter assumption, for the simple reason that in *man* the responses of eating are instinctive, self-rewarding patterns whereas those of food-acquisition are generally not so. Taken exactly, this assumption does not apply without exception even to man. For example, my daughter at the age of five was quite moderate in eating berries when these were provided in a bowl, but she over-ate when left to herself among bilberry bushes. One can actually say that she did not pick the berries in order to eat them, but ate them in order to pick them, and she thus showed an almost inverse relationship between the responses of food-acquisition and those of food-intake as compared to the normal situation. I was able to observe quite similar behaviour in the greylag goose. If I sink plant food in water to a depth suitable for up-ending, it can happen that geese motivated for up-ending will fetch up the plants and (once they are in the beak) chew them and finally swallow the food, although the geese are in a state of satiation in which they would show no inclination to pick up and eat the same plants if they were offered in a bowl. This relationship between the responses of food-acquisition and food-intake, which might always be interpreted as an 'effect of captivity' in the greylag goose and man, is in fact the almost invariable norm for very many animals.[108] With very many predatory animals, for example, where highly-specialized instinctive behaviour patterns of food-acquisition are present, the latter are intensively self-rewarding and form the actual goal of the appetitive behaviour, while the eating responses themselves signify no more than a mechanical continuance of the initiated behavioural chain. Such animals frequently eat insufficiently in captivity because the stimulus situation which represents the real goal of the 'appetite' is entirely lacking. It is an utterly indefensible anthropomorphization to assume that the goal of the appetitive behaviour is represented by analogues of the behaviour patterns which form this goal in man. But it is just this assumption which (explicitly or implicitly) forms the basis of the considerations of almost all those authors who employ the concepts of governing and subordinate instincts. The analysis of the individual case is always lacking, and yet this alone would provide justification (in the particular case investigated) for the statement that one response is subordinate to another and is employed by the animal as a means to an end.

McDougall apparently does not recognize the necessity for such analysis, presumably mainly because from the outset he equates the biological function of the behaviour pattern with the goal of the pattern as set for the animal subject. The phenomena which we have just discussed are apparently completely unknown to McDougall, as is evident from the following – utterly erroneous – statement on p. 101 of his book *Outline of Psychology*: 'It is probable that every instinct is to some extent dependent upon appetite. A predator hunts only when it is hungry. A satiated cat will sometimes allow mice to play upon its tail.' The utter falseness of these statements is evident from what has already been said, and we can summarize the previous considerations to give the correct interpretation: A predator will not hunt when it has completely abreacted its hunting responses and is at the same time satiated. If the predator is hungry, and assuming that the species involved is equipped with fairly advanced intelligence, it will perform the hunting responses 'in order to eat' under certain circumstances, although it actually 'has no desire to hunt'. However, the 'desire to hunt' (i.e. the appetite for performance of the instinctive behaviour patterns concerned) emerges quite independently of the nutritional state of the predator. McDougall should have been aware that in a dog good nutrition (as long as it does not hinder the mobility of the animal through fat-deposition or the like) does not have the slightest influence on the motivation to hunt. An animal of lesser intelligence (e.g. a grebe) hunts only because of an appetite for performance of the response and will not perform the relevant responses at greater intensity when underfed. The animal eats, so to speak, only for the sake of hunting and will soon starve to death if the food is offered in such a way that performance of the species-typical instinctive behaviour pattern of prey capture is rendered impossible.

The thing that is most noticeably absent from the works of McDougall and many authors closely connected with him is the incorporation of specific statements about their conception of the relationship between the instinct and the instinctive behaviour pattern. Even with modern authors, who make use of the word 'drive', we are left in the dark on this point. It must be surprising to everyone that these research workers, who otherwise strive so energetically to limit all statements to objectively experienced matter, do not forego statements about 'the instinct' and instead make the animal *behaviour pattern* and its laws the sole object of their considerations.

Since all of the observational facts discussed above were not known to McDougall, or at least were not at all considered by him, the question arises as to what phenomena in fact provided the observational basis for the assumptions made by him about governing and subordinate instincts and (above all) for the assumption of a specific number of governing instincts. The unitary adaptive function of a group of instinctive behaviour patterns can, as we have seen, give no basis for these assumptions. With the same degree of justification, one could name many further functional categories – as is done, for example, in colloquial language when we refer to the 'instinct for preservation'. Similar concepts have been formed in psychoanalysis. But McDougall did not arrive at his concept of the directive instinct via this pathway; it would be a considerable underestimation of the importance of his work to assume that this were so. What probably led McDougall to assume the existence of a specific number of instincts was (in rough terms) the following: McDougall's great and everlasting contribution was in the recognition of the close relationships which exist between the instinctive behaviour pattern on the one hand and emotions on the other. (The

English expression 'emotion' roughly corresponds to a concept encompassing that of *Gefühle* and *Affekte* and it is best to translate this term into German through these two words.) McDougall sees in these subjective phenomena the experiential correlates of instincts and therefore deduces the number of man's instincts from the number of emotions which can be qualitatively distinguished from one another. Originally, the types of instinct so classified were intended to apply to mammals and man, but they are now widely treated as the only 'first-order instincts' present in the entire animal kingdom.

As regards the basic tenet that instinctive responses are accompanied by subjective phenomena and that each individual response is linked to a specific experiential correlate in the form of a specific emotion, all people with a good knowledge of animals are (consciously or unconsciously) followers of McDougall's theory. Verwey, for instance, defines the instinctive behaviour pattern as a reflex process which is 'accompanied by subjective phenomena'. This is a rather venturesome, but nevertheless excellent, definition of instinct. Heinroth refers to 'Stimmungen' (motivational states) of animals, correlated with specific types of stimulation, and indicates the relationship by making compound words from the name of the instinctive behaviour pattern and the word 'Stimmung' – giving terms such as 'Flugstimmung' (flight motivation), 'Nestbaustimmung' (nest-building motivation) and so on.

These extremely useful new terms provided by Heinroth themselves demonstrate the points at which the animal observer finds that McDougall's theory of instinct is inadequate. The number of types of governing instinct assumed by McDougall is derived from that related to the 'emotions' which can be distinguished in human beings, as has been explained. It is the extension of this conclusion to animals which introduces the very error which Heinroth adeptly avoids with his new terms. Since even the (apparently) least important components in a series of behaviour patterns of unitary function are just as important (as independent mosaic stones) for the function of the entirety as any of the other components and can be elicited quite autonomously, we must – for the sake of consistency – assume that they too are related to qualitatively distinct and independent emotions. In most cases, we must therefore attribute *far more* individual types of emotion to an animal than are known from human beings. In the case of man, we must assume that there has been a process of simplification and dedifferentiation equivalent to that demonstrated in the instinctive behaviour patterns. The terms laid down by McDougall, corresponding to the human emotional framework, are from the outset *too few in number* to describe the subjective processes of animals.

An example of this is the following: The Burmese jungle-fowl and the domestic chicken (*Gallus bankiva*) both possess two distinct warning calls uttered in response to the sight of a flying raptor or to a flightless terrestrial predator, respectively. If the degree of arousal following the raptor warning call reaches a high intensity level, a downwards escape response follows, with the bird heading downwards to the ground and (where possible) *beneath* cover of some kind. This type of arousal is automatically coupled with upward glancing, and at the lowest intensity levels it is only this eye-movement which is elicited. On the other hand, the type of arousal corresponding to the other (terrestrial predator) warning call leads at high-intensity levels to take-off by the chicken, usually leading to refuge in a tree rather than a long-distance flight. We are quite definitely unjustified in assuming that there is a unitary experience for these two entirely-independent responses, unless we wish to fall into indefensible anthropomorphism. The expression 'fear' is certainly insufficient to characterize these two sharply-distinct types of arousal in the chicken. But with the collapse of the assumption of a unitary type of arousal, McDougall's own approach dictates the collapse of the assumption of a unitary 'instinct of escape' which was postulated by that author. As has already been stated, we can in fact speak of such instincts in the plural – but the singular of the term is a mere word without an associated concept.

McDougall's instinct terminology is very useful for classifying groups of responses on functional criteria, since instinctive behaviour patterns ensuring escape from danger, care of the young and so on are of course present in the vast majority of animals. Nevertheless, all of these terms introduce the danger that mere words will be interpreted as concepts as soon as it is overlooked that each of these terms may only be employed in the plural.

III The reflex theory of the instinctive behaviour pattern

The Spencer/Lloyd Morgan interpretations of the instinctive behaviour pattern contrast sharply with those of Ziegler. H. E. Ziegler says of the instinctive behaviour pattern: 'I have defined the difference between instinctive and intelligent behaviour patterns in the following way: the former are based upon innate paths and the latter upon individually acquired paths. The psychological definition is thus replaced with a histological conceptual framework.'

We will ignore the fact that this statement includes an interpretation of 'intelligent' behaviour patterns which does not readily meet with agreement. It incorporates an assertion which could possibly be set up as a working hypothesis, but which at present can be neither proved nor disproved. I should like to point out, however, that the interpretation of the instinctive behaviour pattern as a chain-reflex is almost exclusively confined to authors with radically mechanistic views. This is an interesting – and far from self-evident – fact. Non-mechanistic biologists generally oppose with some emotion the assumption that instinctive behaviour patterns might possibly be explained as reflex processes. McDougall, in particular, always argues against the reflex theory as if such an assumption were the equivalent to equating the organism with a machine. This attitude is explicable on the grounds that the authors rightly opposed by McDougall attempted to carry out this equation in an extremely dogmatic manner. Since the instinctive behaviour patterns quite definitely represent only a *part* of animal behaviour, it would not in fact lead to 'degradation' of the animal to a machine even if the instinctive patterns could be successfully explained on a reflex-physiological basis. This would be just as likely to brand the animal as a machine as (in crude terms) the recognition that a part of the organization of the human organism, such as the elbow-joint, can be explained fairly completely on a mechanical basis would lead to branding of human beings as machines. The fact that the reflex theory is represented by mechanists, whilst the other interpretations are supported by vitalists, has had an extremely unfavourable effect on the discussion.

Before we can tackle the problem of the extent to which the concept which we have formed of the instinctive behaviour pattern is compatible with that of a reflex, we must be quite clear about the meaning which we attribute to the term reflex. It would be erroneous to believe that the word reflex is always used just for one concept with the same content, even though in this case idiomatic confusion has not reached the level attained with the term 'instinct'. For purposes of comparison, it is doubtless advantageous to employ a reflex concept which is defined as *narrowly* as possible, since any equation which is accompanied by an *extension* of one of the two equated concepts is only a pseudo-solution of the problems involved. If we interpret the concept of the reflex as widely as Bechterew, who unhesitatingly terms any motor process occurring in a living organism as a reflex process – including the movements of protozoans, the tropisms of plants and even growth processes – the statement that the instinctive behaviour pattern is a reflex process would be undeniably correct in the formal sense, but it would also be completely valueless. It is self-evident that it must be possible to establish a concept governing both the instinctive behaviour pattern and the generally recognized concept of the reflex. However, if we simply apply the term reflex to this governing concept we are extending the reflex concept. Such a step can be criticized on the grounds that already established limitations of the original, narrower definition – based on reliable experimental results – are disrupted without presentation of substitute limitations. This disruption is definitely not compensated by the acquisition of a new classificatory principle. This criticism is directed exclusively against the general terminology employed by Bechterew. What he has to say specifically about the 'inherited organic reflexes' (his term for instinctive behaviour patterns) is in general terms quite noteworthy.

For the reasons discussed, we will employ in the following discussion of Ziegler's reflex theory the *narrowest* of the practicable reflex concepts. The term 'reflex' will be taken to apply only to a process which is based upon a 'reflex-arc' as its anatomical substrate; the latter consisting of a centripetal and a centrifugal pathway together with an intervening section (large or small) of the central nervous system, which serves as a 'reflex centre' for the transmission of excitation from the centripetal to the centrifugal arm of the reflex arc. Ziegler in fact interpreted instinctive behaviour patterns as composed of such reflex processes with an identifiable anatomical basis.

The most immediate objection to the reflex theory of the instinctive behaviour pattern in its most distinct form is derived from the *regulative phenomena* which can be demonstrated in many instinctive behaviour patterns, particularly the simplest among them. The reader is reminded of Bethe's experiments, which were largely performed for the express purpose of demonstrating the untenability of the theories of nervous pathways and centres. Bethe was able to demonstrate through amputation experiments that the locomotor co-ordination patterns of widely-different animals can exhibit a large number of regulative effects. In animals with many legs, such as crustaceans, the number of possible walking co-ordination patterns is so large that Bethe quite rightly points out the improbability of the assumption that each pattern is based on a specific 'pathway'. It can scarcely be assumed that a crayfish possesses a stored 'reserve pathway' to 'cope with the situation' arising from each of the experimentally-produced combinations of leg amputation! However, one must ask whether the emergence of these 'holistic' regulation processes must necessarily be based upon the response of the organism as a whole and whether one must entirely reject the possibility that highly-complicated systems of reflexes (in the narrowest sense) might be sufficient to explain the observed effects. A decerebrated frog, which of course cannot be regarded as an entire organism and whose responses are interpreted as pure reflex processes by all physiologists, still responds in an integrated fashion with its 'wiping reflex'. The hind leg is always accurately raised to the irritated spot, wherever the stimulus may be applied, and if the leg on the side of irritation is anchored the leg of the other side will be at once employed. Thus, the objections to interpretations based on pathway theories raised by Bethe with respect to the instinctive behaviour pattern also apply to processes which are generally interpreted as reflexes by physiologists.

Proof of an anatomical substrate in the form of a reflex arc has only been established with some accuracy in the very simplest cases. Reflexes whose nervous pathways have been exactly traced by neurologists (e.g the tendon reflex, the abdominal skin reflex, and so on) are in all cases phenomena which are only exactly known from their performance in isolation from the overall activity of the organism. Strictly speaking, then, the explanation according to pathway theory can only be regarded as definitely established in each special case. In fact there is every conceivable transitional stage between such extremely simple events, which are largely open to an explanation on the basis of the pathway theory, and 'reflexes' such as the wiping reflex of the frog, which undoubtedly possess the regulative ability typical of the instinctive behaviour pattern.

Consequently, the objection to the reflex theory of the instinctive behaviour pattern arising from regulative properties is just as applicable to the pathway explanation of many processes generally regarded as reflex patterns. Accordingly, regulative properties cannot be drawn into the attempt to demarcate the instinctive behaviour pattern from the reflex, unless one is inclined to describe processes such as the frog's wiping reflex as 'instinctive behaviour patterns', as distinct from reflexes. In fact, something can be said for a separation of this kind. Very few of the reflexes exactly understood on the basis of anatomical studies can be regarded as adaptive behaviour patterns when performed in experimental isolation. Whatever the extent of their significance to neurologists, they must be regarded by biologists as chance phenomena. Nevertheless they are so obviously closely related to adaptively important processes that, in spite of the present deficiences in dissecting the latter, no physiologist would seriously think of drawing a distinction between the two. Verworn in fact incorporated the adaptive value of the reflex as an important feature in his definition. He states: 'The nature of the reflex depends on the fact that an element perceiving the stimulus and an element responding to the stimulus in a *purposive* manner [my italics] are related to one another by a central junction, such that . . . etc.' The distinction between the instinctive behaviour pattern and the reflex discussed above would be equivalent to defining both concepts according to the *accessibility* of the anatomical relationships. This would be rather pointless, since each new result of analytical research might bring processes previously regarded as instinctive behaviour patterns under the umbrella of the reflex concept.

Finally, I should like to point out that the regulative properties described do *not* present an absolutely convincing argument against the *possibility* of explaining the instinctive behaviour pattern through pathway theory. It seems quite conceivable to construct mechanical models which are capable of imitating the regulative changes in locomotor co-ordination patterns observed by Bethe. A mechanical model of a system of reflexes of this kind would quite definitely have to achieve an extreme degree of structural complexity, but no supporter of the reflex theory acquainted with the facts would think of the instinctive behaviour pattern as anything other than the product of an extremely highly differentiated system of reflex pathways. This also applies to Ziegler. Thus, regulative effects do not, strictly speaking, demonstrate the invalidity of the pathway theory; they are only incompatible with the assumption that an instinctive behaviour pattern is laid down as a single pathway or a small number of pathways.

Although I am here defending the reflex theory of the instinctive behaviour pattern and even (with some reservations) presenting myself as a supporter of this theory, this does not mean that I regard this as a particularly promising working hypothesis. In the study of instinctive behaviour patterns, we rely upon factual evidence derived from a source quite different to that providing the knowledge for reflex physiology. There would seem to be little to be gained from research into the instinctive behaviour patterns using the methods of reflex physiologists, since the immediate loss of the more highly-differentiated instinctive behaviour patterns following the slightest degree of physical damage would make it completely impossible to reach any conclusions on the basis of absence of a function following a vivisectional operation.

An objection to the reflex theory of the instinctive behaviour pattern similar to that raised on the basis of regulative effects is based upon the *variations in intensity* in the performance of instinctive responses, discussed on p. 265. We have already seen that the observed form of one and the same response can vary considerably according to the intensity of arousal of the animal. Since these observed forms grade into one another in a smooth series, one can maintain that they are present in an infinite number. This infinite number can be taken as an obstacle to the assumption of an anatomical substrate for the performance of the response. On the other hand, these scales of intensity themselves are expressed in a compulsory (one is tempted to say 'machine-like') degree of regularity, such that one automatically and involuntarily employs physical comparisons, referring (for example) to 'arousal pressure' and so on.

Another peculiarity of the instinctive behaviour pattern which – without actually providing an objection to the reflex theory – cannot be explained from a reflex nature of instinctive patterns is the lowering of the threshold of the stimulus eliciting an instinctive behaviour pattern, discussed on p. 276. This lowering of the threshold, as we have seen, occurs when the stimuli normally necessary for elicitation of a response are lacking for some time. This eventually leads to stimulus-independent eruption of the instinctive behaviour pattern, a phenomenon which has been termed the 'vacuum activity'. None of these effects can be derived from the stimulus response schema of the reflex and therefore require a supplementary explanation from any adherent of the reflex theory of the instinctive behaviour pattern.

Such an explanation might possibly relate to the relationships which exist between the threshold-lowering for the elicitatory stimuli and the search for these stimuli represented by appetitive behaviour. Quite apart from the purely functional interaction, relationships between threshold-lowering and appetitive behaviour can be sought in the fact that both are presumably connected with the subjective phenomena which accompany every instinctive behaviour pattern and doubtless undergo a considerable increase in intensity following a long period of 'damming'. In addition, with respect to the experimental aspect of the phenomena concerned, I should like to point out that perception is apparently the component of the 'reflex arc' upon which the alteration of the response operates. Goethe says: 'With this potion in your body, you will soon see every woman as a Helen.' Or, to take a more prosaic example, after a period of hunger our salivation reflex will respond even to the odour of a meal which we normally reject with distaste. Eliot Howard says that with variation in the intensity of a response the perceptual field of the animal alters. He quite obviously favours the same interpretation as that which I am attempting to convey here.

The threshold-lowering of the elicitating stimuli is regarded by Craig as a component effect of 'appetite', whose foremost product is a purposive search for the stimuli (i.e. appetitive behaviour). Since Craig classifies this search for stimuli within the concept of the instinctive behaviour pattern, regarding it as a component of the latter, his procedure is a logical extension of his views. In fact *both* operate in a unitary way as regards an increase in the *motivation* of the animal to perform the behaviour pattern concerned. But since we are compelled by our narrower definition of the instinct concept to conduct further analysis of such functionally unitary behaviour patterns, we must sharply separate the threshold-lowering for the elicitatory stimuli as a property of the instinctive behaviour pattern distinct from the stimulus-seeking appetitive behaviour. The latter, even in its most primitive form, must be regarded as belonging within the category of the higher purposive behaviour patterns. The relationships which, as indicated above, possibly exist between the two have no influence on the separation of the two concepts and do not conflict with it.

Even in the examples quoted above, illustrated from the subjective side, the phenomenon under discussion emerges not only as a lowering of the threshold value for elicitatory stimuli but also as a decrease in the selectivity of the organism, as a tendency to respond to not-quite-adequate stimuli. With object-oriented behaviour patterns, this can have the effect that the response, following a long period without the adequate biologically-correct object, will (so to speak) 'make do with' another, not quite suitable object. Just like the biologically-correct performance of the pattern, such performance of an instinctive behaviour pattern in response to a substitute object brings about immediate re-elevation of the abnormally-lowered stimulus threshold, even if the threshold value does not return to the norm. This elevation of the stimulus threshold or, if you like, the increase in selectivity of the perceptor component of the response, has the result that response to a not-quite-adequate stimulus situation will disappear after a comparatively short period of performance. An instinctive behaviour pattern which is repeatedly performed in response to the normal object for some time will thus be elicited only briefly, or only a small number of times, by a substitute object. Lissmann was, for example, able to demonstrate that the fighting responses of male fighting fish (*Betta splendens*) kept in isolation were initially elicited by extremely clumsy plasticine dummies, but rapidly habituated. Habituation occurred more rapidly, the lower the correspondence between these substitute objects and a conspecific combatant. When we removed one offspring from the nest of a night heron and placed it on an open meadow, the adult responded to the abnormal stimulus situation by warming the young bird for a moment and then immediately standing up and departing. Shortly afterwards, the adult defended the youngster *once* against a peacock and then – despite the continuance of the peacock's threat to the young bird – immediately switched its attention from the two and towards the immobile female keeper to beg food from her. When we returned the young bird to the nest, it was greeted for a particularly long period and with intense 'pleasure'. It was at once warmed for some time and it was similarly protected from me with great and undiminishing anger.

I am, generally speaking, no great friend of physical conceptual models for biological processes, since they can all too easily lead to the belief that a process for which one has actually developed no more than a very incomplete model has been completely causally/analytically interpreted. Nevertheless, I believe (with the above reservation) that it is necessary to employ a physical analogue of the relationship between the instinctive behaviour pattern and its elicitatory stimuli during and between performances of the pattern. The analogy of 'arousal pressure' has already been mentioned a number of times, and an animal does actually behave as if some response-specific energy were *accumulated* during periods when a specific pattern is not employed. It is as if a gas were continually pumped into a container in which the resultant pressure continually increases until a discharge is effected under quite specific conditions. I should like to symbolize the various stimuli which lead to the discharge of the 'arousal pressure' as taps which permit the accumulated gas to flow out of the container. The adequate stimulus – in more exact terms the adequate combination of stimulus effects – would correspond to a simple tap which permits the pressure in the container to sink to the level of the external pressure. All other, more-or-less adequate stimuli would correspond to taps which are connected through an obstacle in the form of a spring-loaded valve, which only permits gas to flow out above a certain internal pressure level. Thus, these taps would never be able to completely extinguish the pressure prevailing within the container, and the deficit would be greater, the stronger the spring of the intervening valve (i.e. the greater the difference between the elicitatory substitute stimulus and the normal, adequate stimulus situation). The rapid habituation which the instinctive pattern exhibits with inadequate elicitation can in this way be illustrated in a very fitting – and probably essentially accurate – manner. But this analogue falls down with respect to one important point, since it does not illustrate the vacuum activity, or at least illustrates it only poorly. The elemental, almost explosive emergence of the response, which 'evacuates' the animal to the point of exhaustion, cannot possibly be represented as discharge of pressure through some kind of safety-valve. At the most, it could be envisaged as a rupture of the entire container.

This property of the instinctive behaviour pattern, which is so extremely suggestive of internal accumulation processes, is easily understood in all responses which have to cater for a *requirement* of the animal's body, such as the instinctive behaviour patterns of food and water uptake, and the deposition of faeces, excreted products and sexual products. In many of these cases, the manner in which the internal stimulus develops to form a subjective need through the agency of more-or-less complex systems of indicators is exactly known, as is the mode of reduction, to a greater or lesser extent, of the part played by external elicitation of the response. But it must be emphasized here that with many, possibly *all*, instinctive responses the animal behaves in a completely analogous fashion, i.e. even with responses for which the existence of a simply understood internal stimulus of that kind can be excluded with certainty. Independence from demonstrable internal stimuli[110] is particularly evident in the so-called negative responses – the escape and defence responses of many animals. With these responses, one can provide the best demonstration of threshold-lowering for the elicitatory stimuli, since this occurs almost regularly with captive animals, particularly those which have been reared in human care. I have been able to show that birds reared from an early age exhibit a lowering of the threshold for escape-eliciting stimuli, particularly in those species in which the escape response of the young bird is not elicited by the sight of an innately 'recognized' predator but by warning from the parents, or by their own fright response and consequent flight. Such young birds are consequently deprived of *any* adequate elicitation of their escape response in human care and they are (so to speak) unable to 'rid themselves' of the response. As expected, they exhibit a tendency to respond to the weakest substitute stimuli with an extremely dangerous panic or even to exhibit an abrupt escape response without the operation of demonstrable stimuli. This is extremely inconvenient in the practical task of keeping animals. Many hoofed mammals, particularly antelopes, behave in a very similar manner. In this case too, as Antonius has informed me, it is mainly individuals reared in isolation by human beings which fatally injure themselves against wire fences in blind and unfounded panic.

It has been mentioned that the vacuum activity is generally characterized by a high intensity of arousal, which in the case of some instinctive behaviour patterns may even be of *maximum* intensity, as is the situation with the escape responses discussed above. One might well say that at a low intensity of the need to perform a specific response, accompanied by simultaneous presentation of an optimal external opportunity for the performance of the response, the incipient responses and incomplete patterns described on p. 265ff are observed; whereas at a high internal arousal pressure, accompanied by an insufficient or even completely barren external opportunity, abnormal behaviour patterns of an evidently different type are observed. While in the first case the incompleteness of the pattern prevents attainment of the adaptive goal, it frequently becomes graphically clear in the second case that the fixity of the pattern – which is itself complete – prevents achievement of the biological function as soon as the slightest degree of adaptive change in the behaviour pattern becomes necessary. *These two extreme cases are the clearest proof of the utter lack of any relationship between the adaptive function of an instinctive behaviour pattern and the goal which is actually sought by the animal subject.* In the one case, the adaptive goal of the response is not attained because the internal 'arousal pressure' of the animal – despite the sufficiency of all external conditions – is insufficient to 'drive the animal through' all of the individual components of the behavioural sequence. In the other case, the animal very eagerly performs these behavioural components without omission, although all of the stimuli supposed to be necessary as 'behaviour supports' and (frequently) all of the purely physical conditions, for performance of the response are lacking.

The completeness of performance of the behaviour pattern in the vacuum activity may initially appear to support an interpretation based on pathway theory. However, the lowering of the stimulus threshold, which is to be regarded as the basic process underlying all vacuum activities is something which is entirely foreign to the simple stimulus-response schema of the reflex, and (as has already been explained) this at least requires a special explanation. Just as was the case with the regulative phenomena discussed earlier, the lowering of the threshold is not basically opposed to the reflex theory, but it does show that the latter theory is by itself insufficient as an explanatory principle for the instinctive behaviour pattern.[111]

If the question is now raised as to whether the threshold-lowering effects can be used for a distinct conceptual separation of the instinctive behaviour pattern from the reflex in its narrowest sense, the answer must be a negative one. Every habituation effect can, after all, be interpreted as the same process with a reversed sign. During the recovery phase following repeated elicitation, the tissues serving animal functions of extremely different kinds exhibit the phenomenon of gradually mounting sensitivity and threshold-lowering for the elicitatory stimuli in an unmistakably analogous fashion, although there are considerable quantitative differences. The phenomenon itself is therefore quite definitely not something which is confined to the anatomical substrate of instinctive behaviour patterns.

The most important property of the instinctive motor pattern, which similarly cannot be explained from the stimulus-response schema of the reflex, is the *endeavour to secure elicitation* by means of the behaviour patterns which have already been discussed at length on p. 270 under the heading of *purposive* or *appetitive behaviour*. The fact that the performance of the instinctive behaviour pattern represents the goal of the purposive behaviour does *not* itself speak against a chain-reflex nature of the motor co-ordination patterns. It is simply that it is at first entirely obscure why the animal endeavours to secure elicitation of these chain-reflexes since it does not exhibit such behaviour for *all* of its reflexes. It would never occur to anybody to endeavour to reach a stimulus situation in which his patellar reflex is elicited, simply to secure performance of the reflex. It is in fact an integral part of the concept of the reflex that it should be continuously at the ready, like a laid-up machine, and should only operate when specific key stimuli take effect on the animal's receptors. It is not in the nature of the reflex that the response should (so to speak) make its presence felt, render the animal restless and provoke the animal to *seek* actively for the relevant key stimuli, although this itself does not speak *against* a reflex nature of the eventual performance. In the simplest case, this search is no more than motor restlessness, though this already operates as trial-and-error location. At the opposite extreme, however, the search can be guided by the highest learning processes known in the entire animal kingdom. The emergence of restlessness in the animal – the *impetus* for directed or undirected seeking for a specific stimulus situation, in which the innate releasing schema of the requisite response is first brought to elicitation – is exactly what I would describe with the word *drive*. (I am completely aware that this concept of drive is even less customary than the concept of the instinctive behaviour pattern which I am employing.)

There are without doubt two factors which directly provoke the animal to search for the stimulus situation for elicitation of a given instinctive pattern. The first factor is the 'drive', as it has just been defined, and the second (according to previous experience) is represented by the pleasurable sensation accompanying performance of the instinctive behaviour pattern. However great the obstacles to providing a causal explanation for these subjective phenomena, it is an obvious step to attribute a teleological significance to them – of course in terms of survival of the species. The animal is 'driven and attracted' to perform the motor co-ordination patterns necessary for survival of the species. We have already discussed (p. 275) the manner in which the functional urge operates as a bait in the acquisition of instinct-conditioning intercalation patterns, ensuring that the subject proceeds along the path laid down for survival of the species. Without this double motivation of all appetitive behaviour, there would be insufficient guarantee for the requisite extent of performance of instinctive patterns. Indeed, it can be said that without this any animal species would be condemned to rapid extinction.

Although on closer observation neither the phenomenon of regulative capacity nor that of threshold-lowering for the elicitatory stimuli has proved to be of use for conceptual distinction between the instinctive behaviour pattern and the reflex, the endeavour to secure elicitation, with performance of the pattern serving as the goal of the appetitive behaviour, is certainly applicable as criterion for a definitive separation of the instinctive behaviour pattern as a special category of reflex processes. At first sight, it may seem unsatisfactory that a definition of the instinctive behaviour pattern should be dependent upon subjective phenomena. For this reason, it is emphasized once again that the concept of purposive behaviour can be defined entirely objectively following Tolman, in full accord with the principles of ethology. A definition of the instinctive behaviour pattern as 'a reflex pattern whose elicitation is sought after' is a more precise form of the interpretation given by

Verwey: 'If reflexes and instincts can be at all separated, it is on the basis that the reflex is performed mechanically, whereas the instinctive behaviour pattern is accompanied by subjective phenomena.' However valuable it may be to be able to define the term purpose from the standpoint of objective ethology, I believe that it would be deliberate ignorance of an important fact if we were to fail to point out that it is actually the *subjective* concomitants of the instinctive behaviour pattern which represent the *immediate goal* of appetitive behaviour.

I am well aware that definition of the instinctive pattern as 'a reflex pattern whose elicitation is sought after' is not entirely free from philosophical difficulties. Combination of the concept of seeking, which is essentially a subjective process despite its objectively determinable features, with the physiological concept of the reflex incorporates some of the naïvety of Descartes' view that the pineal gland is the point of attack of psychological factors upon bodily processes. But this itself clearly characterizes the philosophically-difficult, yet consequently important and possibly even revealing, part played by the instinctive motor pattern in the theory of animal and human behaviour. This may even lead to more exact formulation of a question which the natural philosopher must pose for the biologist.

Summary

The impetus and justification for this critique of virtually all of the customary interpretations of the instinctive behaviour patterns and for the definition of a new concept of the instinctive behaviour pattern was almost entirely provided by *new observational material.* These observations are indeed far from an exclusive product of my own investigations, but the material is nevertheless derived from a group of animal observers so well known to me and so closely aligned in their approach that I am presumably justified in assuming that these facts have never before been taken into account in the formation of a concept of what the instinctive behaviour pattern actually is. I should not omit to mention that the vast majority of these observational facts have been known to me far longer than the theories which are here criticized with their help, and that the views presented in this paper had already been basically established in my mind – even if the scientific formulation was lacking – before I had even heard the names of the great theoreticians of instinct.

Since every one of these observational facts, which are incompatible with so many customary views, appear *repeatedly* in this paper and are discussed in different places with respect to different aspects of their interpretation, I believe that the clearest summary of the tenets of this paper is to be provided by repeating these facts in compressed form. This will not be done in the sequence in which they occur in the paper but in an order corresponding to my assessment of their importance and significance.

In this process, first place must be accorded to *instinctive behaviour patterns which fail to perform their adaptive function.* Both those which remain incomplete owing to insufficient internal response intensity (p. 266) and those failing to perform their biological function because of insufficient external conditions (p. 277) can be observed with great frequency in captive animals. It was this which convinced me quite early on that the biological function of the response and the purpose set to the animal subject have nothing to do with one another and may definitely not be equated. The extreme case of the non-adaptive performance of an instinctive pattern because of inadequate external conditions – the vacuum activity performed without an object – exhibits truly photographic similarity to normal performance of the motor actions involved, leading to attainment of the biological function of the pattern. This demonstrates that the motor co-ordination patterns of the instinctive behaviour pattern are hereditarily determined down to the finest detail. The vacuum response permits us to study the instinctive behaviour pattern in what may be called 'pure culture' in captive animals reared in isolation. This response must be regarded as an irrefutable basic phenomenon which demonstrates the invalidity of a school of thought which attempts to derive all animal and human behaviour from conditioned reflexes – a school which still attracts supporters in America. The similarity between the motor patterns of the vacuum response and the normal, biologically-adaptive response prevents us from the outset from interpreting the instinctive behaviour pattern as a form of purposive behaviour. It is simply not tenable that proof has ever been brought for changes occurring in an instinctive behaviour pattern in relation to a specific goal recorded by the animal subject.

As a second basic fact, I should like to remind the reader of the *relationships of the instinctive behaviour pattern in the zoological system* discussed on pp. 281–285. These show us that the instinctive motor co-ordination pattern behaves exactly like an organ as regards phylogenetic variability and that it can and must be similarly understood on a comparative systematic basis. The relationships of the instinctive behaviour pattern in the phylogenetic system show quite plainly how pointless it is to make statements about 'the instinct' and demonstrate that our conclusions can only relate to inherited motor patterns, to instinctive *behaviour patterns.* These patterns themselves can only be dealt with in the bounds of more-or-less restricted units of the zoological system.

Both facts – the completeness of the motor patterns in biologically-unadaptive performances and the organ-like relationships of instinctive behaviour patterns in the zoological system – must, by their very nature, make us sceptical about reports of adaptive variability of the instinctive behaviour pattern mediated through individual experience. On the basis of our investigations of phylogeny (p. 281) and our experiments on ontogeny (p. 262) of instinctive motor patterns, we can state that it is probable that in all cases in which there have been reports of apparent adaptive change of an instinctive behaviour pattern mediated by individual experience there has been confusion with *maturation processes*.

A further similarity to organs in the ontogeny of the instinctive behaviour pattern, the interpretation of which I will gladly leave to others, is represented by the peculiar process of acquisition which has been termed *imprinting* (p. 279). This process, on the basis of its dependence upon conspecific material, its attachment to a short-lived developmental state and (above all) its irreversibility, exhibits an unmistakable parallel to the process of inductive determination known from embryogeny.

Finally, I must draw attention to *intercalation patterns* incorporating instinctive and purposive behavioural components (p. 268). The fact that innate, instinctive components and purposive, adaptively variable components can follow immediately on from one another in a functionally unitary behavioural sequence is of great importance in two respects. In the first place, exact analysis of such behavioural sequences has prevented us from assuming that smooth transitions between the instinctive behaviour pattern and purposive behaviour exist, and we have thus not fallen into the trap encountered by so many authors who have formed a too broad definition of instinct, incorporating appetitive behaviour. Secondly, observation of the ontogenetic development of a particular type of intercalation pattern – the instinct-conditioning pattern – has led to the demonstration of a fact recognized by Wallace Craig some time ago. He recognized that the performance of the instinctive behaviour pattern is the goal of the purposive behaviour and represents the goal presented to the animal subject (pp. 269, 310). This provides the only possibility for conceptually separating the instinctive behaviour pattern, as 'a reflex pattern whose elicitation is sought after', from other 'pure' reflex processes (p. 311).

It may appear that this relatively small number of genuinely new conclusions is not sufficient justification for establishing a concept which is so basically distinct from the customary view. But I can see no danger in formulating my opinions in the clearest possible manner, as long as it is not forgotten that they still to some extent represent pure working hypotheses which might have to be altered at any time in the face of new facts.

I do hope and believe, however, that I have convincingly achieved one thing – the demonstration that investigation of the instinctive behaviour pattern is not an area for highly-complex philosophical speculation but a field in which (at least for the time being) only experimental investigation of individual cases is decisive.

Taxis and instinctive behaviour pattern in egg-rolling by the Greylag goose (1938)

Introduction

The experiments which we conducted with greylag geese last season, to investigate the pattern of rolling eggs back into the nest-cup, are far from complete. In addition, our experimental animals were to a large extent related to domestic stock. Although some control experiments have already been carried out on pure wild-type Greylags, giving completely consistent results, we intend to carry out series of experiments on pure-blooded geese next year, placing particular emphasis on a detailed film-analysis of the observational material reported here. Nevertheless, we believe that there are two good reasons for publication of this paper in its present form. In the first place, discussion of two papers recently published by Lorenz (1937a and b) has uncovered a number of misunderstandings which can best be eradicated by discussion of our findings on the egg-rolling response of the Greylag goose. For this reason, we wish to see our findings published as soon as possible, as a sequel to the two papers mentioned. Secondly, we hope that discussion of our present results will provide us with additional knowledge, which would benefit the experiments to be conducted in the years to come. At this juncture, gratitude is expressed for permission from the *Deutsche Reichstelle für den Unterrichtsfilm* to make use of photographs from a film made in their commission by cand. zool. A. Seitz (Vienna) in Altenberg.

I Theoretical considerations of the taxis and the instinctive behaviour pattern

Whatever concept one chooses to associate with the word instinct. it will always be necessary to consider the existence of specific 'motor formulae' which are innately determined in an invariable form in the individual. These 'formulae' can represent characters of great taxonomic value, typifying a species, a genus or even an entire phylum. Since these motor patterns fairly closely represent the nucleus of what was interpreted in earlier descriptions of animal behaviour as the operational effect of the instinct, we have employed (or, more exactly, retained) the term *instinctive behaviour pattern*[112] to refer to them. The form of an instinctive behaviour pattern exhibits remarkable independence from all receptor process – not just independence from 'experience' in the broadest sense of the term but also from stimuli which impinge upon the organism *during* performance of the pattern. In this feature, instinctive patterns are quite distinctly demarcated from *taxes* or *orientation responses*, with which they share a purposive character (in terms of survival of the species) and independence from individual learning effects, and with which they are frequently bundled together under the concept of 'instinct'. In Drisech's definition of the instinct as 'a response which is complete from the outset', the conceptual delimitation covers both the motor pattern of innately-determined form *and* every innate orientation process, without taking account of the basic and far-reaching difference between the two. In order to clarify the conceptual framework of our work, it is necessary to briefly discuss the relationships and differences between the taxis and the instinctive behaviour pattern.

1. THE TOPOTAXIS

Most of the processes which we refer to as *taxes*, and particularly *topotaxes* (after A. Kühn) include – *in addition to* orienting responses – fixed instinctive behaviour patterns and 'employ' specific motor patterns, particularly those of locomotion. Nevertheless, the main motor component of the topotaxis – and *its only unique characteristic* – remains the *turning motion* of the animal's whole body or one of its parts (e.g. the eyes or the head), *which is quantitatively determined by the external stimuli*. This is not the case with Kühn's *phobic response* or phototaxis, and the relationship exhibited between this type of response and our instinctive behaviour pattern in terms of the fixity of the motor formula and its independence of the nature and direction of impinging stimulus permits us to distinguish it quite clearly from the topotaxis. Even if the analogies between the phobic response and the instinctive behaviour pattern are only external in kind and the two are little related in causal terms (e.g. as regards the dependence of the instinctive behaviour pattern upon the central nervous system), we do not wish to distinguish the phobic response as markedly as the topotaxis from the instinctive behaviour pattern in establishing our working hypothesis.

The turning motion (in its widest sense) leading to the spatial orientation of the animal has been termed the 'oriented turning response' by K. Bühler. This extremely broad concept includes every alteration of the motor status of the organism relating it to the spatial parameters of the environment, thus covering both the simplest 'towards' or 'away' turning motion and spatial orientation to external stimuli controlled by the highest forms of intelligence. Since the 'away' turning movement of the phobic response, which is *not* quantitatively controlled, is also covered by this concept, we must conclude that in this paper we are interpreting under the concept of the orienting turning motion only a fraction of the oriented turning responses categorized by Bühler. We are exclusively concerned with the quantitatively stimulus-controlled turning movement which is uniquely characteristic of the *topic* response. Since we are exclusively concerned with the topic response in the following comparison of taxis and instinctive behaviour pattern, we will confine the term taxis to this response for the sake of brevity. In the last analysis, it is only a question of convenience as to which concepts are defined, and the definitions selected here in order to permit the briefest possible presentation of the factual evidence are by no means intended as a critique of established concepts.

Even if some motor formulae which are themselves rigidly fixed play a part as elements of the overall movement pattern of the taxis, it is the *overall form* of the taxis which is dependent upon the quantitatively stimulus-controlled turning movement. The adaptation of the overall movement pattern to spatial environmental conditions, which is an important feature of the taxis, is itself ensured by the *norm of the response* to specific external stimuli. Without discussing the purely philosophical question of whether the animal subject steers towards these stimuli or is passively steered by them, it can be stated that we are here objectively concerned with *moulding of an adaptively purposive motor pattern by stimuli of external origin*. In the simplest, and best analysed case, this process is quite definitely *a reflex in the true sense of the word*. But even with orientational responses which are derived from the so-called higher psychological functions, the activity of the central nervous system is essentially a process of evaluation and reaction to external stimuli and is thus a reflex-like process, at least in functional terms.

2. THE INSTINCTIVE BEHAVIOUR PATTERN

The instinctive behaviour patterns also exhibit relationships to the reflexes, in that they are similarly elicited by particular external stimuli which are often very specific. However, closer investigation shows that it is *only the releasing mechanism* and not the subsequent performance of the motor formula which represents a genuine reflex. The overall form of the motor pattern, once elicited, is apparently independent not only of external stimuli but also of the animal's receptors generally. Instinctive behaviour patterns are not innate response norms like the taxes – they are *innate motor norms*.

However improbable it may at first sight appear that finely co-ordinated and extremely adaptive behavioural sequences of an animal can occur without interaction with the receptors, just like the movements resulting from spinal sclerosis (i.e. that they are *not* constructed from reflexes), there are several important reasons for assuming this to be the case. E. von Holst has been able to provide entirely convincing evidence, from deafferentiation studies of the central nervous systems of greatly different organisms, that *automatic, rhythmic stimulus production processes* occur in the central nervous system, *where the impulses are actually co-ordinated*, such that the impulse sequence transmitted to the animal's musculature arises in an adaptively complete form without interaction with the periphery and its receptors. The centrally co-ordinated motor sequences so far investigated are almost exclusively concerned with locomotion, and it is just these locomotor patterns which have been previously almost unanimously regarded as chain-reflexes, in which each movement was supposed to elicit the next only by a detour through the peripheral receptors.[113]

Without a trace of doubt, the motor patterns investigated by von Holst must be regarded as typical instinctive behaviour patterns in the sense of the concept defined here. This does not mean, however, that the converse must also apply – i.e. that all of the instinctive behaviour patterns which we have previously described must depend upon exactly similar processes in the central nervous system. The fact that the receptors (in the broadest sense of the term) do not play a part in the production of the adaptively-functioning motor pattern even in more highly-differentiated instinctive behaviour patterns is strongly indicated by a number of observational facts and experimental results which were described in detail in an earlier paper (1937). Even if it may seem premature at this time to equate the motor automatisms investigated by von Holst with the instinctive behaviour pattern, considerable emphasis should be given to the fact that two of the most important and most characteristic features of the instinctive behaviour pattern, which present virtually insurmountable obstacles to any other explanation, can be easily interpreted on the assumption that they are based on automatic, rhythmic production of the elicitatory stimuli and upon central co-ordination of the impulses. These two features are *lowering of the threshold* for the releasing stimuli and the consequent *vacuum performance* of the instinctive behaviour patterns in situations in which even the physical/mechanical prerequisites for adaptive function are absent.

Even in cases where the instinctive behaviour pattern is elicited by an unconditioned reflex, which normally responds only to fairly (or even extremely) specific combinations of external stimuli, it can under certain circumstances be observed that the *form of the elicited motor pattern is independent of that of the elicitatory stimuli.* If these stimuli are absent for a period of time longer than that to be expected under natural conditions, the instinctive behaviour pattern soon exhibits *reduced selectivity* of elicitation. The animal will then respond with this same motor pattern to other stimulus situations which are no more than generally similar to the adequate situation. The longer the period of 'damming' of the response, the smaller the requisite degree of similarity between the available stimuli and the normally specific elicitatory stimuli. In cases where the stimuli eliciting a particular instinctive behaviour pattern are to some extent open to quantitative determination, it is observed that the threshold value of the response to be elicited sinks continuously with the length of the inactive period. Since both of these phenomena are obviously based upon only *one* process in the central nervous system, I have included them under the concept of 'threshold-lowering' in the two papers already mentioned. Threshold-lowering can in very many instinctive behaviour patterns (perhaps basically in all) literally progress to the limiting zero value. Following a more-or-less extensive period of 'damming', the entire motor sequence can be performed *without* demonstrable operation of an external stimulus, and this has been referred to as the 'vacuum activity'. Threshold-lowering and vacuum activity present a double support for the assumption of automatic, rhythmic stimulus-production and for central co-ordination of the impulses involved.

The phenomenon of threshold-lowering indicates the existence of internal accumulation of response-specific arousal (von Holst assumes that stimulatory substances are involved), which is continuously produced by the central nervous system and reaches a level proportional to the period of absence of discharge resulting from performance of the instinctive behaviour pattern concerned. The higher the level of accumulation reached, the more intensive is the eventual eruptive performance of the instinctive behaviour pattern and the harder it is for the central nervous factors governing the instinctive behaviour pattern to maintain inhibition against such performance in the 'incorrect' situation. These considerations, which are reproduced here in grossly simplified form, were developed before von Holst's observations were known. We now know from von Holst's work that the operation of the continually-active automatic stimulus production processes is in actual fact persistently inhibited by the activity of higher (or more central) inhibitory components of the nervous system and that elicitation of the response signifies no more than removal of central inhibition. When left to themselves, the peripheral components (e.g. a spinal cord preparation in the case of von Holst's experiments) will continuously perform the motor pattern in accordance with the automatic rhythm. If the spinal cord preparation is then provided with a substitute for the inhibitory effects normally produced by the activity of higher centres – in the form of supplementary stimuli – one observes a phenomenon which Sherrington described under the term 'spinal contrast'. The greater the duration and intensity of operation of the inhibitory effect, the more violent the eruption of the inhibited motor pattern when inhibition is ceased. This correspondence between the automatic rhythms investigated by von Holst and the instinctive behaviour pattern (as it is here understood) goes much further and covers minor details which will not be discussed here.

To the same extent that the threshold-lowering and elevation of intensity which we can observe with virtually every instinctive behaviour pattern clearly supports the assumption of a stimulus-producing automatism, the phenomena which we observe during the eventual stimulus-independent eruption of the vacuum activity clearly indicate central co-ordination of the impulses which are transmitted by the automatism.[114] It is of undoubted importance that a motor sequence frequently performed virtually '*in vacuo*' resembles down to the finest detail the motor pattern observed during the normal, adaptive response. This is particularly obvious when the behaviour pattern performs mechanical work in the normal situation. In other words, the motor pattern is normally performed against an opposing force which is missing with the vacuum activity. (This will be considered in detail later on.) The complete independence of the form of the motor pattern from the conditions and stimulatory properties of the natural environment can only be explained on the assumption that the impulses for the individual muscular movements are transmitted in an already co-ordinated form and sequence from the centre, and that they (so to speak) 'experience nothing' of the prevailing environmental conditions. Otherwise, it is impossible to explain how a starling, for example, can perform the motor sequence of catching, killing and swallowing small insects in the complete absence of such animal prey; or to explain how a humming-bird can attach 'non-existent nest-material fibres' to a twig with a wonderfully co-ordinated weaving movement, as was recently observed by Lorenz at the zoological garden in Berlin.

Even though such vacuum performances of highly-differentiated instinctive behaviour patterns quite conclusively demonstrate the independence of the form of the motor pattern from all external receptor processes, we unfortunately do not possess an object on which we can exclude the co-operation of proprioceptor reflexes with the same degree of certainty as von Holst achieved with the spinal cord of the fish. For this reason, the part played by proprioceptive receptors in the performance of highly-differentiated instinctive behaviour patterns is still extremely obscure. Nevertheless, it can be observed that even extremely coarse mechanical effects which directly and forcibly cause temporary inhibition of the instinctive motor sequence or force the body of the responding animal into abnormal postures produce no alterations in the further performance of the response. Such coarse mechanical disruptive effects would automatically lead to alterations in the form of the subsequent parts of the motor pattern if the proprioceptive registration of each individual component movement were decisive for the form of the next in the manner postulated in the old chain-reflex theory of the instinctive behaviour pattern. We have so far never seen such alteration. After cessation of the mechanical pressure, the instinctive behaviour pattern – if it continues at all – continues in a completely typical manner. Examples of this will be provided later.

For all of these reasons, it would seem permissible to construct a working hypothesis on the assumption that the co-ordination of the individual movements of the instinctive behaviour pattern is independent of *all* receptors and is not influenced by them.[115] This may be a completely gross simplification of the factual evidence and it may signify a too extensive restriction of the concept of the instinctive behaviour pattern, particularly since we know from experiments conducted by von Holst that an impulse emanating from a centrally co-ordinated automatism may be superimposed upon one emanating from a reflex to give *one* muscle contraction. Nevertheless, provisional disregard of this fact would seem to be justified in view of another fact which complicates the analysis of animal behaviour in general and of the instinctive behaviour pattern in particular. This complicating fact, which justifies an extensive simplification of the factual evidence as long as the provisional nature of this step is continually borne in mind, is based on the following circumstances:

3. INTERCALATION OF TAXIS AND INSTINCTIVE BEHAVIOUR PATTERN

An intact higher organism in its natural environment *almost never performs a centrally co-ordinated motor pattern in isolation*. Instead, the organism usually 'does several things at once', following (for example) an orienting response *during* the performance of an instinctive behaviour pattern. Nothing is farther from the truth than the consistently revived statement, usually employed in the distinction of the instinctive behaviour pattern from the 'reflex', that the instinctive pattern represents a response of 'the organism as a whole', in contrast to the reflex, which involves the activity of only 'one part of the organism'. As far as one can really maintain that only a part of an organism is involved in a response, one can confidently state that this is the case with the instinctive behaviour pattern as well. The automatic rhythm and the central co-ordination of the instinctive behaviour pattern render it far more independent than the reflex, for the simple reason that the instinctive pattern is consequently independent of the receptors. The automatism of the instinctive behaviour pattern is a virtually closed system. It is subject to the systemic control of the central nervous system only with respect to the timing and degree of disinhibition. Despite the independence of the automatism of the instinctive behaviour pattern and its refractory nature with respect to the highest controlling factors of the central nervous system even in man (i.e. with respect to the 'ego'), its sphere of influence in the motor apparatus of the higher animal is always very limited. The automatism always transmits its impulses to quite specific groups of muscles and never (apart from a few exceptions) simultaneously to the entire body musculature. Therefore there is always the possibility that other muscles and groups of muscles can simultaneously perform movements which are brought about by quite different processes in the central nervous system.[116] Consequently, in the study of the instinctive behaviour pattern itself we shall have to select predominantly those processes in which the centrally co-ordinated impulse sequence of the instinctive behaviour pattern is transformed into visible and describable movements without disruptive side-effects derived from other nervous processes. Alternatively, we must select processes in which the operation of non-instinctive motor processes can be easily analysed and eliminated from considerations of the instinctive behaviour pattern proper.[117]

Certain non-instinctive movements, which can be easily analysed, accompany *every* instinctive behaviour pattern performed by higher animals: The tropotactically-controlled orienting response of equilibriation proceeds continuously even when instinctive behaviour patterns are performed. With the sole exception of the copulation response of the rabbit buck, we know of no instinctive behaviour pattern in which a normally gravity-oriented organism loses its orientation and falls over! But even instinctive behaviour patterns for which the movements of tropotactic orientation to gravity are the only accompanying orienting responses represent definite exceptions. In the search for an instinctive behaviour pattern which could be clearly represented in curvilinear form with some degree of accuracy, for the purposes of a comparative and genetic investigation, it was not at all easy to find a centrally co-ordinated motor pattern which was not overlain by simultaneously active orienting mechanisms! The final choice fell upon the courtship motor patterns of certain duck species. These patterns are performed more or less in the sagittal plane by the duck whilst swimming in an upright posture, and it is possible to prepare a fairly complete reconstruction and graphic representation of the movement from a ciné-film taken from an exactly lateral position. From this projection, the individual, simultaneous orienting responses (the lateral movements of the equilibriating tropotaxis) can be completely ignored. A second orienting mechanism operates even with some similar instinctive behaviour patterns of the same birds, serving a biologically identical function. In such cases, the courting drake always orients his movement in the gaze direction of a conspecific female, such that the optically stimulatory markings displayed in the motor performance concerned are fully exposed to the courted duck.

The accompanying orienting responses naturally carry far greater significance for instinctive behaviour patterns whose adaptive value depends upon a *mechanical* interaction with certain environmental objects than for those which operate entirely optically (or partially acoustically) as stimulus transmitting patterns. Of course, only a few of these mechanically-operative motor patterns are functional *without* an orienting response to bring the organism into spatial relationship with the object of its instinctive behaviour pattern. The tactic movement 'towards' or 'away from' can rarely be dispensed with in the case of mechanically operative instinctive behaviour patterns, unless the animal is to interact mechanically with a homogeneous medium. With plankton-eating animals, even the food may represent a homogeneous medium under certain circumstances, and it is therefore among such inhabitants of open aquatic environments that we find the greatest independence from taxes existing among free-moving organisms. In the world of solid objects, however, no animal with the power of locomotion can maintain itself successfully without orienting responses, since even the most differentiated and finely-adapted system of centrally co-ordinated motor patterns cannot function unless the animal provides the correct spatial orientation of each individual instinctive behaviour pattern towards the object.

This alteration of position of the animal, which is always governed by a taxis, by an orienting response in the broadest sense, can be completed *before* the performance of the instinctive behaviour pattern, so that a purely tactic and a purely central co-ordinated motor sequence are seen in temporal progression. A typical case of such progression from a taxis to an instinctive behaviour pattern is provided by the snapping response of the sea-horse (*Hippocampus*). By means of an orienting response of undoubted complexity, this fish orients itself such that the prospective prey is situated in the sagittal plane of its head, obliquely above and in front of its mouth at a specific distance and in a specific direction. It can take some time before this is achieved. The sea-horse will often follow a small crustacean backwards and forwards for a period of minutes, twisting and turning as far as its rigid armour will allow. Only when the spatial orientation described has been properly established is the necessary stimulus situation for elicitation of the muscular response pattern or snapping presented. An orientation response of this type, preceding the performance of an instinctive behaviour pattern, provides the simplest case of what we call 'appetitive behaviour' following Wallace Craig.

Psychologically, it is doubtless the most important and remarkable property of the instinctive behaviour pattern (and at the same time the most difficult property to explain on a causal basis) that the purely automatic performance – which itself lacks any orientation towards a goal sought by the animal subject – may itself represent *in toto* a goal of this kind. The organism exhibits an 'appetite' for performing its own instinctive behaviour patterns. Disinhibition and performance of the instinctive behaviour pattern are accompanied by certain subjective phenomena, and both man and animals actively *strive* to enter stimulus situations in which these processes will occur.[118] We attribute such subjectively rewarding experience of the instinctive behaviour pattern to any organism possessing such patterns – not merely by way of analogy, but because we regard this subjective experience effect as one of the most important processes ensuring survival of the species. As has been particularly emphasized by Volkelt, subjective experience is not a chance side-effect (an 'epiphenomenon') of physiological processes! Without the 'sensual pleasure' which presumably represents the experiential aspect of every instinctive behaviour pattern, performance of the pattern would only take place when the organism entered the elicitatory stimulus situation purely by chance. It is undeniably the subjective experience effect which makes the instinctive pattern an attractive goal. The incorporation of subjective phenomena in the casual chain of adaptive, physiological process presents the greatest philosophical difficulties; in fact, it must be regarded as the central feature of the mind–body problem. It is particularly remarkable that these pleasurable sensations, which are largely proprioceptive, should accompany motor processes whose form and co-ordination characteristically arise without participation of the proprioceptors.[119]

The adaptively variable 'appetitive' behaviour directed towards elicitation of an instinctive behaviour pattern can, in the simplest case, be represented by one of the most common and best-analysed topic responses. However, there are all conceivable transitional stages between such responses and the highest known feats of learning and insight. In the Lorenz publications already mentioned (1937a and b), detailed consideration was given to the remarkable fact that, from the standpoint of objective ethology, no distinction can be made between the simplest orienting response and the highest form of 'insight' behaviour.

Apart from taxes which precede the elicitation of the instinctive behaviour pattern, *procuring* elicitation and thus clearly exhibiting an appetitive character, there are also taxes which *continue to function during* the performance of the instinctive motor pattern, as has already been demonstrated with the example of topotactic orientation to gravity. With instinctive behaviour patterns of mechanical function, they perform the task of maintaining the spatial orientation of the animal to the object established by the appetitive orienting response. The fact that this spatial relationship, in psychological terms, frequently corresponds to a subjectively pleasurable stimulus situation automatically means that one can very often interpret these continuing responses together with the orienting mechanisms of the instinctive behaviour pattern as a kind of appetitive behaviour. A grazing cow, for example, continuously holds its head in a position dictated by orienting responses, such that the necessary spatial relationship between the mouth and the grass is maintained. On the objective side, this relationship renders the purely instinctive motor patterns of the jaw, tongue and lip muscles mechanically operative and ensures their adaptive function, while on the subjective side the animal is presented with a pleasurable stimulus situation whose preservation represents a rewarding goal.

Obviously, such simultaneous co-operation of taxis and instinctive behaviour patterns is extremely common in unitary motor sequences of higher organisms. Such 'simultaneous intercalation patterns', as we shall call these composite behavioural units composed of simultaneously operative heterogeneous factors, can equally obviously attain a virtually unlimited degree of complexity. This can present almost insurmountable difficulties in analysis employing the few methods of investigation available. However, the comforting degree of optimism necessitated by the size of the task involved is brought closer by one fact: Just as there are cases in which we find pure instinctive behaviour patterns free of taxes, or instinctive behaviour patterns and purposive behaviour in an easily analysable sequence, there are also cases in which particularly favourable circumstances permit us to separate the instinctive behaviour pattern and the orienting response during a bout of stimulus activity.

II Aims of this study

The task of the research discussed in the following account was the investigation of the co-operation of an instinctive behaviour pattern and a simultaneously active orientation response, using a particularly simple case which was also especially suitable for analysis for supplementary technical reasons. The simplest imaginable case of such co-operation is provided when the centrally co-ordinated impulses of a motor pattern operate in *one* plane, while the movements produced by the taxis operate in a *perpendicular* plane, thus causing receptor-controlled deviations from the plane of the instinctive behaviour pattern. In order to illustrate a process of this kind, one may construct a simplified model. Let us consider a skeletal component which is rotatable around a ball-joint, something in the manner of a sea-urchin spine, and is moved by two perpendicularly-oriented pairs of antagonistic muscles, one pair operating vertically and one pair operating horizontally. Let us further assume that this skeletal component must oscillate through a particular spatial plane (e.g. vertical) to provide the necessary survival value. This oscillation of the skeletal organ can very well be operated by a centrally co-ordinated automatism, which transmits its impulses to the approximately vertically oriented pair of muscles. But in order to ensure that the movement is performed exactly in the vertical plane, a taxis controlled by a gravity receptor must be incorporated. Without affecting the spatial orientation of the animal's body as a whole, this taxis arrangement can ensure (through the orienting mechanism) that the second, horizontally contracting pair of muscles perform contractions which cause the skeletal component to be displaced from the plane of oscillation of the central automatism just enough to produce the exactly vertical movement necessary for survival of the species. This special case, in which an instinctive movement occurs exactly in one plane and is steered – like a horse between two reins – by perpendicularly arranged antagonistic muscles controlled by an orienting response, presumably occurs quite frequently. Similar relationships seem to be present in human chewing actions, the frictional movements of copulating male mammals and in other behavioural acts.

The mechanism outlined above appears to apply quite particularly to a motor pattern which can be observed with brooding birds of many different species. The adaptive function of this motor pattern is the return of eggs which have rolled out of the nest-cup. The behaviour pattern incorporates movements which can be immediately recognized as balancing actions, and can therefore scarcely be explained as anything other than products of an orienting response, and also movements which quite clearly exhibit all the characteristics of a true instinctive behaviour pattern. Since these movements also occur simultaneously and at a right angle to one another, we were convinced that we had found a particularly suitable object for the study, and attempted analysis, of taxis and instinctive behaviour pattern. When the Altenberg geese began to brood in the spring of 1937, we did not neglect the opportunity to conduct relevant observations and experiments.

The analytical approach governing the observations and the arrangement of experiments was based on a very simple consideration: If, as has been assumed, the taxis is a *response norm*, whilst the instinctive behaviour pattern is a *motor norm*, it must naturally follow that, in a case where the taxis and the instinctive behaviour pattern operate through separate muscles of the animal, absence of the steering stimuli must be accompanied by absence of the movements dependent upon the orienting response. Therefore, in the 'vacuum activity', the instinctive behaviour pattern must be performed *alone*, without the normally simultaneously functioning orienting response. In addition, alteration of the stimuli characteristic of the overall situation must lead to similar changes in the movements controlled by taxes, since the latter are dependent upon the former. These changes must take the form of adaptation to the new spatial conditions, whereas the instinctive behaviour pattern – because of its independence of receptor-control – must always be performed in exactly the same way, regardless of whether the performance occurs in the complete absence of an object, with a substitute object greatly deviating from the norm or with the biologically-adequate object. In addition, any alteration of the spatial relationships which requires even the smallest adaptation of the form of the motor pattern must lead to complete functional failure of the instinctive behaviour pattern. Finally, it was an evident step to exploit the action-specific fatigue effect characteristic of the instinctive behaviour pattern, in an attempt to cut out the central automatism through exhaustion, in order to see how far the appetite for the normally elicitatory stimulus situation and the associated orienting responses would be altered.

III Observations of the egg-rolling movement

The facts derived from pure observation without experiments will take first place, not only because they represent 'involuntary' experiments free of many of the sources of error of planned experiments, but also because they temporally preceded all experimental results in the collection of data.

The normal adaptively functional performance of the response takes roughly the following form: When the brooding goose sights an egg lying not too far away outside the nest-cup, the egg-rolling movement definitely does not emerge immediately as a 'decisive' action. Instead, the goose – on first sighting the stimulus-transmitting object – briefly gazes at the egg and then looks away. When the goose's gaze returns to the egg, it is fixated for a longer interval, and it is possible that a mild movement of the head towards the egg may be performed at this stage. After a brief series of intention movements of rapidly-increasing intensity, but of highly-variable length, the goose will stretch its neck

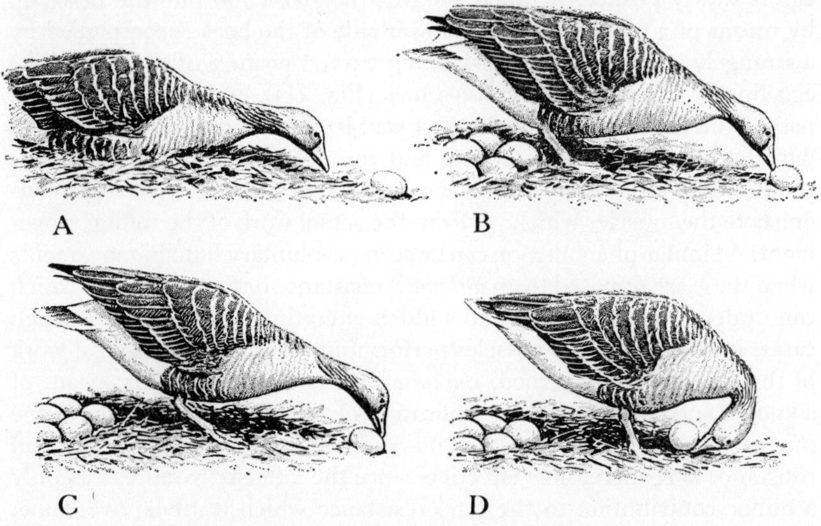

Fig. 1. *Normal performance of the egg-rolling movement*

towards the egg lying in front of its head, without moving the rest of its body (Fig. 1A). The goose frequently remains squatting in this position for a number of seconds, 'as if spell-bound', until rising slowly and hesitatingly from the nest (without altering the posture of the head and neck) and walking towards the egg with the peculiarly careful gait characteristic of all locomotion in the neighbourhood of the nest-cup. The premature stretching of the neck towards the egg gives the impression that the bird cannot localize the egg properly in space and 'hopes' to be able to reach it without having to get up from the nest. This impression, which is doubtless erroneous, is strengthened when the bird carefully approaches the egg, counting every step. This aversion to moving away from the nest-cup is based on the very good reason that the instinctive behaviour pattern is only really 'appropriate' and capable of transmitting the egg into the nest in a single performance when the goose does not move farther than the wall of the nest. When the goose has approached the egg sufficiently, the egg is first touched with the tip of the beak at the point on its surface nearest to the goose (Fig. 1B), and the slightly-opened beak is presumably pushed against the egg. The lower side of the beak is then passed slowly forwards over the upper surface of the egg, maintaining continuous contact, until it has extended beyond the egg and approaches the substrate on the side furthest from the goose (Fig. 1C). At this point, the bird's entire body is seized with a peculiar tension; the neck becomes taut and the head begins to quiver quite noticeably. This delicate quivering motion persists while the egg is slowly pushed or rolled towards the goose and into the nest-cup by means of a movement of the lower side of the beak accompanied by a strangely stiff and awkward-looking curved posture of the neck. The egg finally arrives on the goose's toes (Fig. 1D), and if it has by then passed the highest point of the nest wall it will roll into the nest-cup to join the other eggs. The tension and the delicate quivering motion is produced by a relatively strong accompanying innervation of the antagonists to the muscles which perform the actual work of the rolling movement. A similar phenomenon can be seen in voluntary human movements when these are opposed to an *unknown* resistance or to a resistance which can undergo unpredictable and sudden alterations in intensity. In such cases, in addition to the muscles performing the actual mechanical work of the movement concerned, *the antagonists* are activated. By means of a known and controllable intensification of the internal resistance to the movement, the unpredictable and sudden variation of the external resistance is rendered less effective, since the latter now represents only a minor contribution to the total resistance which must be overcome. This does indeed lead to a far greater consumption of muscular energy than is actually required for performance of the required external work, but stray movements are reduced to a fraction of their expected extent. This kind of control of muscular power, restrained by wasteful antagonistic activation, is the only means of stabilizing the form of the intended movement in the face of variations in the external resistance, which occur too rapidly for timely operation of receptor-controlled braking of the stray movement. On the basis of this necessity to do without receptor-control, we can explain the accompanying innervation of antagonists, the resultant tension of the part of the body employed and the frequently evident quivering movement observed with the instinctive behaviour pattern. The centrally co-ordinated motor pattern as such is not influenced by the receptors and therefore, in cases where mechanical work is performed against external resistances, the behavioural action *always* operates against a (so to speak) unknown resistance. In all cases where the resistance may vary, the pattern will require internal control of muscular power to a far greater extent than any voluntary pattern of movement. We do in fact find the described conditions of tension and antagonistic quivering in many mechanically-operative instinctive behaviour patterns. A particularly clear case is the lateral pushing movement with which herons and related birds insert twigs into the nest structure. A further indication of accompanying innervation of antagonists is provided by the fact that with instinctive behaviour patterns which have to overcome considerable resistances in their normal performance, the absence of resistance in occurrences of the vacuum response does not produce a noticeable alteration in the pattern or speed of movement.

Apart from the motor pattern described, which is performed more or less through the bird's sagittal plane, there are other motor acts which can be observed. These are *lateral* movements of the head and beak,

Fig. 2. *The hollowing movement*

which are normally restricted in extent and evidently perform the task of preventing the egg from slipping to the right or left and rolling back past the beak as it is rolled up the nest wall. *The egg is balanced on the underside of the beak.* This is particularly obvious when the path of the egg leads very sharply upwards so that a large part of the total weight is carried by the goose's bill. The extent of the movements immediately increases when the egg threatens to become unbalanced and can only be prevented from rolling away by application of a more extensive compensating movement. In such cases, the balancing nature of the lateral movements is immediately obvious.

After the conclusion of this egg-rolling movement, and always *before* the appearance of the settled brooding posture, the goose performs a specific instinctive behaviour pattern which will be referred to as the *hollowing movement*. The elbows are slightly extended such that the wing shoulder is lowered and pushed forward to press against the anterior edge of the nest. At the same time, the goose pushes itself forward with alternating backwards shuffling movements of the feet, so that the breast and the wing shoulders exert a strong pressure on the anterior edge of the nest-cup, whilst the feet themselves push nest-material towards the edge of the nest situated behind the bird (Fig. 2). This movement is employed alone for production of the nest-cup at the beginning of nest-building, and the same movement is used by the bird to remove the protective layer of down when returning from a brooding pause. Adoption of the normal brooding posture apparently never occurs without prior performance of this instinctive behaviour pattern.

From this normal, adaptive performance of the response itself, one can conclude fairly reliably that two fundamentally different motor processes are involved. An entire range of immediately evident characters of the instinctive behaviour pattern proper, such as the described tense condition accompanied by antagonistic quivering, the reliable photographic repetition of the movement, a number of comparative zoological features and other phenomena, permit the extremely safe assumption that centrally co-ordinated motor processes are present. On the other hand, the lateral movements of the beak which balance the egg are completely dependent in direction, extent and form upon the tactile stimuli which emanate from the egg in the course of its varying deviations to the right or left. These movements must therefore be regarded as the products of a taxis. A number of further facts, though derived from observations conducted on the basis of the analytical approach described rather than from experiments, support this assumption.

The most important of these observations, which strictly speaking includes all the results which later experiments were to provide, is the following: The egg rolled towards the nest with the motor pattern described will by no means always arrive at its goal. It frequently slips away from the beak as it is pushed up the nest wall, dropping back on more than half of the observed occasions when the wall was steeply constructed. But when there is a failure of this kind, the motor sequence is not always broken off – it is frequently continued in a peculiar manner. The tucking movement of the head and neck continues exactly as if the movement were really transporting an egg beneath the goose's abdomen. *But in this vacuous completion of the movement the balancing lateral movements of the head are absent.* The continuing movement, which proceeds exactly along the sagittal plane of the bird standing in the nest-cup, eventually leads to contact of the beak with the eggs remaining in the nest-cup. This tactile stimulus appears to elicit a new egg-rolling movement. At least, the contact almost always leads to particularly intensive and thorough rolling of the entire clutch, whereas this intensity is not usually observed when the egg rolled from outside really does reach the nest-cup. This gives the impression that the goose has been left 'un-satisfied' following the short stretch of the movement performed without the egg and consequently 'enjoys' with an enhanced appetite the tactile contact of the beak against the eggs. After rolling the clutch around, the goose settles onto the eggs, performs the hollowing response already described (Fig. 2), and – as long as there are at least a few eggs to be felt beneath the abdomen – will remain settled in completely contented fashion until the runaway egg is spotted afresh. To use an involuntary anthropomorphization: the goose seems to exhibit amazement at seeing 'yet another' egg outside the nest. At least there is no doubt that the goose is not aware whether the egg-rolling pattern has transported the egg into the nest-cup or not. Following an interval of varying length and renewed summation of the stimuli emanating from the egg, a new rolling response is performed.

The interpretation of the portion of the movement performed without the egg as a vacuum response may be countered with the supposition that the goose simply moves its beak along the shortest path to the nearest available egg after losing the egg it was rolling. For this reason, Lorenz intends next year to procure more extensive vacuum activities than those so far observed by artificially arranging for the rolled egg to disappear as rapidly as possible (e.g. by using a pit trap or by suddenly whisking away a light dummy with a thin thread). The responses will be recorded on film so that the form of the motor pattern in vacuum responses can be directly compared with that in the normal performance. We believe that a comparison of this kind will demonstrate the correctness of the assumption made here.

The fact that the balancing lateral movements disappear in the vacuum activity shows that these movements are dependent upon the tactile stimuli derived from the egg and therefore represent orienting responses. This does not necessarily mean, however, that the remaining part of the motor pattern lacks control through taxes. But if such additional, simultaneous responses are present they definitely play no more than a minor rôle. In any case, during the egg-rolling movement the goose's beak does *not* follow all of the minor vertical displacements which the egg undergoes because of unevenness of the substrate. The beak contacts the egg at its upper face when it is rolling through a small hollow and low down when the egg happens to pass over a prominence. Probably, the only process which orients the rolling movement roughly parallel to the substrate, to the pathway of the rolling egg, is the stretching of the neck towards the egg observed *before* the egg-rolling movement is observed. There is therefore no orienting response operating simultaneously with the instinctive behaviour pattern, but a preceding telotaxis belonging in the category of appetitive behaviour (in its narrowest sense). The motor pattern can also exhibit coarse irregularities in its parallel orientation to the substrate, under certain circumstances. The egg-rolling pattern is apparently quite specifically tailored for the situation in which the performing goose is standing exactly on the edge of the nest. Under natural conditions, this is doubtless usually the case. In the first place, the goose is subject to a considerable inhibition against leaving the nest (i.e. against proceeding further than the nest wall); in the second place, this is usually sufficient for the goose to reach an egg which has not been deliberately placed some distance away, but has been allowed to roll freely down from the edge of the nest, so that the spatial relationship between the egg and the goose is doubtless closest to that under natural conditions. Since a goose nest is always located in the middle of thick-stemmed vegetation, the egg always rolls just to the edge of the area covered with nest material, such that a fairly constant distance from the centre of the nest is established. Under these conditions, the motor pattern in the sagittal plane appears well adapted. The form of the pattern automatically determines that the vertically-operating component of the pressure from the beak is strongest at the point where it is most necessary, namely towards the end of the movement, where the egg has to pass the steepest gradient on the nest wall. The nest of our experimental goose was composed of pine needles (i.e. an extremely unnatural form of nest-material) and the wall was far lower and less steep than that of all the nests which we have seen under natural conditions. In Fig. 1D, one can almost see that the motor pattern is actually adapted for a somewhat steeper nest wall. Genuine disruption of the rolling movement actually occurred when the egg was *nearer* than expected under natural conditions. The goose, in this case, did not move forward but remained stationary over (or even behind) the eggs, so that the strongest lift exerted on the egg by the beak appeared at a point on the pathway where the substrate had already begun to decline again, because the peak of the nest wall had already been passed. The egg was then often raised into the air, only to fall from some height onto the rest of the clutch after several seconds of continuous juggling on the underside of the beak. This frequently led to cracking of the eggs.

According to these observations, there would definitely appear to be no taxis-controlled movement in the sagittal plane of the egg-rolling response, apart (of course) from the extension of the head towards the egg, which is to be regarded as appetitive behaviour. This latter movement only quite generally determines the angle of inclination of the neck and back of the goose and it occurs in temporal separation from the centrally co-ordinated motor sequence. Nevertheless, the possibility of finding other taxes which function simultaneously with the instinctive pattern must be examined in more detail. If it is at all possible, this will be attempted next year through preparation of ciné-film of the egg-rolling movement taken exactly from the side, with the profile of the path varied as much as possible in a number of successive shots. Such film material would permit graphic reconstruction of the curve of the movement, so that any influence of tactile stimuli from the egg upon the movement can be clearly demonstrated.

IV Experiments

This year's experiments were confined firstly to investigating the dependence of the balancing lateral movements upon tactile stimuli and secondly to demonstrating the fixity and receptor-independence of the sagittal movement which must be present if a pure instinctive behaviour pattern is involved. We at first looked for objects which would elicit the goose's rolling response and were nevertheless so different in form from a goose's egg that the stimuli necessary for the thigmotactic orienting response would be greatly modified or entirely absent. In addition, we looked for objects which would require receptor-controlled modifications of the sagittal movement and would literally cause a breakdown if the movement were independent of the receptors and purely centrally co-ordinated. This search for substitute objects had the result that we discovered a number of things concerning the conditions necessary for elicitation of the rolling response. This combination of hereditarily-recognized characters of the object – the so-called innate schema of the egg – will be briefly discussed.

1. THE INNATE RELEASING SCHEMA

The term innate releasing schema applies to the hereditarily-determined readiness of an animal to respond to a specific combination of environmental stimuli by performing a specific behaviour pattern. There is an innate receptor correlate for a stimulus combination, which (in spite of its relative simplicity) characterizes a biologically-important situation with sufficient exactitude to link the appropriate response firmly to the conditions of that situation and to prevent inappropriate elicitation by environmental stimuli exhibiting chance similarities. The elicitation of motor processes by the reaction of an innate schema corresponds in every detail to an *unconditioned reflex* and is allied to the simplest unconditioned reflexes by a continuous series of intermediate forms. However, this only applies to the releasing mechanism itself and it is not permissible to draw conclusions about the released or inhibited motor pattern. The latter can in fact be of an entirely different character. The reaction of an innate schema can just as easily elicit a taxis-free instinctive behaviour pattern or appetitive behaviour leading to performance of an instinctive pattern, and the appetitive behaviour can be represented by a simple orienting response or by the highest forms of purposive behaviour. Similarly, a pure taxis leading to the converse situation of a refractory resting state, rather than to an elicitatory stimulus situation, may be elicited by an innate schema.

In the case of the egg-rolling response, the innate schema of the situation 'egg outside the nest' elicits appetitive behaviour in the form of the previously-described orienting response of stretching the neck. The stimulus parameters representing the characters of 'the egg' are so simple and restricted in number that one is amazed that they characterize the egg to a biologically adequate extent. It is a well-established fact (Koehler and Zagarus, Tinbergen, Goethe, Kirkman, etc.) that many brooding birds, when performing the egg-rolling response, will accept objects which bear very little similarity to the egg of the species concerned. Herring-gulls will brood on polyhedrons and cylinders of any colour and of virtually any size, and black-headed gulls appear to be even less selective. At first, our Greylag geese behaved in a basically similar way, but they later exhibited a remarkable change in their behaviour determined by individual learning. It is also possible that pure-blooded individuals might react in a manner different from that of the hybrid animals investigated. It is a particular feature of innate releasing schemata that *loss of characters* can occur as effects of domestication, and this leads to a considerable broadening of the object-schema. The observations must be repeated with pure-blooded geese in order to determine whether such an effect operated with our experimental animals.

The purely optical elicitation of the appetitive orienting response of stretching the neck is in fact linked only to the character of *surface continuity* of the object. Any smooth object presenting an approximately even contour will arouse the attention of the goose and will provoke neck-stretching. However, a tactile character operates a moment later. Even before the goose stretches its beak over the object, the degree of hardness is determined by lightly pushing the beak against the object (Fig. 1B). A rounded toy chicken made of white rubber will clearly awaken appetitive behaviour for the rolling response. The goose rises and stretches out its beak towards the object, but after gently tapping with the beak the interest immediately disappears. A yellow children's balloon blown up to the size of a goose's egg is treated in the same way. A hard-boiled, peeled Muscovy duck egg is tested with the beak and then greedily eaten. In our experience, the colour of the object plays no part whatsoever in the goose's behaviour, and this provides a certain contrast to gulls and plovers. Size has no effect over such a wide range that one is scarcely able to decide whether failure of the response with too large or too small eggs is due to limits imposed by perceptive mechanisms or by purely mechanical factors. On the other hand, the requirements imposed with respect to surface continuity of the object at the beginning of the appetitive behaviour are considerably reinforced in the succeeding components of the response. The performance is particularly prone to immediate blockage by optical or tactile perception of any projecting edges or appendages on the object. A hollow wooden cube opening on one face is immediately rolled by the goose, but this persists only until the opening comes to point upwards. The rolling response then ceases as the goose begins to peck intensively at the exposed rim of the opening. A white toy chicken carved from a piece of wood the size of a goose's egg was not rolled but nibbled wherever projections (such as eyes, beak and feet) had been glued on to the egg-shaped body. Similarly, a large cardboard Easter egg was nibbled where the paper frills projected from the decorated seam. The right-angled edges of a cube or the edges of a plaster-case cylinder, on the other hand, are not noticed and do not disrupt the egg-rolling response.

This intensive response to all projecting appendages is not a chance phenomenon. The described form of the response is only elicited when the nibbled prominences are attached to an object *which otherwise agrees well with the egg schema* and arouses the appetite for rolling and brooding. This is doubtless a quite specific response which is presumably similarly characteristic of all Anatids and guarantees the adaptive function of *removal of broken eggs*. Experiments with various duck species and Egyptian geese have shown that such eggs are 'dismantled' from the edges of the crack and devoured, before the other eggs in the clutch are soiled by leaking egg content and consequently handicapped in respiration. Of course, this response must be inhibited prior to the onset of hatching, since the goose would otherwise kill the emerging chicks. It so happened at this time that pure-blooded Greylag goose eggs, which had been brooded by a hen, were just about to hatch. For technical reasons (regarding the requisite oiling of the down feathers of the goslings by the plumage of an adult goose), we intended to arrange for the actual hatching process to occur under our experimental goose. This proved to be utterly impossible, however, since the brooding goose immediately began to nibble greedily at the edges of the pipped areas on the eggs, responding at an evidently high intensity level. The goose pecked so uninhibitedly that the edge of the shell was at once pressed inwards, and blood began to flow from the torn egg membranes. The goslings would definitely have been killed, one and all, if the eggs had not been rapidly removed. When the goose's own offspring hatched only two days later, however, the response of nibbling edges and prominences had completely disappeared. We were not even able to elicit the response by directly presenting the goose with the pipped openings. The goose did in fact incline her head towards the peeping egg in an aroused and attentive fashion and the egg was even touched with the slightly opened beak, but no actual pecking resulted. The same movement of the head and beak is often exhibited by a guiding mother goose towards very young goslings, in which case the impression of an affectionate gesture is given. It only remains to provide experimental backing for our supposition that acoustic stimuli emanating from the hatching egg provoke response-reversal in the mother, so that even the slightly-pipped egg is already treated 'as a gosling'. This supposition is supported by the fact that pipped eggs are evidently no longer turned, since in the undisturbed clutches of hatching goose and duck eggs the pipped areas are virtually always uppermost and remain so. In addition, the mother no longer treads upon the eggs, as she does without hesitation in earlier stages of development (see also Fig. 5), and she does not even lower her full weight onto the clutch. Instead, there is already an indication of the squatting position which later serves to warm the young goslings. Unfortunately, we did not investigate whether a markedly pipped egg lying outside the nest-cup would still be rolled or not.

An attempt to elicit the rolling response with a freshly-dried, but still helpless, gosling gave an unusual result. Surprisingly, the goose initially responded in exactly the same way as at the onset of the normal egg-rolling response. She stood up, bent towards the gosling and touched it tentatively with her beak. On a few occasions, we even observed the mother reaching over the gosling with her beak. However, a retrieving, pushing movement never followed; instead, the goose always withdrew her head 'disappointedly'. This gave the definite impression that the goose *intended* to transport the gosling into the nest and attempted to achieve this with the rolling movement, but that the response was disrupted by the poor correspondence between the gosling and an egg. The goose was obviously aroused. A few seconds after the failure of the response, the mother once again stretched her neck towards the peeping gosling. However, since the relevant experiments with pure-blooded greylag geese have not yet been conducted, we will not discuss this behaviour at this juncture. It is possible that the continual backwards and forwards motion of the mother's head might eventually lead a scarcely mobile young gosling back into the nest, but it is equally possible that the experimental goose was subject to domestication-induced disruption of a specific response to a helpless chick situated outside the nest-cup.

Our experiments on the innate releasing schema of the egg-rolling response came to a remarkable close long before the end of the brooding phase of our experimental animals. After a pause of several days in the middle of the brooding phase, we attempted to induce the tamest of our geese (which had been used for almost all of the previous experiments) to roll a number of objects of different kinds, but we obtained no response. It soon emerged that the egg-rolling response, from this point onwards, *could only be elicited by genuine goose eggs*! All of the objects previously described as substitute objects were ignored and even an intact, small chicken's egg evoked no response. On the other hand, the egg-rolling response was performed with goose eggs immediately and without any particular tendency towards habituation; so our suspicion that the intensity of operation of the central mechanism had been reduced was shown to be unfounded.

2. EXPERIMENTS TO SEPARATE TAXIS AND INSTINCTIVE BEHAVIOUR PATTERN

The wide range of variation in substitute objects permitted by the crude and simple character-schema of the egg provided us with the opportunity of presenting appropriate objects of specific size and form in order to produce situations which facilitated the dissection of the entire behaviour pattern into tactic and instinctive components.

Our first task was experimental verification of the hypothesis that the lateral movements of the beak which balance the egg during the egg-rolling pattern are directly evoked by tactile guiding stimuli, as was deduced from the observations discussed on pp. 331–332. For this purpose, we needed to find a substitute object which – unlike the rolled egg – would not exhibit persistent lateral departures from the path and would thus fail to give tactile stimulation of the lateral margins of the underside of the beak. We first attempted to induce the goose to roll a plaster-cast cylinder along a smooth pathway by presenting the object on a broad plank resting on the nest wall. This experiment proved to be unsatisfactory, to the extent that the concomitant disturbance produced by the unfamiliar (and therefore fright-inducing) board weakened the goose's response and led to rapid habituation. In addition, the goose frequently failed to contact the cylinder in the middle (as would have been the case with egg-shaped objects). The cylinder was grasped so far to one side that pronounced lateral departures from the path occurred and elicited even more marked lateral head movements. These movements were nevertheless noticeably distinct from those usually observed. Since the straying movements of the cylinder were slower and less frequent than those of the egg, the relationship between each straying motion of the object and the compensating lateral movement of the beak was far more distinct than in the normal performance.

Exclusion of the balancing lateral movements proved to be much simpler and far more extensive when the goose was presented with an object which would slide rather than roll. A small, extremely light wooden cube, which remained stationary at any point on the nest perimeter and exhibited no tendency to slide back or to roll in a particular direction, proved to be the best object for our purposes. This was probably partially attributable to the purely mechanical fact that the cube usually tipped over so that one of the relatively broad, flat faces came to lie across the two rami of the underside of the beak (Fig. 4, p. 343). This naturally meant that the cube would tend to remain in the plane of movement, and it did in fact emerge that the cube was always pulled directly into the nest *without the slightest lateral movement*.

On the basis of these two experiments, we believe that it can be taken as established fact that the lateral movements of the beak are directly elicited by tactile stimuli produced by sidewards straying of the rolled object.

As a counterpart to this conclusion, it was necessary to establish whether the movement in the sagittal plane (which has been described as a pure, centrally co-ordinated instinctive pattern) is indeed free from modification by supplementary orienting responses serving to produce more detailed adaptation to the spatial relationships in any given individual situation. However improbable such modification may seem on the basis of the observations already discussed, it is nevertheless good practice to attempt to induce some slight modification of the innate co-ordination of the sagittal neck and head movement. The largest substitute object used in our experiments (the large cardboard Easter egg already mentioned) seemed to be suitable for this purpose. Since the diameter of this object required entirely different neck postures to those necessary for a genuine goose egg, it was obviously a promising step to observe the operation of the rolling movement under the resulting altered conditions. The size and curvature of the object determined a disruption right at the beginning of the movement. The action of reaching over the egg (as seen in Fig. 1C) is initially followed by a movement of the head alone, in which the head is inclined at an acute angle towards the neck, at first without altering the extended neck posture. This typical initiation of the sagittal movement will push a genuine goose egg towards the nest – but not the dummy egg used in the experiment. The beak is pressed against the surface of the substitute object at such an awkward angle that the head slips away, and we were forced to intervene slightly in order to permit the response to continue (Fig. 3A). All features of the motor pattern exhibited invariable adaptation to a much smaller object in other details of the response as well. Even before the pronounced flexure of the neck (Fig. 1D) could occur, the dummy egg *jammed* between the goose's beak and thorax so that the posture shown in Fig. 3B remained unaltered through 1½ metres of our film-strip. The response was then broken off and the goose adopted an upright posture. Characteristically, this event was never followed by the 'hollowing movement' typical of the 'satisfactory' performance of the response. Instead, the goose remained standing in the nest 'in discomfort'. Such forceful disruption of an instinctive behaviour pattern is apparently experienced as something extremely unpleasant, since the experimental goose exhibited 'disinclination' towards the large Easter egg (expressed

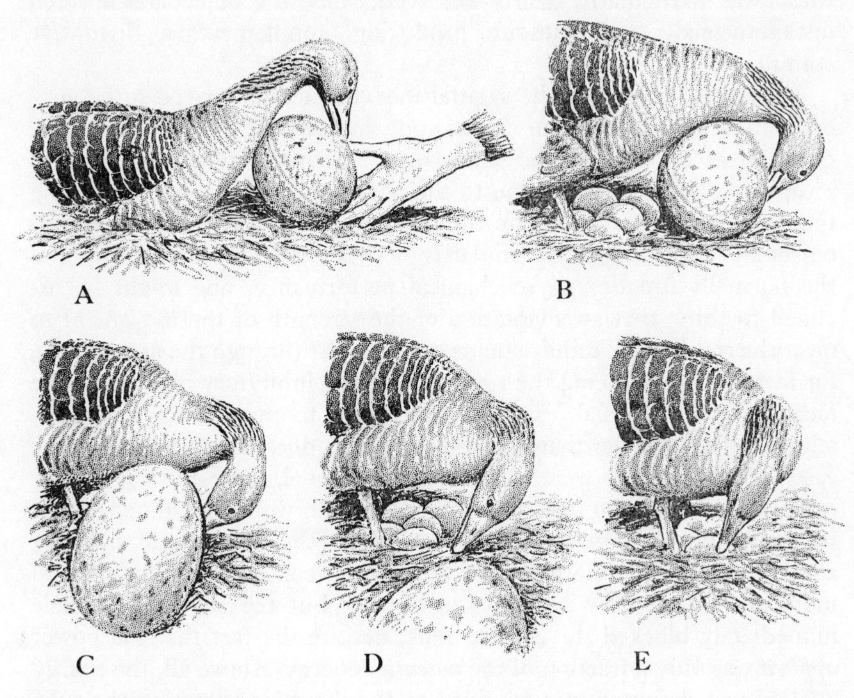

Fig. 3. *Explanation in text*

as lowered response motivation and rapid habituation) earlier and more intensively than towards other substitute objects. The goose never hit upon the idea of transporting the egg into the nest by walking backwards *without* the pronounced flexion of the neck.

In a number of our experiments, including that providing illustrations for Fig. 3A–E, the goose reached a subjectively far more satisfying solution: The dummy abruptly slipped out from between the beak and the thorax, like a cherry-stone pressed between two finger-tips. In Fig. 3C, it can be clearly seen how the beak follows the egg to the right 'in the attempt' to balance it. Four frames later in the film, the sagittal movement can be distinctly seen progressing *in vacuo*. In Fig. 3D, the head is still displaced somewhat to the right following the ultimate, extremely pronounced thigmotactic movement, but thereafter the movement of the beak is not related to the object, which is lying far out in the foreground. The movement is subsequently followed by the intensive rolling of the clutch described on p. 333 (Fig. 3D), and the goose eventually settles and performs the 'hollowing movement'. In all cases where this abrupt slipping effect of the large, but quite light, cardboard egg robbed the response of its object, the progression of the pattern *in vacuo* was particularly clearly observed, since the object disappeared instantaneously and without producing supplementary disruptive stimuli.

Although the *form* of the sagittal movement thus proved to be generally invariable in all observations and experiments, there remained the question as to whether the *power* of the response might be open to receptor-controlled adaptation to the obstructing resistance. Following the described observations on the associated innervation of the antagonists and on the detailed similarity between the vacuum activity and the normally-functioning mechanical performance, one might be inclined to think that an adaptation of the strength of the movement to meet the mechanical requirements might occur through the antagonists, for instance by reducing the resistance of the inhibitory muscles in the face of greater external resistance. According to our findings with substitute objects heavier than a goose's egg, this does not appear to be the case. These findings were initially unintentional, since the egg-rolling response failed in the face of resistances which we had assumed would be overcome as a matter of course. With the plaster-cast cylinder already mentioned, the power of the beak sweep was just sufficient to move the object over a smooth substrate, but the slightest obstacle immediately blocked the action. This, despite the fact that the power *applied* was only a fraction of the *potential* energy. Above all, the energy applied was definitely not confined to the direction in which the substitute object was rolled; the interplay of antagonistic muscles already described was just as obvious as ever before. This weak and unsuccessful wavering of the response with an object which is only a little overweight itself gives the most convincing impression of unintelligent, machine-like behaviour, at least for the initiate who is aware of the amazing power which a greylag goose's neck can otherwise exhibit. For example, playful performance of the grass-plucking pattern can produce a pull which will drag a heavy oak chair across a rough floor or whisk away the tablecloth from a table laid for several people. The fact that only an inordinately small fraction of this capacity can be employed in the rolling response for rolling a slightly overweight object supports the assumption that the energy employed in the sagittal movement is not accessible to any appreciable receptor-control.

With substitute objects which are lighter than a goose egg (e.g. the wooden cube mentioned several times previously), the movement in the sagittal plane showed virtually no difference from the vacuum activity. Towards the end of the movement, where the goose egg may normally be lifted and balanced in mid-air (see p. 334), the cube was regularly swept upwards and often pressed against the goose's thorax so that it was jammed in that position for a moment by the beak (Fig. 4). The ease and regularity with which the light cube was lifted created the definite impression that the tension of the muscles concerned in the sagittal movement had been calculated once and for all to cope with the normal average weight of a goose egg. On the other hand, this lifting of the cube itself awakens the suspicion that even in the muscles performing the

Fig. 4. *Retrieving a cube*

sagittal movement there are minor reflex processes operating alongside the centrally co-ordinated impulses. In the completely vacuous performance of the movement, in which the beak has nothing whatsoever to carry, one would expect the head to lift even further than in the carriage of the cube, and this is not the case. Instead, the beak is held noticeably low in the vacuum activity and even scrapes the substrate in some cases (Fig. 3D). The tactile stimulation from the rolled object presumably does not elicit an orienting response in the true sense of the term but produces a *tonus effect* in the muscles which incline the neck and (particularly) the head. A reflex tonus effect of this kind would of course be dependent upon receptors, but it could only be interpreted as an orienting response if it were to emerge that the intensity of the effect depended upon the weight of the rolled object. But this would appear to be improbable on present evidence. On the basis of the observations reported here, we incline more to the view that the power employed remains the same with both light and heavy objects, even if it may be slightly greater than the power involved in the vacuum activity.

3. APPETITIVE BEHAVIOUR FOLLOWING OVER-SATIATION OF THE INSTINCTIVE PATTERN

In conclusion, we should like to discuss two further observations which do in fact provide direct experimental evidence for the separation of taxis and instinctive pattern, but which are most important with respect to a different problem. From different quarters, compulsive elicitation of the 'instinctive' behaviour pattern (i.e. with the orienting responses often included under the same concept) has often been emphasized as an important characteristic. The instinctive behaviour pattern as defined here is, of course, not elicited with the same compulsive quality as a knee-jerk reflex or the like, since elicitation of the former always signifies *disinhibition* and since performance of the pattern is dependent upon central arousal states. Exhaustion of the central mechanism renders the normally elicitatory stimuli utterly inoperative. Such exhaustion appears in an action-specific form extremely rapidly with many instinctive behaviour patterns, long before the organism as a whole or the participating effectors are exhausted. Some instinctive behaviour patterns, such as the injury feigning performed by a number of birds to distract predators from the nest-site, can only be repeated a few times in succession. Even if the elicitatory stimulus situation persists, the behaviour pattern is not repeated. In the case of injury feigning, the bird presents an indifferent activity to the potentially elicitatory human observer and begins to search for food, to preen itself, etc. This is far from the case with the taxis; an orienting mechanism can usually be set in operation on a virtually limitless number of successive occasions. A beetle turned on to its back will right itself with a persistence greater than that of the experimenter. Similarly, an ant which is displaced from its path of locomotion will consistently re-orient itself with respect to the direction of the sun.

We must now ask whether this typical resistance to fatigue is also found with orienting responses which operate as appetitive behaviour, bringing the organism into the appropriate spatial relationship to the object for performance of the corresponding instinctive behaviour pattern. This is not generally thought to be the case; instead it is widely believed that the ease of elicitation of the appetitive behaviour, even where this is represented by a simple taxis, vanishes with satiation through the performance of the instinctive behaviour pattern. J. von Uexküll states: 'The effector signal always extinguishes the receptor signal – the behaviour pattern is thus concluded. The receptor signal is either objectively extinguished, for example when it is derived from a food object which is eaten, or it is subjectively wiped out when a state of satiation is reached – the sensory filter is closed.' But it is our impression that this extinction of elicitatory signals is not always complete, particularly when the elicited appetitive behaviour is a relatively simple orienting response. One can provide many examples in which an instinctive behaviour pattern can no longer be elicited after repeated performance, whilst the preceding orienting response can still be elicited. It is more or less the rule that an over-satiated organism will still attend to the stimuli normally eliciting a behaviour pattern (e.g. by orienting its gaze or performing some other orienting action), even when the instinctive behaviour pattern itself has been exhausted by repeated performance.

It is a remarkable fact that an orienting response which, by its very nature, is only functional as appetitive behaviour should be open to elicitation even in a case where the actual appetite (in terms of active purposive behaviour on the part of the subject) has been extinguished. Such elicitation has a distinctly compulsive, reflex-like appearance. It is also remarkable – and probably fundamentally important – that such reflex-like elicitation of such 'appetite-less appetitive behaviour' is evidently accompanied by subjective displeasure. Even we human beings, after satiation, will turn away in disgust from even the most delicious dishes and push the plate out of reach. In brief – we withdraw from the stimuli emanating from the object of the instinctive behaviour pattern just concluded. This itself is clear indication of a reflex-like, compulsive response to the stimuli concerned. It would not be necessary for us to withdraw from these stimuli if they had really been 'subjectively extinguished' by satiation (to use von Uexküll's terminology) and if we were not compelled by further operation of these stimuli, despite our lack of appetite, to an unpleasurable repetition of our response. Even if the above-mentioned quotation by von Uexküll is generally applicable, there are definitely a number of exceptions. It is possible to cite many cases in which an organism will not become completely indifferent to the stimulus situation eliciting appetitive behaviour at any level of satiation and will instead respond with a withdrawing response immediately after extinction of the approaching orienting response.

The described persistence of the introductory orienting response after exhaustion of the instinctive pattern was easily observable in the egg-rolling response of the greylag. The instinctive behaviour pattern concerned habituates rapidly. This is not surprising, since the tempo of central stimulus-production and the associated maximum frequency of performance of the pattern are adapted to the natural environmental conditions, under which the eggs are certainly not dislodged with the frequency pertaining in our experiments. It was completely unnecessary for us to conduct special fatiguing experiments; without our intervention we became more familiar with fatigue effects of the pattern than we actually desired. The differences in the behaviour elicited by a genuine goose egg and the oft-mentioned large Easter egg were conspicuous and probably significant. Since the observations concerned were conducted shortly before the point where the goose began to rigorously ignore all substitute objects, the paradoxical nature of the behaviour to be described is probably due to the fact that the goose was acquainted through personal experience with the goose egg as a satisfactory object for rolling and brooding, whereas the dummy had been experienced as a substitute object which did not provide full satisfaction.

Even after utter exhaustion of the instinctive pattern, the tactic

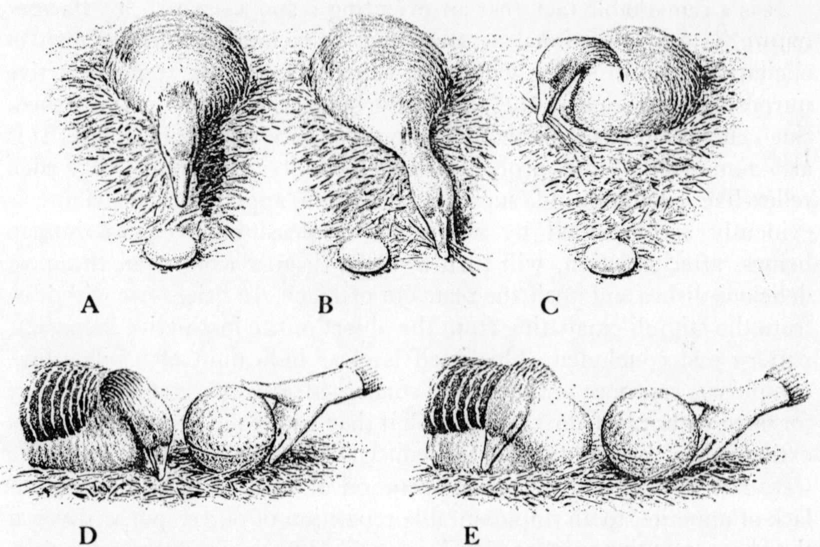

Fig. 5. A–C: *The 'gathering pattern'*. D and E: *Submissive gesture*

movement towards the egg persisted. The orienting stretching of the neck (Fig. 5A), or at least an attentive glance towards the egg, could be elicited virtually at any time. The stimuli emanating from the egg will apparently not allow the goose to settle down; the neck is repeatedly stretched towards the egg until the goose eventually performs a type of behaviour which we regard as being of great theoretical interest. In the place of the habituated, refractory egg-rolling pattern, the neck-stretching movement was abruptly followed by *a different instinctive behaviour pattern*. Immediately after stretching the neck towards the egg, the goose performed the '*gathering pattern*', which is otherwise employed in nest-building. This motor pattern is exhibited by all Anatids and serves the function of collecting and consolidating nest-material at the nest-site (Fig. 5B, C).

Such substitution of a biologically inappropriate instinctive behaviour pattern for the expected pattern has been recognized in birds in a number of different situations. Birds which are greatly alarmed by the presence of a predator close to the nest (or offspring) will often interpolate, between the threatening movements and the well-known guiding-to-safety responses, behaviour patterns of food-seeking, preening and even settling down to sleep. Lapwings disturbed on the nest may rise and perform pecking movements; Corvids which are not quite confident enough to attack will peck with all their might at the branch on which they are sitting; and so on. Apparently, these 'erroneously' erupting instinctive patterns can be produced in *two* different ways. In some cases, such as that of nest-defending passerines which abruptly follow a small number of mock limping responses (or the like) with preening patterns, there would appear to be much justification for Howard's assumption that after exhaustion of the specific motor pattern appropriate to a given situation the receptor-controlled arousal state will flow over 'into other channels' and lead to the eruption of 'some other' instinctive behaviour pattern.[120] In other cases, such as that of a Corvid pecking at its perch, we are inclined to assume that two incompatible instinctive behaviour patterns (in this example, patterns of attack and escape) block one another so that the general arousal is channelled off into a third pattern. A further possibility for the production of similar phenomena would be the following: From some mammals we know of a functionless mode of performance of instinctive behaviour patterns which nevertheless exhibits some degree of 'insight'. Everyone is familiar with the 'begging' motion of scraping with the fore-hoof which Equids – from the carthorse to the wild stallion – will exhibit towards a human food-donor. Dogs which no longer dare to bite after a painful experience with a wasp or a hedgehog will regularly begin to dig in the direction of the arousing object. The individual scratching movements usually do not touch the object, so that a hedgehog rolled into a ball may eventually stand more or less on a pedestal, surrounded by a trough. In intelligence tests, monkeys will sometimes throw sand in the direction of the goal; and so on. In all of these cases, it is highly probable that an extremely primitive form of insight is involved, to the extent that 'something must be undertaken' in a particular direction.

It may be an overestimation of the powers of intelligence of the Greylag to assume that the behaviour described above is closest to that of the mammals mentioned, but this was our immediate impression when observing the process in action. It seemed as if the goose were 'unable to bear' the sight of the egg lying outside the nest-cup without doing something about it. This subjective description is probably less naïve than may appear at first sight, since human beings experience similar phenomena in just the same type of situation, where innate releasing schemata and instinctive behaviour patterns play a major part. We believe that the goose attempted to cope with the unsettling and unpleasantly intrusive stimulus situation following the failure of the egg-rolling response by the purposive 'application' of another accessible instinctive behaviour pattern. In fact, we feel justified in assuming that it was by no means a chance effect that the motor pattern which emerged happened to be one which similarly transported something towards the nest. A major feat of intelligence of this kind – the most impressive which we would attribute to a bird – itself demonstrates the animal's pronounced dependence on innate motor patterns.

Whereas a Greylag goose egg lying alongside the nest induced our experimental goose to perform the described behaviour after extreme exhaustion of the egg-rolling response, the cardboard egg had no effect at all once the response had suffered the least habituation. The goose then continued to brood in an entirely settled fashion in sight of the dummy egg. However, if the Easter egg was forced upon the goose (as shown in Fig. 5D), the beak was drawn in (as shown), as if to avoid any tactile contact with the egg. If the egg were pressed even closer, the goose turned her head right away (Fig. 5E). The head movement shown is not that of 'gathering' of nest-material but represents the so-called 'submissive gesture' which a brooding goose always exhibits when roughly harassed on the nest by a higher-ranking conspecific and does not quite dare to chase the intruder away. Even a highly-intrusive approach by a human being (e.g. in stroking or scratching the brooding goose) was not met with this gesture, so such an approach was obviously a less unpleasant stimulus for the goose than the forcible introduction of the literally 'unappetizing' dummy egg. Although the form of this movement is also innately determined, the motor pattern in this situation gave the observer an extremely strong recollection of the emotional state which well-bred children have often summarized with the following words in the face of cake-foisting aunts: 'Thanks awfully, but I feel sick already!' It is doubtless significant that this behaviour appears most prominently in response to an object which is inappropriate for the instinctive behaviour pattern.

Summary and conclusions

The subject of this study was the manner in which an instinctive behaviour pattern (i.e. a receptor-independent, centrally co-ordinated motor pattern) can co-operate with one or more receptor-controlled taxes and combine with them to produce a functionally unitary and adaptive behavioural sequence when performed by the animal concerned. The motor sequence which the Greylag goose (*Anser anser* L.) employs to return a stray egg to the nest-cup incorporates an instinctive behaviour pattern and both simultaneously and separately operating taxes. Oriented stretching of the neck represents an introductory orienting response, producing the stimulus situation in which the instinctive behaviour pattern is elicited and simultaneously providing the necessary spatial relationship which is a prerequisite for adaptive performance of the pattern. In other words, this is a typical case of 'appetitive behaviour' in Craig's terminology. The instinctive behaviour pattern which is then performed consists of ventrally-directed bending of the head and neck, such that the egg lies against the lower side of the beak and is pushed towards the nest. From the very beginning of this purely sagittal pattern of movement to the end, another orienting response is simultaneously operative. By means of thigmotactically-controlled lateral movements, the egg is balanced on the underside of the moving beak so that the major direction is maintained.

Our assumption that *the movement in the sagittal plane is a pure instinctive behaviour pattern* is based on the following findings:

1. The motor pattern exhibits the phenomenon of *vacuum performance*, which characterizes independence of the instinctive behaviour pattern from receptor-control (p. 333).

2. The form of the motor pattern is always the same. It proved to be impossible to produce receptor-controlled adaptations of the pattern to modified spatial relationships. Neither the characteristics of the pathway along which the egg rolls, nor the form of the rolled object produce recognizable modifications of the motor pattern. In cases where it was attempted to forcibly produce changes through mechanical factors (using a very large object – p. 340), the motor pattern *jammed* and was discontinued.

3. The power with which the sagittal movement is performed remains *constant* within extremely narrow limits. The tactile stimuli deriving from the object do have a minor tonus effect on the muscles performing the movement, but this effect would appear to remain at a constant level even if the weight of the object varies. Thus, as soon as the object is only a little overweight the motor pattern fails.

4. The motor pattern exhibits the property of response-specific *habituation* characteristic of the instinctive behaviour pattern as opposed to the taxis.

The assumption that the *lateral balancing movements* – in contrast to the motor processes occurring in the sagittal plane – are *orienting responses* controlled by tactile stimuli is based on the following facts:

1. The lateral movements do not appear in the object-less vacuum performance of the sagittal movement (p. 332).

2. The lateral movements are lacking during rolling of objects which do not exhibit lateral straying from the path of movement (p. 340).

3. With objects whose lateral departures differ from those exhibited by the goose egg, the lateral movements exhibit complete *adaptation to the manner of movement of the object* (p. 339).

In addition, the following supplementary findings should be mentioned:

1. The appetitive behaviour in the form of orienting stretching of the neck is elicited by any object which exhibits surface continuity and a more or less cohesive outline; the size of the object is immaterial over a wide range (Figs. 3, 4).

2. If the object possesses any protruding corners or appendages, the rolling movement is not continued. Instead, a different response is performed – the object is broken up and eaten. This response, which serves for the removal of broken eggs, is presumably extinguished by means of acoustic stimuli before the chicks are due to hatch.

3. In the course of the brooding phase, rough recognition of the object of the egg-rolling response, which is initially determined only by a few innate characters, is supplemented by acquired recognition of characters, producing enhanced selectivity in the performance of the response.

4. Following extinction of the instinctive behaviour pattern through the easily-procured habituation of the central automatism, *the introductory orienting response can still be elicited.*

5. Following exhaustion of the rolling response, the goose attempts to eliminate the stimulus situation presented by a goose's egg located outside the nest-cup by *employing a different instinctive behaviour pattern.*

6. The goose turns away to *withdraw* from the stimuli emanating from a potentially elicitatory object which has proved to be unsatisfactory.

Inductive and teleological psychology (1942)

In delivering this response to the critique of my theories proffered by Bierens de Haan, I am not concerned with defending my work or with converting him to adopt my views. My only concern is that there may be other workers who are similarly unaware of certain basic differences between inductive scientific method and studies of nature based entirely on teleological principles. In fact, one might think that this theme had already been given sufficient treatment after H. Weber's publication on environmental phenomena in this journal, providing a succinct and lucid text dealing with the usage of anticipatory explanatory principles (1939). Only the fact that concepts such as the instinctive principle as used by Bierens de Haan are still being applied long *after* Weber's publication justifies a renewed consideration of this matter. In this article, I shall employ Bierens de Haan's grandiose, all-purpose explanatory principles on the one hand and my own modest and all-too-fragmentary observational and experimental results on the other as concrete examples for a renewed discussion of the difference between teleological 'holistic considerations' and inductive scientific research.

The extent to which Bierens de Haan has failed to grasp this distinction is evident from his statement that the major difference between our two methods of study lies in the different concepts which we associate with the word instinct. For this reason, I should like to correct this basic error (which is evident in the very title of his article) at once: The dispute is not concerned with the concept of animal instinct but with the settlement of two far more penetrating questions. The first question is: Is it permissible for a student of nature to be satisfied with the unearthing of adaptive purpose and to confine himself to 'holistic considerations' of organic systems, or is he bound by the laws adhering to the internal structure of natural sciences to search for causal explanations? Secondly, there is this question: Is it logically and methodologically permissible to agree that it is the justifiable duty of the biologist to search for natural causal principles and then to introduce *in parallel* anticipatory explanatory principles (e.g. that of 'entelechy', vitalistic phantasy or instinct) which are designed to explain 'everything'? Or does this, as H. Weber suggests, represent the greatest danger which has ever threatened free scientific research? These two questions will be treated separately.

I Basic facts about finalistic, holistic analysis and causal analysis in psychology

The attempt to reach greater understanding of psychological processes by means of causal analysis automatically establishes a precondition which biologists regard as axiomatic. However much this may be disputed by some metaphysicists, there is the undeniable precondition that all 'pure' psychological processes are simultaneously neurophysiological processes. Different expressions have been coined to describe this fact, which is presumably accepted as a basic tenet by Bierens de Haan as well, since he himself states that psychological events are 'one facet' of the organic unit. Other authors speak of a mind–body polarity, or of a correlation between psychological and physiological processes; or they refer to the psychological aspect as an 'epiphenomenon' of neurophysiological events. All of these terms are misleading to the extent that the literal formulation incorporates the old dualism of body and soul, which is just what we are trying to avoid. However, it is just as erroneous to adopt the other, 'panpsychological', interpretation, according to which physiological processes and psychological events are 'the same'. Not all living processes are simultaneously psychological in nature – in fact, by no means all neurophysiological processes are of this type. Only a few, quite specific events have this dual nature; the relationship between psychological and physiological events is entirely comparable to that between living processes and physico-chemical events. By no means every physico-chemical process represents a life process, though every event in a living organism is a physico-chemical phenomenon. In an analogous manner, every psychological event falls under the governing concept of living processes, while the converse does not apply.

This delineation of the concepts of inorganic, organic and psychological processes, which itself corresponds to a quite concrete evolutionary succession, provides the scientist with a specific series, along which unknowns must be traced back to established facts. It is fundamentally unjustifiable, for example, to interpret the reflex as an 'extreme case of an instinct' (Buytendijk, 1939) or to regard as an 'explanation' the intrinsically inaccessible statement that any orienting response of a lower invertebrate represents a feat of 'insight' on the part of the animal concerned. By contrast, there is every justification in attempting to trace orienting responses back to reflexes or to interpret human spatial insight as a complex interplay of innate orienting responses, as Kühn has done. However far one must re-trace to arrive at an established element and however hopeless attainment of this goal may appear, the *direction* must be continually borne in mind if the factual evidence from psychological, physiological and physical research is even to be fused into a logical framework of causal interpretation. This direction itself imposes certain *requirements* regarding the scientist's knowledge. A general biologist must be able to understand physics and chemistry, since the subject of his research is no less an aspect of physics and chemistry than the problems investigated by physicists and chemists. The fact that the material considered is, *in addition*, in some way distinct and 'higher' does not free the biologist from the necessity to acquire an exact knowledge of the 'lower' inorganic foundation of organic processes. Nobody would ever think of writing a book about the general basis of biology without possessing such knowledge. Exactly the some considerations – possibly a stronger form – apply to the dependent relationship between psychological research and certain physiological foundation-stones. Consequently, the task which Bierens de Haan set himself, in writing his treatise *Die tierische Instinkt und sein Umbau durch die Erfahrung* ('The Animal Instinct and its Transformation by Experience') as a 'pure' psychological work (i.e. to the deliberate exclusion of physiological facts), is from the outset just as pointless as the attempt to write a book about 'pure biology' without considering (or even acquiring) chemical and physical knowledge. In general, it can be said: A narrow delimitation of a field of research – as expressed through the word 'pure' – is only scientifically legitimate when conducted *in the direction of a more complex field*. It is possible to study physics and chemistry which is 'pure' as regards biology, but the converse would be nonsensical. It is possible to study physiology without paying the slightest attention to psychological facts, *but the converse is impermissible.*[121]

Of course, what has been said does not mean that biology and, particularly, psychology can only become 'exact' natural sciences when the object of study has been traced to its chemical/physical elements. Any system, whatever its complexity, can and must provide an object for scientific research. Even if we are to study a complex phenomenon in animal or human behaviour and have no idea of the physiological foundations (quite apart from appreciating the underlying physico-chemical processes), we cannot ignore the necessity to observe the behavioural process concerned, to describe it and to conduct analysis as far as is possible. In doing this, analysis must always proceed *from* the whole *to* the element and never *vice versa*. It is not an inherent property of C-, N-, O- or H-atoms that they should give rise to human beings or oak-trees. None of the properties of these atoms dictates that these particular end-products should emerge and even the most detailed knowledge of all the properties of the elements would be fundamentally insufficient to permit us to synthetically derive the organic systems which they produce. On the other hand, even the highest organisms exhibit important properties which are necessary derivatives of the nature and form of the component elements.

The direction of research from the complex to the simple is also determined by the *history*, of organic material. A discrete, concrete causal process has passed from simple to more complex systems, such that the simple component always provided the causal basis for the more complex component and not *vice versa*. All living organisms have a history, and a genuine understanding of the basis for their present form can, quite categorically, only be reached through a historical understanding of the evolutionary process which produced that particular form and no other. The answer to the question as to *why* a mammal should have a hyoid bone in a particular form includes, alongside other specifications, the statement: '*Because* it has evolved from aquatic vertebrates with functional gill-arches.' The above-mentioned, idealistically-conceived gradation of the processes to be subjected to causal analysis derives a concrete backing from the factual evidence of the historic passage of causal chains.

Apart from the search for causal relationships, there is another possible question relating to the interaction of components of an organic system. To take the example of the hyoid bone of a mammal (e.g. an ant-eater), one can also ask *for what purpose* should this bone exhibit a particular form. The logical answer would then be: '*So that* the animal can extend its tongue far into the tunnels of a termite nest when feeding.' Understanding of such purposivity will be sought in all cases where a particular structure of an organ indicates a specific function. The question '*for what purpose*' is a feature peculiar to biology, since inorganic processes do not incorporate purposive system-maintenance. But the finalistic question is by no means the only object for consideration, as is assumed by Bierens de Haan in his theory of instinct. We do indeed generally attempt to settle the question regarding the adaptive function of a process or structure *before* analysing the causes, but this is mainly because knowledge of the adaptive function is a precondition for the understanding of causal origins. For example, understanding of the origin of all differentiation processes in which natural selection is involved is dependent upon detailed knowledge of the adaptive value of these processes.

The logical and methodological relationship between causal and finalistic analysis is thus quite clear and simple. In fact, one is scarcely able to understand why these two approaches should ever have been confused. Nevertheless, it repeatedly occurs that the question 'why?' is answered with a 'so that . . .' reply, as if the demonstration of adaptive function itself dealt with the question regarding natural causation. Particularly in psychology, even today, there are many theories which blindly confuse causal and final analysis. McDougall's school, despite all of the progress which was made in fighting the unscientific explanatory monism of the Behaviourists, aided the proliferation of this logical error because the founder of the school defined 'the instinct' exclusively on grounds of purposivity. This neglect of causal analysis later led to the enormously exaggerated conclusion that questions regarding causal relationships are superfluous in this field. Anybody who follows Buytendijk (1939) in accepting an anticipatory principle, such as the 'vitalistic phantasy' of Palagyi, as an adequate explanation or accepts as law Bierens de Haan's creed (1940): 'We can recognize instinct, but we do not try to explain it,' is not really a natural scientist. One-sided teleological consideration of natural phenomena in fact regularly leads to a deep-seated resistance to analytical research, since far-reaching holistic considerations, which are utterly satisfying for people with no urge to understand causal factors, can only be disrupted by causal analysis of individual details. For instance, when faced with our (subsequently repeatedly experimentally justified) attempt to dissect certain behaviour patterns into rigid, innate motor processes and processes acquired through individual conditioning, Bierens de Haan disarmingly makes the following appeal: 'Are we to believe that behavioural elements remain sharply distinct, like the ivy embracing the oak, without undergoing fusion? If so, what has happened to the holistic nature of the animal behaviour pattern?'

This we counter with the question: What has happened to the holistic nature of an ascidian after experimental embryologists have provided incontrovertible evidence that *each* cell from the two-cell stage of the embryo will produce *half an ascidian* if artificially separated? Is the 'holistic character' of the normal species-typical end-product of the development of such a 'mosaic' embryo any different from that of the end-product of a regulative developmental process? Is it not the case that ascidian eggs produce entire ascidians just as sea-urchin eggs produce entire sea-urchins, although we know that these two end-products are achieved in entirely different ways? Teleologists complain that we cannot see the wood for the trees, but at the same time they seem to believe that the existence of the wood as a 'holistic' living community is somehow threatened by recognition of the fact that wood just happens to consist of trees as well as other components. Even in cases where the composite, mosaic co-operation of components is so convincingly demonstrable as with the ascidian embryo or with the many composite animal behaviour patterns composed of innate and acquired motor components, the organic interaction of the components is neither questioned nor destroyed through analytical knowledge. No analytical natural scientist will ever lose sight of the fact that ascidian eggs always produce entire ascidians.

Considerable intellectual one-sidedness is necessary for anyone to regard the finalistic approach as decisive in studying living processes and to desire simply to rule out the search for natural causes. One is tempted to believe that causal categories are simply absent from the '*a priori*' conceptual equipment of these natural philosophers, so that the attempt to explain to them the significance of the causal approach is doomed from the outset. Nevertheless, I shall attempt to explain the general necessity for causal analysis through an example which should be particularly accessible to a finalistic thinker: Let us ask the question '*For what purpose* does the human being possess a pronounced innate drive to acquire causal understanding?' This question can be answered with a parable: A man is driving a car through the countryside. The purpose of this process is that the man is to give a lecture in a distant town. (The successive finality of this act is immaterial in this context.) The human being is there 'to give a lecture' and his car, which directly serves the same finalistic goal, is there 'for driving'. The man in the car is simply basking in the consideration of this wonderfully holistic conglomeration of mutually-correlated finalities. He can only marvel at the 'vitalistic phantasy' of the car manufacturer, and it is far from his intention to consider the mutual causal relationships of the 'sub-entireties' of his vehicle, which he refuses to regard as 'component phenomena'. At this point, an entirely usual phenomenon occurs: the motor begins to run irregularly and finally ceases to operate. The occupant of the car is at once most forcibly struck with the fact that the '*causa finalis*' of his journey will not make the car go. To use a favourite phrase of mine: the finality 'will not stand up to tugging'. If the man does not succeed in recognizing the general *causes* of the normal functioning of his motor and the specific cause of the fault, the entire finality of his journey is void. What has happened to the 'holistic factor'? It is to be found essentially in the elucidation of the causality of the fault (i.e. discounting an improbable chance success)! The finality of the journey will of course underly the activities of the amateur mechanic as a 'factor' providing the motivation and governing the intensity of his behaviour, but experience tells us that the man would be far better off *not* thinking of this goal at first. He should concentrate his entire mental activity on the intermediate goal of finding the cause of the fault. The success of his investigation will then usually provide him with the means of pursuing his former goal without further ado.

The relationship between finality and causality within the framework of an organic body is no different from that in the parable. The vitalists, who are apt to complain that the natural scientist concerned with causal analysis 'degrades the living organism to the level of a machine', regularly overlooks the fact that cars, steamers or radio sets can never be found in a 'wild form', like the primæval ox or the Przewalski horse. The above comparison, which is repeatedly made by the vitalists, would only be applicable if machines (like these animals) were able to lead an existence independent of man and yet (in contrast to the wild animal) lacked an intrinsic finality. But machines are in fact subject to the finality of an organic system (man), just like any given human organ.

Just as finality is unable to span a gap in the causal chains constituting the functional system of a machine, it is quite incapable of freeing any given organ function from its dependence on the laws of natural causality. This is a fundamental feature of *all* organic processes, even if these are processes whose finality is utterly clear, whilst the causality is temporarily completely inaccessible to study. An example of the latter is the phenomenon of organic *regulation* (which is always treated by vitalistic teleologists as being completely independent of causality), or – to take a 'purely psychological' process – *causal thinking in man*, which is the most regulative and purposive of all organic processes on this planet. The 'freedom' and lack of structure in these processes is only an illusion produced by the virtually indefinable complexity and detailed composition entailed in the interaction of the component elements, and these features can accordingly be disrupted by specific damaging effects just as easily as any machine process. The finality of the whole system is no less dependent upon the causality of its organs than is the travelling lecturer in the parable upon the function of his car motor. It is simply that the regulative function of causal analysis, as applied to organic systems, is presented with an infinitely difficult task, not just because these systems are far more complex in structure than the most complex machines, but also because they – unlike machines – are not produced by man, so that we have only a very incomplete impression of the causality and derivative history of these systems. A doctor who is faced with the task of restoring the overall function of an aberrant organic system therefore has many more difficulties than the car mechanic. But even the holistic-regulatory activities of the doctor are just as dependent for their success upon his study of causes and they are just as independent of the degree of 'urgency' applying to the endangered finality. It is fundamentally immaterial to the relationship between the finality and the process of acquiring causal understanding to ask how the disrupted holistic system came to exist. The question as to whether the system was constructed by an engineer, or created by a god, or whether it owes its origin to a natural process involving mutation, selection and many other processes, is of no importance for the functioning of human causal understanding, which can master a system of this kind and literally restore its wholeness after certain disruptive effects.

Thus, the final significance of human causal research is that it operates as our most important regulative factor and *provides us with the means of controlling natural processes.* It is completely immaterial whether these processes are external and inorganic in form (e.g. lightning or storm) or internal and organic (e.g. bodily ailments or 'purely psychological' defects in human social behaviour patterns). Without causal understanding, it is never possible to follow purposive aims, and conversely causal research would itself be functionless if the human research worker did not have purposes for his activity. The attempt 'to search for causes in our surroundings as far as we can and to follow the chain as far as it can be traced according to the laws which we have discovered' (Kant) is therefore not 'materialistic' in the philosophically moral sense, as is only too readily assumed by the teleologist, but represents the most intensive existing service to the ultimate finality of all organic processes. This activity, when it is successful, provides us with the *power* to intervene in an assisting and regulating manner where values are endangered and where the purely teleological observer is only capable of holding his hands in his lap and mourning the evaporating finality of the collapsing 'whole'.

II The difference between the anticipatory explanatory principle and the inductive concept

Bierens de Haan states that a theoretical system should be able to arrange facts 'elegantly' and that this should satisfy our thirst for explanation. 'Elegantly' doubtless signifies the external form of a *mathematical proof* (i.e. a fundamentally *deductive* process), whilst 'satisfy' obviously means that our *entire* thirst for explanation should be stilled once and for all by the system concerned. In actual fact, one can find the occasional mathematically ('elegantly') deduced system, based on pre-defined concepts, which may be held to fulfil this requirement. For this reason, Bierens de Haan understandably regards the working hypotheses of inductive natural scientists (concerned with the laborious task of studying the ramifications of real phenomena) as the ultimate in 'inelegance', since he completely erroneously judges them by his own standards and assumes that they are grandiosely trying to establish a system which will provide the ultimate solution to all problems. His entire critique of my working hypotheses is exclusively based on this opinion, without consideration of inductively-determined facts. I shall illustrate this with a number of concrete examples.

The first example is provided by his critique of the separation of appetitive behaviour and endogenous instinctive motor patterns, which was carried out by Craig and myself (1937). This is intermingled in an extremely revealing manner with a critique of my observations on the co-operation of instinctive motor patterns and individually conditioned acts (1935), and one can only explain this on the basis of his interpretation that I believed that these two extremely modest scientific advances had provided me with 'the principle which will explain everything satisfactorily'. A typical case of such judgement of others according to his own standards is evident from his statements on p. 102 of the article concerned: 'He first invented this "appetitive behaviour" in 1937, however. Previously (1932, 1935), he had already established other terms, at a time when he had discovered the solution in the form of "innate-conditioned intercalation patterns" or "instinct-conditioning intercalation patterns". These terms immediately struck me as being suspect . . . etc., etc.' With the best will in the world, this can only be interpreted to mean that Bierens de Haan, on reading my discussion of the intercalation of endogenous and acquired motor patterns, attributed to me the opinion that I could solve all the problems of innate behaviour with this new 'explanatory principle' and assumed that I had later abandoned this opinion 'in favour of' the concept of appetitive behaviour which had been 'invented' for this specific purpose. This assumption is just as justifiable as the statement that geneticists later abandoned Mendel's Laws in favour of the chromosome theory of heredity (Boveri-Sutton hypothesis). The two concepts are in no way mutually contradictory – each provides discrete information which does not reduce the validity of the other. In the shrike, the motor pattern employed for impaling captured insects on thorns is demonstrably innate. It is equally demonstrable that 'recognition' of the thorn used for the purpose, together with the purposive application of similarly innate taxes for locating the thorn and for establishing correct orientation of the impaling movement, must be acquired by *individual conditioning*. The factual evidence for this is established, and we aptly refer to such co-operation of innate and conditioned motor patterns as 'intercalation'. The attempt to deny on deductive grounds the presence of a 'gap' in the innate behavioural sequence – which must be filled by a conditioning process – is (to say the least) unscientific.

We can lead on to the concept of appetitive behaviour through the observation that intercalation patterns of this kind would never arise if the animal did not in some way actively strive towards performance of its innate motor patterns. 'Something' induces the shrike to *seek*, by trial-and-error, the situation in which the impaling movement can be properly performed. The correct performance of the motor pattern must be subjectively experienced as a 'reward', since it *conditions* the shrike to search for the thorn just as the pieces of meat given by the lion-trainer will condition a circus lion to perform certain conditioned acts. (This is literally what I said in my first publication on intercalation patterns.) The fundamentally important feature of Craig's concept of appetitive behaviour is represented by just this recognized fact – that the goal which the animal subject strives to attain is not the prosecution of the adaptive purposivity of its 'instincts' (as Bierens de Haan still chooses to believe) but the satisfying performance of the instinctive behaviour pattern itself. I can give Bierens de Haan a very precise answer to his question as to what a wolf 'desires' when hunting its prey: Its first aim is to *shake its prey to death*. The opinion that the wolf hunts with the same motives as a hungry human being performing an unpleasant profession is a pure anthropomorphism. The motor pattern of shaking prey is definitely the most emotive and pleasurable motor act in Canids, and it is therefore much more decisive in the motivation of hunting than subsequent devouring of the prey. Even a completely-satiated domestic dog retains an unaltered, irrepressible appetite for performance of this particular motor pattern. That is why this pattern is frequently performed with a substitute object – take, for example, a dachshund or a terrier shaking his master's slippers 'to death'. Of course, this by no means signifies that a hunting carnivore does not experience 'lust after the prey'; but in this case Uexküll's statement that animal events are a question of *action* is particularly applicable. For the wolf, the prey arouses a lust for 'shaking to death', just as a female is an object for mating activities, or – to take our 'appetite' in its strictest sense – a rosy apple is an object 'for biting into'. If Bierens de Haan, adhering to the concept of holistic animal behaviour, still chooses to support the interpretation that the hunting wolf is driven by a generally directive instinct (compelling the wolf to avoid starvation, to ensure its own nutrition and thus ensure survival of the species), and thus to assume that the adaptive finality of the behaviour is the immediate goal of the animal subject, then we are far from understanding what really drives the animal to act. It was Craig who recognized that performance of the instinctive behaviour pattern *for its own sake* is sought by animal and human subject alike. The fact that such a pattern is physiologically fundamentally different from all purposive/plastic behaviour patterns (Tolman's 'purposive behaviour') has been demonstrated quite convincingly by von Holst. These two factors together underly a relationship between instinctive behaviour patterns and appetitive behaviour which is actually 'antagonistic', in rough analogy to the relationship between two muscles exerting opposite effects around a particular joint. A more clearly-defined interpretation of the two concepts is, in actual fact, a genuine extension of Craig's conclusions. As Bierens de Haan would have discovered by reading my earlier article in this journal more carefully, my definition of these concepts was carried out in close co-operation with Craig and is virtually a *product* of such co-operation. Craig himself, whom I am pleased to regard as one of my most respected teachers, is in agreement down to the last detail with my more narrow and precise definition of the two concepts. Bierens de Haan is quite unable to put himself in the place of a natural scientist in the face of new results which necessitate alterations to his previous working hypotheses. This is because he is only familiar with the deductive method based on anticipatory explanatory principles, with which any mistake is to be regarded as a culpable 'error' on the part of the author. With the inductive method, on the other hand, it is not a deficiency but a success of the working hypotheses when new facts necessitate alterations of the previously employed conceptual system. I have scarcely been so thoroughly pleased as by the fundamental alterations of my old concept of appetitive behaviour recently necessitated by the results obtained by M. Holzapfel and (particularly) Baerends. These authors employed new factual evidence – and not deductive speculation – for criticism of my former concepts.

Bierens de Haan adopts exactly the same attitude towards von Holst's findings as towards the factual evidence of intercalation patterns and appetitive behaviour; he dismisses this keynote of modern behavioural research with a footnote veiled in misunderstanding. One must first of all be quite clear about the following developmental succession in our knowledge regarding endogenous-automatic motor patterns: For years, the best and most knowledgeable animal observers (Whitman, Heinroth, Craig, Howard and many others) had repeatedly puzzled over the fact that a given behaviour pattern, obviously adapted to a quite specific situation or object, can be performed *independently* of the relevant stimuli. Every good animal observer knows that some motor patterns (e.g. those of nest-building in swans and geese, the gathering gesture of the plover and the substrate-hollowing or stone-cleaning patterns of some Cichlid fish) may even appear *more frequently* as 'vacuum activities' than in situations where the adaptive function concerned is properly performed. This applies particularly to captive animals. My own detailed observations of the stimulus situation eliciting a non-adaptive, non-functional performance of this kind demonstrated that there is a significant relationship to the *time* which has elapsed since the last occasion when the instinctive pattern concerned was elicited. The longer the period since the last elicitation, the greater the ease with which the motor pattern will emerge, i.e. the greater the permissible departure from the biologically adequate stimulus situation. In the extreme case the response will be performed *without* external stimulation. Without any preconceived ideas, description of these regular phenomena involved expressions (such as 'damming of the response', 'internal pressure', and so on) which, on a purely analogous basis, suggested the idea of *accumulation processes* involving an endogenously produced response-specific arousal factor. It must be remembered that all of the observers at that time adhered, explicitly or implicitly, to the belief that *the reflex* is the basic element of all innate motor patterns, and that the latter must be interpreted as chain reflexes. I myself still adhered to the Ziegler theory of instinctive motor patterns in its customary form in 1937, merely making the modest observation that threshold-lowering and vacuum activity required a supplementary explanation, since the nature of the reflex does not incorporate spontaneous production of excitatory energy, and since vacuum responses (which are performed in the complete absence of external elicitatory and steering stimuli) are not dependent upon stimuli to the extent that would be expected of reflex chains. Similarly, Lashley has shown with rats how little some innate motor patterns are affected by disruption of sense-organs and afferent pathways, and he himself has stated that observation of this fact was actually entirely unexpected. At this stage in the development of our knowledge, the bomb of von Holst's results exploded: The 'reflex' is not the only 'element' of neural processes; it is one of the most important functions of the CNS *to produce its own stimuli*. For example, a completely deafferentiated spinal cord of an eel produces automatic, rhythmic stimuli which are centrally co-ordinated, such that the transmitted impulse produces swimming movements in their complete, adaptive form, *without the participation of afferent neurons*. The occurrence of 'spinal contrast' (Sherrington) and other phenomena indicate that centrally co-ordinated automatisms are probably dependent upon *material* production of a response-specific, accumulatory source of energy. These results, derived on the basis of exemplary and convincing experimental procedures, provided a sweeping solution to the remaining problems – which were incompatible with the reflex theory. We were abruptly able to understand why the instinctive behaviour pattern (unlike a reflex) does not remain inactive for an indefinite period, waiting for the elicitatory stimuli, but – so to speak – intervenes on its own accord. There is not only a gradual lowering of the threshold value for the elicitatory stimuli; the instinctive pattern itself comes to represent an unspecific stimulus, a *drive* (as defined by Portielje, Kortlandt and

Baerends), which is why there is such a thing as appetitive behaviour. We could at once understand why the vacuum response is independent of guiding external stimuli, why (for instance) a captive starling will trap and eat non-existent flies with the same, photographically corresponding motor pattern as used for real predation. All of the phenomena which had remained as incomprehensible paradoxes at once became self-evident, even theoretically postulated, products of a single, clearly-recognized basic process. It is possible to judge how little Bierens de Haan had traced the development of our knowledge about endogenous-automatic instinctive patterns (which is sketched only briefly here) to be able to dismiss these basically important and utterly reliable experimental results with the following words in his footnote: it was 'not quite apparent' why I 'had unnecessarily constructed my hypotheses on such an unreliable foundation'. At another point, he wrote that one must wait and see whether scientists might later discover that the central nervous automatisms described by von Holst were reflex chains after all, since it was so easy to overlook external stimuli. This almost gives the impression that he has not read the paper concerned. He would otherwise have been forced to appreciate that in the nerve cord of an earthworm which has not just been deafferentiated but completely isolated, or in the spinal cord of a decapitated eel with transected sensory roots, intact co-ordinated reflex chains simply cannot function. It can scarcely be assumed that Bierens de Haan would be unable, despite careful perusal of the paper concerned, to understand the importance of these fundamental facts for psychology and physiology. And what is to be made of the fact that Bierens de Haan employs as *criticism* of my papers things which are included in these papers themselves? For example, in discussing the subjective processes accompanying vacuum responses, he writes: 'it is very difficult to determine what an animal perceives, or thinks it perceives', and that there are perhaps hallucinatory processes which induce the animal to perform the response, such that the latter appears to be 'vacuous' to the observer, but not to the animal subject. This criticism is in the first place understandable from the repeated false assumption that physiological explicability of a process excludes a 'psychological concomitant', and secondly on the grounds that Bierens de Haan has not read my discussion of this question. In fact, in 1937 I wrote that the damming of response-specific energy alters the *perceptive field* of the animal or human subject so far that a normally inadequate object will be subjectively experienced as adequate, or – in the extreme case – experienced as a pure hallucination. 'With this potion in your body, you will soon see every woman as a Helen', or, when the process is taken far enough, as an hallucination in empty space. It is this very aspect which renders the physiology of endogenous instinctive behaviour patterns so important for perceptive and experiential psychology.

Overall, Bierens de Haan's attitude towards our conclusions regarding the physiology and psychology of endogenous-automatic motor patterns is exactly the same as towards our analysis of instinct-conditioning intercalation patterns and appetitive behaviour: He is endowed with anticipatory, finalistic understanding; he deduces from this an 'elegant' pseudo-explanation and ignores the factual evidence which conflicts with it.

In a somewhat different manner, his criticism of our work on orienting responses illustrates the inhibitory effect exerted by anticipatory explanatory principles on the progress of analytical research. I hold today to the opinion I expressed in 1937, that one cannot draw a clear distinction between obviously reflex orienting responses (such as that of a slight detour made by an animal around an obstacle lying between it and its prey) and the most complex, 'methodical', insight-controlled behaviour of human beings. This is because even the latter behaviour is almost always based upon operations involving concepts of spatial dimensions, which are ultimately composed of tactic responses. This provoked Bierens de Haan to attribute to me the opinion that there is no qualitative difference between man and the amoeba, and that a person suicidally leaping into the water is suffering from nothing more than an overdeveloped positive hydrotaxis. Once again, Bierens de Haan is extrapolating from himself to others. In the first place, he blandly overlooks the fact that we have, of course, never maintained that we can interpret all human behaviour on the basis of the few psychophysical processes which have so far been subjected to more-or-less complete analysis. In contrast to that author, we have a healthy respect for unanalysed residues. Secondly, he employs the word taxis in a form which we have never employed, since he includes the *goal* of the orienting response in his definition. We have avoided this, because of the danger that the characterization of the process thus produced might appear to be an intended explanation. Bierens de Haan seems to be unaware that the taxis concepts formulated by Kühn are not defined according to the adaptive ultimate success of the response concerned, but according to the causal, physiological mechanism producing the response. In the light of his limited accuracy in reading analytical papers, it is possible that the terms applied to Kühn's taxis categories may have led him astray. The terminology (in contrast to the definitions of the concepts) is derived from the most frequent adaptive function which a particular type of taxis performs. For instance, 'tropotaxis' does not mean a turning movement of the animal concerned, but an orienting response based on a balance of excitation; 'telotaxis' does not mean following of a particular goal, but a taxis in which a stimulus 'fixates' on a specific site of the receptor; etc.

Finally, I must emphasize that the taxis concept which we employ is not broader than the original taxis concept formulated by Kühn (as is assumed by Bierens de Haan), but *narrower*. This restriction arose from recognition of the fact that every one of Kühn's taxes incorporates locomotor movements which are quite definitely not reflexes, but endogenous automatisms as understood by von Holst. Consequently, the concept of the taxis was limited to the small, quantitatively receptor-controlled movements to the right and left, or up and down, which determine the *direction* in which the elicited locomotory automatisms will lead or orient the organism. The 'positive Americotaxis' of a HAPAG steamer is based firstly on the automatic function of the propelling machine, which can only be retarded or disinhibited, and secondly on the steering. The intensity of the motor function depends both upon the position of the disinhibiting control valve and upon endogenous steam production, while the special motor co-ordination involved is not under the control of the helmsman. Apart from inhibition and disinhibition of the machine, the steering consists only of minor impulses to the right or left, which are so calculated by the helmsman – according to external stimuli such as those deriving from compass, stars or the sight of a landmark – that the desired course is followed. The automatism of locomotory movement and the taxis controlled by external stimuli operate in just the same way in any fish or other organism swimming in a directed manner. This restriction of the taxis concept, which automatically emerged through the enrichment of our knowledge by von Holst and the consequently expanded possibilities for analysing orienting movements, was utterly approved by Kühn as a matter of course. Tinbergen, employing extraordinarily shrewd experimental techniques, has carried out the analysis of oriented movements which I had put forward as a theoretical suggestion. This was conducted firstly with the egg-rolling movement of the greylag goose, as a joint study together with me, and later (with Kuenen) on the directed gaping movement of young blackbirds and thrushes. In this case, too, a regrettably inelegant complexity and mosaic-like multiplicity of interacting causes emerged. This result compares extremely unfavourably with Bierens de Haan's assumptions (which are so satisfying to his own desire for explanations), in that they are not only apparently unsuited to a definitive conclusion of our hypotheses, but actually provide stimulation towards a completely unpredictable glut of further investigations. (This must appear even worse to Bierens de Haan!)

The investigation of the elicitatory and directive stimuli for the gaping movement of young blackbirds and thrushes carried out by Tinbergen and Kuenen will serve for discussion of another concept, which – according to Bierens de Haan – is the vaguest and most obscure in my entire 'conceptual system', namely that of the *innate schema*. Bierens de Haan maintains that I myself was not quite clear what this concept was intended to convey; but a large number of other authors – particularly Tinbergen and Kuenen – have been contrastingly able to understand without difficulty the clearly demarcated (even if enigmatic) phenomenon which I have characterized with this term. Instead of satisfying themselves with the explanation that an infallible instinct tells the blackbird nestling that it must orient its opened beak towards the parent's beak, in order to obtain food, Tinbergen and Kuenen presented the nestlings with a wide range of objects, some of them 'extremely unholistic'. The following results emerged: With two rods presented at the same height, the nestlings will gape towards the *nearest*; with two rods at the same distance, towards the *highest*; and with two rods of different size (within certain limits), towards the *smallest*. Tinbergen and Kuenen were able to establish the latter relationship mathematically using an ingenious experimental set-up based on a *quantitative* balance between the effectiveness of the characters 'higher' and 'smaller': Two adjoining discs of different size were slowly rotated with respect to one another so that the smaller disc was first higher and then lower. As long as the characters 'higher' and 'smaller' coincided the nestlings of course gaped towards the smaller disc. When the smaller disc had been rotated to lie quite a bit lower than the larger, the character of relative height dominated over that of relative size and the gape-response was switched to the upper edge of the larger disc. The relative size of the smaller disc with which this switch is observed latest (i.e. with which the maximal directive effect operates on the response of the young bird) proved to be 1 : 3. If one constructs a model in which the three relative characters mentioned are combined, the result is a rounded structure with a smaller, similarly-rounded appendage located on its upper anterior face. Thus, one produces a greatly-simplified 'schematic' reproduction of the head and body of a bird, a 'schema' which could be constructed purely on the basis of the young bird's responses even if it were not aware of the normal biological situation eliciting directed gaping and we had never seen a parent bird feeding 'with its head'. In other words, the young bird has a pre-formed, simplified 'schematic' correlate to a biologically relevant stimulus situation, permitting the nestling to respond in an adaptively functional manner without the participation of learning processes. The characters constituting an innate schema are not always relative, as in the case of the example given. Sometimes, a colour plays the most important part, sometimes a rhythmic pattern of movement and sometimes all three things together. For example, in order to elicit the courtship responses of a female stickleback, the dummy corresponding to the male must be red on its anterior ventral aspect (colour character plus relative character) and also perform certain movements (Tinbergen, 1939). According to Bierens de Haan, all of these investigations are unnecessary, since we already know that an instinct aids an animal in any given case to recognize the correct object for its behaviour. According to him, the instinct is 'the characteristic psychic predisposition on the basis of which certain sensations, perceptions or memories are followed by certain feelings and emotions, while this process of recognition and feeling itself gives rise to a specific drive and certain aspirations which are realized in behavioural acts. Conversely, specific perceptions and feelings are awakened and influenced by these aspirations. More briefly: The instinct is the physic predisposition which couples a specific feeling to a specific recognition and a specific aspiration to the feeling awakened by the recognition and also renders both recognition and feeling dependent upon the aspiration' (*Die tierischen Instinkte und ihr Umbau durch die Erfahrung*, 1940). This may represent perfectly apt speculation about the subjective processes which accompany the elicitation of appetitive behaviour and instinctive motor patterns through innate schemata. What we want to know, however, is *why* a specific drive follows a specific perceptive effect, and, above all *how* we are to understand the 'recognition' of the appropriate object on a causal basis. We are not satisfied by the explanation that 'the instinct' (which we may 'recognize' but not explain) gives rise to 'a specific perception'. We are still unable to understand why a female mallard which is isolated from conspecifics from the egg onwards and reared with pintails will have nothing to do with a male pintail which is available, but will immediately 'recognize' a mallard drake at first sight (or, more exactly, will respond with extremely intensive courtship activities). What is even less explicable is that the instinct – which is allegedly not affected by causal trifles – is *not* capable of providing the mallard drake with the same 'recognition' of the sexual partner under the same conditions. Strangely enough, a mallard drake which has grown up with pintails does not 'recognize' a conspecific female as a sexual partner, but responds instead to pintails with courtship and copulatory behaviour, though the mallard drake significantly does not distinguish between the sexes of the foreign species and attempts to tread both drakes and ducks indiscriminately. Surprisingly, however, the mallard drake will 'recognize' male conspecifics as his own kind and joins up with them, rather than courting pintail drakes, to perform the social courtship ceremony. *Thus both sexes possess only the innate response to the bright signal colours and conspicuous motor patterns of the drake, but not towards the less distinctive plumage of the duck.* This would be a remarkable functional limitation for the behaviour of an 'instinct', whereas it is an evident assumption to make that the 'Prachtkleid' (display plumage) of the drake in some way transmits stimuli to both sexes, inducing copulatory inhibition in males and courtship elicitation in females. This would explain why the mallard drake reared in isolation with pintails mounted male pintails, since the latter lacked the innate copulation-inhibiting mallard drake plumage. Seitz has investigated the same phenomenon in the mouth-brooding fish *Astatotilapia strigagena*. In this case, too, the male is very brightly coloured and is characterized by specific instinctive motor patterns; he responds innately to conspecifics of the same sex, but not to the inconspicuous female, with specific behaviour patterns. We do not yet know what an innate schema is, but we do know that there is a specific *restriction: The innate schema cannot respond, like acquired Gestalt perception, to a complex quality integrated from a large number of characters; it is always dependent upon a small number of distinguishing characters.*[123] This fact explains the wide distribution of systems which serve exclusively for the transmission of specific, simple stimulus combinations and which I have termed 'releasers' (1935). The provisionally enigmatic nature of the innate schema, its remarkable 'stimulus-filter effect' (which determines that *only these and no other stimuli* can produce elicitation) and the remarkable contrast of its function to that of the Gestalt phenomenon (as in the case of the gaping of young blackbirds, where the directive schema is based upon the sum of three relative characters, each of which has the same effect as all together, but at a lower intensity – Seitz's 'stimulus summation effect') all contribute to make the innate schema one of the most stimulating and attractive objects for experimental research. It is therefore not surprising that a rapidly-growing flock of investigators (Tinbergen, Kuenen, Kraetzig, Goethe, Seitz, Baerends, Lack, Peters and many others), and more recently the entire newly-founded Institut für vergleichende Psychologie here in Königsberg, should be occupied with these basically important problems, which are also relevant to human psychology – particularly for certain phenomena in human '*a priori*' ethics. Of course, the concept of the 'innate schema' is determined exclusively according to the stimulus-situation selecting *function* of this remarkable receptor apparatus; but *we will not forget this for a moment.* It would not surprise us in the least if closer analysis should lead to division of the 'innate schema' into two or three causally (i.e. physiologically) completely distinct processes, which require new concepts. We in fact have some idea of the manner in which this will happen. Thus, we know quite exactly what the concept of the schema is intended to convey. Unfortunately, we can safely promise Bierens de Haan that the results of the planned research will become still more 'inelegant' with increase in our knowledge. This is not our fault, however, but a result of organic creation, which determined the particular form of the emerging organisms.

I feel that the examples of intercalation patterns, appetitive behaviour, endogenous automatisms and the innate schema are sufficient to demonstrate the difference between a purely finalistic approach and inductive behavioural research. Detailed examination of Bierens de Haan's article will yield a number of others, but I will confine my further comments to one last case. His critique reaches its peak in a comparison between the practical applications of his conceptual system and mine. His final comparison is based upon our differing capabilities for explaining *bird migration*; I cannot explain it, whereas he can – in an utterly satisfactory manner: In autumn, the instinct drives the bird to seek out its winter quarters and infallibly shows the bird where to go. 'The explanation that migration is a taxis is of no further use to us.' With regrettable backwardness, the *Kaiser-Wilhelm Gesellschaft zur Förderung der Wissenschaften*, ignoring these facts, continues to support two well-equipped bird observatories, which stubbornly follow the belief that the assumption that there are taxes involved in migration (though it is indeed no explanation) represents a *working hypothesis*, which will one day lead to the discovery of the stimuli according to which a migrating bird sets its course.[124] (This is still a complete mystery.) Just as stubbornly, they are investigating the part played by endogenous stimulus-production processes in bird migration (Putzig, 1939). It has been known for some time that migratory restlessness and the drive to fly long distances appear even in captive birds and that free-flying tame Greylag geese, though not flying away in a directed manner, will 'fly out of their soul' the increased production of the instinctive motor patterns of migratory flight, in excursions lasting several hours or even days. It is therefore an entirely logical step to attempt to analyse the phenomenon of migration on the basis of the hypothesis of an interaction between endogenous instinctive behaviour patterns and guiding stimuli.

In summary, I should like to make the following observations: Like every anticipatory explanatory principle, Bierens de Haan's concept of instinct has the result that any adherent is fundamentally cut off from natural science. This is not my personal opinion, but that of every natural scientist. Bierens de Haan really gives me too much credit in stating, in the introduction to his article, that he was aware from the outset that a settlement between his conceptual system and that of 'the group around Lorenz' would have to come sooner or later. I am quite unqualified to claim to be the central member, or even a particularly prominent representative, of 'the group' against which Bierens de Haan has set his face, namely the circle of all relatively disciplined representative thinkers of inductive natural science. In historical terms, he ought to refer to 'the group around Galilei', or – if he only means the restricted field of research into innate animal and human behaviour patterns – to 'the circle around Heinroth and Whitman'. Rejection of the anticipatory explanatory principle by every thinker who has clearly recognized the nature of inductive method is far from new. John Dewey has formulated this principle very well in his book *Human Nature and Conduct*. In particular, he says of the old concept of instinct, as still used by Bierens de Haan:

> It is maintained that fear is a real phenomenon, as is anger, the competitive spirit, the mania to dominate others or to subject oneself, maternal love, sexual lust, the need for social contact and jealousy; and each of these things has a corresponding act. Of course they are real phenomena. Just as real are the sucking action of a vacuum, the rusting of metals, thunder and lightning and steerable airships. But science and human intervention never advanced as long as human beings gave themselves to ideas of particular powers which were supposed to explain these phenomena. Human beings have actually tried this path, but it only led to pseudo-learned ignorance. They talked of a repulsion of nature against emptiness, of a power of combustion, of an 'inner tendency' to do this or that and of lightness and heaviness as forces. It emerged that these 'forces' are only the old phenomena in new robes, that they have been transformed from their particular, concrete form (in which they were at least something real) to a generalized form, in which they are no more than words. In this way, they changed a problem into a solution, which then provided simulated satisfaction.

The bad effects of anticipatory explanations on experimental research was evident even in Kant's time. In his *Bestimmung des Begriffs einer Menschenrasse*, Kant says of the introduction of non-causal 'factors': 'If I were to admit to just one single case of this type, it would be as if I were to introduce a ghost-story or sorcery. The barriers of reason are at once broken and madness penetrates thousand-fold through the gap.' Bierens de Haan, the admirable experimental investigator, is presumably unaware of the weapons placed in the hands of every enemy of free research by Bierens de Haan, the 'teleologist', with his prohibitory statement: 'We can recognize instinct, but we do not explain it.'

Notes

1 *P. 2, l. 15.* Thus, when first mentioning the phenomenon of 'imprinting' I already state that the demonstration of a true imprinting process in one species does *not* justify the assumption that it also occurs in another, however closely related. That the following responses of Mallard ducklings or Greylag goslings are fixated on the object by typical imprinting is in no way contradicted (as R. Hinde, P. Bateson and others seem to assume) by the fact that this is not true of young Coots or domestic chickens.

2 *P. 5, l. 13.* It must be remembered that, at that time, the phenomena of central co-ordination functioning without participation of proprioceptors were as yet unknown.

3 *P. 8, l. 24.* First realization of the innate releasing 'schema' or mechanism.

4 *P. 11, l. 20.* The 'crop' in Corvidae is a pouch between the tongue and the bottom of the mouth cavity, not, as in pigeons, a widening of the oesophagus.

5 *P. 32, l. 39.* Two years after the conclusion of this paper a trio of jackdaws built one nest with two cups. The foundation was built by the male and each cup by one female, the eggs being distributed between the two.

6 *P. 37, l. 5.* This is the first record of a rudimentary instinctive motor pattern. Further records by Ahlquist, I. Eibl-Eibesfeldt and H. H. Haas.

7 *P. 37, l. 31.* This is my first mention of threshold-lowering due to 'damming' of a response.

8 *P. 41, l. 26.* Amply substantiated since.

9 *P. 47, l. 2.* This was only true of the first brood of my jackdaws. In later broods made by birds which had lived in full liberty for years there definitely emerged a mechanism of distributing food equally among the nestlings, which normally are of equal size when fledging.

10 *P. 55, l. 6.* The inference of this paragraph is of great general importance to the evaluation of all observations made on captive animals. It is a very simple rule: if the captive animal does show a complicated behaviour pattern of recognizable survival value, the observer is fully justified in the conclusion that the behaviour pattern observed is a constant property of the species in question. On the other hand, the absence of a behaviour pattern in a captive animal can always be caused by poor health of the individual and permits no deductions concerning the behaviour of the species in the wild.

11 *P. 57, l. 18.* At the time of writing I was still a strict adherent of classical Sherringtonian reflex theory.

12 *P. 61, l. 7.* Here is a typical case of founding a correct theory on an incorrect observation. In fact, young shrikes which are only slightly physically impaired have to learn to direct the motor patterns of impaling towards the thorn. Later experiments by G. Kramer, U. v. Saint-Paul and myself showed that, in absolutely healthy young birds, the orientation to the thorn is entirely innate. In fact, the first presentation of a thorn causes the inexperienced bird to search for a substitute object for impaling (for example a dry butterfly-wing), while, conversely, the presentation of a large insect causes the bird to search for thorns. Many cases are known in which motor co-ordinations are entirely innate, while the responses guiding them in space must be individually acquired. The dropping-out of a response due to lack of general fitness in a captive bird will be discussed on pp. 35, 36. However, the abovementioned observation on Red-backed Shrikes, made independently by Oskar Heinroth, Gustav Kramer and myself, hitherto represent the only known case in which a *pathological defect* of innate behaviour mechanisms was demonstrably *compensated* by learning.

13 *P. 65, l. 36.* The following are typical examples of imprinting; see also p. 124 ff.

14 *P. 67, l. 24.* This statement, made in 1932, may seem rather assertive. However, I still regard it as entirely true. My reasons for this are given in detail in my book *Evolution and Modification of Behaviour*.

15 *P. 68, l. 23.* This is a clear example of threshold differences caused by habituation, as opposed to those effected by exhaustion or damming-up of action specific potential.

16 *P. 71, l. 25.* This is to be explained not by late maturation, but by the restitution of a behavioural mechanism which had been pathologically disturbed in the young bird.

17 *P. 72, l. 13.* This is still true of instinctive motor patterns, though their histological basis is certainly not provided by a chain-reflex as assumed at the time of writing.

18 *P. 73, l. 11.* This statement will be found corrected in the second volume. It is not possible to give a definition which differentiates between behaviour guided by a taxis and that governed by insight. Taxis of all kinds are, of course, functional in lower animals.

19 *P. 73, l. 24.* The fixed motor patterns here discussed are, of course, not in actual fact dependent on chain-reflexes (see p. 318 ff.).

20 *P. 74, l. 8.* At that time I already suspected the presence of echolocation, later demonstrated by Dykgraf, Griffin and others, because my bats would avoid all sorts of unexpected obstacles, but flew straight into a plush-upholstered armchair which did not reflect an echo.

21 *P. 75, l. 23.* See the picture of ravens in the paper on pair-formation in the second volume.

22 *P. 76, l. 32.* This has been confirmed since by a number of investigations, particularly those of M. Konishi. He showed that in some passer-

ines the form of the song is phylogenetically programmed in the form of an auditory template and not in the form of an innate motor pattern. Individuals deafened at an early age failed to produce the species song, while birds with undisturbed hearing developed a normal song even when reared in isolation in sound-proof chambers. Warning calls, flight calls, etc., develop normally even in deafened individuals because they are based on innate motor patterns.

23 *P. 77, l. 8.* This is exactly what Konishi showed to be the case.

24 *P. 77, l. 40.* Gwinner has since found that in the raven there are certain other vocalizations with a communicative function which can be altered by learning and actually form 'dialects' in different raven populations.

25 *P. 78, l. 5.* The conception of the releaser will be extensively discussed in the next paper.

26 *P. 78, l. 32.* I am speaking here of the motor pattern which is released in the conspecific.

27 *P. 79, l. 12.* This observation is of importance in so far as to the best of my knowledge, it is historically the first case in which traditional knowledge was clearly demonstrated to exist in an animal species.

28 *P. 82, l. 2.* See note 12.

29 *P. 84, l. 4.* This is the first reference to be found in my papers showing a dawning appreciation of the fact that the releasing mechanism is something physiologically different from the motor pattern which is set off by it.

30 *P. 85, l. 22.* The inference of this important fact has already been given in note 10.

31 *P. 91, l. 12.* This view has been developed into a theory by A. Kortland, who assumes that chimpanzees had already reached a higher degree of 'humanization' than they show at present, when they were forced to retire into the woods by the competition of more advanced pre-hominids.

32 *P. 93, l. 15.* The possibility of a motor pattern developing, on the basis of species-specific structure, by trial-and-error learning is a theoretical consideration of great interest. The proof that this has ever been realized in an animal species is, however, still lacking.

33 *P. 93, l. 39.* This speculation, though amusing, is epistemologically correct. If one concedes subjective experience to the animal at all, one has to assume that it is concomitant with specific patterns of behaviour.

34 *P. 95, l. 15.* Read: innate motor patterns.

35 *P. 96, l. 5.* There are certain exceptions from this rule, as R. J. Andrew has pointed out (see also p. 227 in this volume).

36 *P. 97, l. 42.* The description of these phenomena is one of the initial steps towards the discovery of Alfred Seitz' law of heterogenous summation.

37 *P. 99, l. 43.* I wish to emphasize here the perseverance with which I held to Sherringtonian reflex theory.

38 *P. 101, title.* The German word 'Umwelt', as used by Jakob von Üxküll has the connotation of the subjective world of any living creature. When

we translate it here simply by 'environment', it should be kept in mind that this conception encompasses only those features of the animal's objective environment which are responded to by the subject.

39 *P. 101, l. 25.* See note 20.

40 *P. 104, l. 15.* This statement is not quite correct. The function of Gestalt perception achieves quite as much, though it has nothing to do with insight.

41 *P. 107, l. 11.* This has since been extensively investigated by J. Nicolai.

42 *P. 108, l. 27.* The German word 'Kumpan' denotes a concept somewhat different from that associated with the word 'companion'. The German word has a slightly derogatory meaning; a 'Kumpan' is not the companion of your soul but a fellow who shares your pleasure in hunting, drinking, frolicking, etc. (Saufkumpan, Jagdkumpan, etc.).

43 *P. 110, l. 22.* This applies literally to all experimentation on animal behaviour, even in those cases in which learning processes are the subject of research. See my paper 'Innate Bases of Learning'.

44 *P. 110, l. 32.* See notes 11, 17, 19, 37, 52, 89.

45 *P. 116, l. 19.* This is not quite correct. In greylags and other geese, it frequently occurs that the young of last year are actively driven away by the parents when these begin to nest again.

46 *P. 116, l. 38.* See notes 11, 17, 19, 37, 52, 89.

47 *P. 117, l. 22.* Experimentally proved later by Doppler.

48 *P. 119, l. 11.* See note 12. The principle, however, has since been experimentally verified by Eibl-Eibesfeldt.

49 *P. 120, l. 37.* This is one of the few fundamental errors of early ethology. Even if a learned behaviour pattern replaces the function of a disappearing fixed motor pattern, this learning process itself necessarily has its own phylogenetically programmed basis.

50 *P. 122, l. 6.* Bethe's results, though incompatible with chain-reflex theory, find a natural explanation on the basis of Erich von Holst's results concerning endogenous stimulus generation and central coordination.

51 *P. 122, l. 40.* See notes 11, 17, 19, 37, 52, 89.

52 *P. 123, l. 19.* The phenomena of threshold-lowering and vacuum activity are referred to here for the first time. They could not be more clearly described, but the perspicacious reader will notice that, at the time, my bias in favour of the chain reflex theory permitted me only to notice their incompatibility with purposivistic theory, though they are equally, or even more, incompatible with the reflex theory.

53 *P. 124, l. 13.* Some instances of sexual imprinting in mammals have since become known (Leyhausen).

54 *P. 126, l. 22.* Communication effected by vocalization between the newly-hatched bird and its mother plays an important rôle in directing the infant's imprinting to the right object, as demonstrated by H. Fischer, P. Klopfer and others.

55 *P. 132, l. 15.* The problem of such imprinting on a species rather than on an individual has been thoroughly investigated by F. Schutz.

56 *P. 133, l. 4.* The word 'schema' suggests, incorrectly, that the animal possesses a sort of image-like perceptual template. The only case in which this has proved to be correct concerns the innate 'knowledge' of the species-specific song in some passerine birds (M. Konishi). Otherwise, the innate perception pattern has proved to be a mosaic of extremely simple receptor correlates activated by specific key stimuli. The total of these correlates which, by their joint effect and in agreement with Seitz' law of heterogenous summation, activate the response, is now called the innate releasing mechanism (I.R.M.). There is hardly one I.R.M. in existence which is not further developed and made more selective by learning in individual life. See W. Schleidt.

57 *P. 133, l. 18.* See also W. Schleidt.

58 *P. 139, l. 5.* This of course represented an answer to the duckling's 'lost piping' and had the effect already commented on (p. 126).

59 *P. 142, l. 28.* Many more examples of this will be discussed in the Anatinae paper in the second volume.

60 *P. 143, l. 32.* The statement that the curlew hatchling is unimprintable because of its specific information has since been corrected by Alfred Seitz, who succeeded in experimentally procuring the stimulus situation necessary for this process.

61 *P. 145, l. 29.* The small number and the relative simplicity of the key stimuli which are innately responded to are explicitly discussed in W. Schleidt's paper.

62 *P. 146, l. 7.* See note 61.

63 *P. 160, l. 24.* One curious exception from this rule is represented by the curlew as demonstrated by Otto von Frisch. Newly-hatched birds of this species respond innately to the silhouette of a flying bird by crouching. As the impinging of this stimulus-situation practically always coincides with that of the parent's warning call, the young bird's crouching response is quickly conditioned to the latter. In the naïve hatchling, the warning call usually has no visible effect, though it can be demonstrated to lower the threshold of the response to a flying bird.

64 *P. 162, l. 37.* European Blackbird, *Turdus merula.*

65 *P. 165, l. 25.* This is an important example of threshold lowering in a response which would be classified as an aversion by Wallace Craig.

66 *P. 175, l. 33.* W. and M. Schleidt have shown in the turkey, and E. v. Holst (by brain-stem stimulation) in the domestic hen that these specific postures are motivated by the conflicting responses of escaping from an enemy and of staying with the chicks. Not being able to run away, the bird shows what H. Hediger has termed a *critical response.*

67 *P. 181, l. 34.* This waning of a response is the effect of two independent processes. One is the exhaustion of action-specific potentiality (A.S.P.), in other words the effect which is the physiological counterpart to threshold lowering. The other process is stimulus-specific habituation, often called 'sensory adaption' by physiologists. Both processes can be experimentally differentiated (M. Schleidt; L. Franzisket).

68 *P. 184, l. 3.* This has in fact been found to be the case in geese. A snow goose which had lost one of her four half-grown goslings, which we had removed for treatment, searched for it so persistently as to endanger the remaining three.

69 *P. 186, l. 8.* This has hardly changed during the intervening thirty years.

70 *P. 192, l. 4.* This is the first reference to the problem of the displacement activity and, at the same time, to the phenomenon of ritualization.

71 *P. 194, l. 39.* This attempt to distinguish certain types of pair-formation is only partly successful. The lizard type as described here is only characteristic of the Iguanidae described by G. K. Noble. The lizards of the genus *Lacerta* behave quite differently, as they possess typically female releasers. The cichlid type of pair-formation also proved to be a special case only and occurs, I am sorry to say, not in cichlids but only in herons and some other territorial and colony-nesting birds. In cichlids I had missed, at the time of writing, the fleeting stage of rank inferiority in the female. (See Seitz and particularly Oehlert.) What is said about the labyrinth fish type is more or less correct.

72 *P. 196, l. 1.* Exactly the same is true of the blue throat in *Lacerta viridis* (G. Kitzler). In this reptile, the largest female reacts with submission to the smallest male, provided the latter is in possession of a blue throat.

73 *P. 197, l. 23.* This statement by Noble and Bradley is a typical error arising from generalization. What the authors say here is not even true of all lizards, let alone of all birds.

74 *P. 197, l. 34.* The explanation of this was given in 1958 by B. Oehlert. In the individual which happens to be rank superior, female sexuality is suppressed while, conversely, a dominated individual is unable to activate the patterns of male sexual behaviour.

75 *P. 198, l. 26.* This was a complete error. I had simply missed the period of female submissiveness which, however short, is as indispensable for the pair-formation in these fish as it is in labyrinth fishes.

76 *P. 206, l. 4.*

> Ein Jüngling liebt ein Mädchen, die hat einen andern erwählt,
> der andre liebt eine andre, und hat sich mit dieser vermählt.
> Das Mädchen heiratets aus Ärger den ersten besten Mann,
> der ihr in den Weg gelaufen, der Jüngling ist übel dran.
> Es ist eine alte Geschichte, doch bleibts sie immer neu,
> und wem sie just passieret, dem bricht das Herz entzwei.

77 *P. 208, l. 24.* Geese, for instance, do not fit into these categories at all. The complicated mechanism of their pair-formation has been investigated by H. Fischer.

78 *P. 210, l. 22.* This is an important assertion which I still hold to be essentially correct. Of course, what is meant here by 'physiological processes' are the cumulative effects of hunger, thirst, or the tension of empty organs. The important problem which still awaits solution is the question as to what physiological processes underlie those instinctive activities whose rhythmical recurrence is not explained in this manner.

79 *P. 211, l. 6.* See also note 10.

80 *P. 222, l. 39.* This has been confirmed by later observations by H. Fischer and particularly by those made by E. Gwinner on the ravens.

81 *P. 232, l. 17.* These birds, however, do defend tiny territories around the nests, according to K. Immelmann. The same is true of colony-nesting Estrildinae like *Taeniopygia castanotis.*

82 *P. 243, l. 40.* As explained in note 56, this conception can here be equated with that of the innate releasing mechanism (I.R.M.).

83 *P. 244, l. 4.* This is the first clear statement concerning the selective function of the innate releasing mechanism, which acts as a filter impermeable to irrelevant stimuli.

84 *P. 245, l. 42.* Learning is here equated to classical Pavlovian conditioning. If one defines learning in a more embracing manner as any adaptive modification of behaviour, imprinting represents one very peculiar type of learning.

85 *P. 246, l. 40.* All these characteristics of imprinting have since been thoroughly demonstrated by the work of Schutz, Immelmann, Hess, Schein, Fabricius and others. The conflicting results which seemed to contradict the peculiarity of imprinting were attained by authors investigating birds in which imprinting does not occur (Hinde, Bateson). See also note 1.

86 *P. 248, l. 33.* Actually, priority for this must be accredited to Charles Otis Whitman who, unknown to Heinroth and myself, applied this method more than ten years earlier.

87 *P. 256, l. 15.* The existence of phylogenetically programmed neural and sensory mechanisms influencing behaviour is still violently denied by some anthropologists who cling to the doctrine that man is entirely instinctless and that all human behaviour is learned (F. M. Ashley Montagu, 1968).

88 *P. 261, l. 2.* This opinion is still held by some behaviouristic psychologists and even by some English-speaking ethologists.

89 *P. 261, l. 20.* This certainly is not true any more. I call attention to the stubbornness with which, at that time, I still adhered to the reflex theory.

90 *P. 261, l. 30.* Though I might today formulate this opinion in a less aggressive manner, I have to confess that I still hold to it absolutely.

91 *P. 272, l. 4.* 'Instinctive behaviour' is here equated to the fixed instinctive motor pattern.

92 *P. 272, l. 41.* This narrowing-down results ultimately in the conceptualization of the fixed motor pattern as a behavioural element different from the taxis and other phylogenetically programmed behaviour mechanisms. (See also p. 318.) Orienting mechanisms, though wholly dependent on phylogenetically programmed mechanisms, are here treated as being 'non-instinctive'.

93 *P. 273, l. 30.* The subjective purpose is, of course, not only the arrival of the releasing stimulus situation, but also the consummatory act itself with all the 'rewarding' and objectively reinforcing proprioceptions

which it entails. Furthermore, this is not, as I thought at the time of writing, the only type of purposive behaviour in existence. In the equally important second type, the stimulus situation which constitutes the purpose has no other effect than to switch off appetitive behaviour. Monika Meyer-Holzapfel has chosen for this type of purposive action the term 'appetence for quiescence' (Appetenz nach Ruhezuständen). Wallace Craig terms this type of behaviour 'aversion'. According to Hull, most or all reinforcements are ultimately due to this type of mechanism; in other words, he ignored the first type, as I did the second.

94 *P. 275, l. 2.* Again the fixed motor pattern is meant.

95 *P. 275, l. 9.* This description anticipates, in part, what N. Tinbergen and G. P. Baerends later found to be a hierarchical organization of instinctive activities.

96 *P. 277, l. 24.* In fact, Wallace Craig's great discovery implies much more: every fixed motor pattern is in itself an autonomous purpose as well as an autonomous reinforcement of precedent behaviour.

97 *P. 279, l. 8.* There are, in reality, more than two separate physiological mechanisms involved in the process. In 'Innate Bases of Learning', an attempt at their analysis is made.

98 *P. 280, l. 19.* According to E. Hess, the sharply-defined maximum of imprinting occurs in the mallard during the 17th hour after hatching.

99 *P. 280, l. 32.* This number of observations has since been vastly enlarged, see note 84.

100 *P. 282, l. 17.* To the best of my knowledge, the credit for first having applied truly comparative method in the study of behaviour is due to A. and E. Peckham.

101 *P. 283, l. 31.* The few exceptions have already been pointed out, p. 227.

102 *P. 284, l. 34.* This is an error based on the assumption that intermediate forms which connect two taxonomical characters unconditionally imply their homology. Here is one of the few paradigmal cases in which it is not so: the pre-flight movement of *Anas* is a true intention movement derived from the motor patterns of jumping into the air. The motor pattern of *Anser* is derived, by ritualization, from the displacement movement of bill-shaking. *Alopochen* actually shows a highly ritualized mixture of the two.

103 *P. 286, l. 9.* I am glad to say that the programme here outlined has since been abundantly fulfilled.

104 *P. 286, l. 17.* As already pointed out (see note 102) miscibility of motor patterns does not unconditionally prove their homology.

105 *P. 289, l. 23.* As already pointed out, this is only half the story (see note 92).

106 *P. 291, l. 6.* I now think that it is definitely misleading to call this emergence of 'new gaps' a *reduction* of instinct. As P. Leyhausen has shown, the cutting-up of one behavioural chain into a number of independently available parts implies an *increase* of the number of motivations which become autonomous in the process.

107 *P. 292, l. 5.* This statement may be charitably interpreted as a mention of other types of purposive behaviour. I do not think this is justified; I honestly believed, at the time, that the sequence of appetitive behaviour–I.R.M.–consummatory act represented the only possible functional combination of the behavioural elements then known to me.

108 *P. 295, l. 6.* Again, this is a dawning of the idea of hierarchical organization of instinct, as clearly analysed by Tinbergen and Baerends.

109 *P. 298, l. 13.* More striking examples of such a 'sliding hierarchy' of motivations have been demonstrated by P. Leyhausen in the behaviour of cats.

110 *P. 308, l. 42.* After these considerations, it would seem more than obvious to draw the inferences mentioned in the Introduction to this volume. I keep stressing the fact that I was not 'inventing' spontaneity in order to explode the reflex theory but, on the contrary, for many years pig-headedly did my best to save it.

111 *P. 310, l. 13.* See note 109.

112 *P. 317, l. 4.* The nomenclature conference at the ethologists' congress in 1949 decided on the term 'fixed motor pattern' for this concept.

113 *P. 319, l. 19.* At this point, the perspicacious reader who has suffered through all the inconsistencies of the author, ought to heave a sigh of relief.

114 *P. 321, l. 19.* These phenomena, including that of the 'spinal contrast' of Sherrington himself, are all the more convincing arguments for E. v. Holst's theory, since they are collected by people who were not yet aware of it.

115 *P. 322, l. 19.* Gray, Lissman and others have shown that, in certain cases, an influx of stimulation from one or two posterior roots is necessary to keep the automation going. Even in these cases, however, the afferent input has no influence on the co-ordination.

116 *P. 323, l. 20.* Later studies of mixed motivation have shown that this is actually the rule and that it is actually an exception when an animal's behaviour is motivated by a single automatic pattern.

117 *P. 323, l. 29.* Even central nervous mechanisms which are completely innate, in the sense of being entirely programmed in the phylogeny of the species, are here termed non-instinctive. This rather awkward (and later abandoned) terminology was motivated by my striving to emphasize as clearly as possible the unique properties of the fixed motor pattern.

118 *P. 325, l. 20.* What is not yet clearly stated here (though it ought to have been) is that the consummatory motor pattern, once it is set off, demonstrably acts as a reinforcement of the precedent appetitive behaviour.

119 *P. 325, l. 40.* The problem here raised is thoroughly discussed and, I think, solved in 'Innate Bases of Learning'.

120 *P. 347, l. 18.* This is the first, unclear description of what N. Tinbergen and A. Kortland called a *displacement activity* while G. F. Makkink called it '*sparking over*'.

121 *P. 353, l. 31.* There are some American psychologists of great repute who do not agree with this principle, which I still believe to be obligatory in all science.

122 *P. 359, l. 36.* See note 12.

123 *P. 368, l. 17.* See note 56.

124 *P. 369, l. 23.* These problems have since been thoroughly investigated by G. Kramer, W. Braemer, F. Sauer and others, who discovered that birds are able to orient much on the same principles as human navigators.

Bibliography

ADLER, M. J. *The difference of man and the difference it makes*, New York, 1967

ALLEN, A. A. Sex rhythm in the Ruffed Grouse (*Bonasa umbellus* Linn.) and other birds, *The Auk*, **51/2**, 1934

ALLEN, F. H. The role of anger in Evolution, with particular reference to the colours and songs of birds, *The Auk*, **51/4**, 1934

ALVERDES, F. *Tiersoziologie*, Leipzig, 1925

ALVERDES, F. Die Ganzheitsbetrachtung in der Biologie, *Sitzungsbericht der Gesellschaft zur Förderung des ges. Naturwiss. zu Marburg*, **67**, 1932

ANDREW, R. J. The origin and evolution of the calls and facial expressions of the primates, *Behaviour*, **20**, 1–109, 1963

ARDREY, R. *The territorial imperative*, New York, 1966

BAERENDS, G. P. Aufbau tierischen Verhaltens, in Kükenthal: *Handbuch der Zoologie*, **8**, 10 (3), 1–32, 1956

BATESON, P. P. G. An effect of imprinting on the perceptual development of domestic chicks, *Nature*, **202**, 421–2, 1964

BECHTEREW, W. *Reflexologie des Menschen*, Leipzig and Vienna, 1926

BETHE, A. Plastizität (Anpassungsfähigkeit) des Nervensystems, *Bethes Handbuch der normalen und pathologischen Physiologie*, **15/2**, 1045–1130, Berlin, 1931

BETHE, A. Plastizität und Zentrenlehre, *Bethes Handbuch der normalen und pathologischen Physiologie*, **15/2**, 1175–1222, Berlin, 1931

BIERENS DE HAAN, J. A. Der Stieglitz als Schöpfer, *Journal für Ornithologie*, **80/1**, 1933

BIERENS DE HAAN, J. A. Probleme des tierischen Instinktes, *Die Naturwissenschaften*, **23/42**, 43, 1935

BINGHAM, H. Size and form perception in *Gallus domesticus*, *Journal of Animal Behavior*, 1913

BOWLBY, J. Maternal care and mental health, *World Health Organisation, Monograph Series* 2, 1952

BOWLBY, J. The nature of the child's tie to his mother, *International Journal of Psychoanalysis*, **39**, 350–73, 1958

BOY, H. L. and TINBERGEN, N. Nieuwe feiten over de sociologie van de zilvermeeuwen, *De Levende Natuur*, 1937

BRADLEY, H. T., *see* NOBLE, G. K.

BRAEMAR, W. A critical review of the sun-azimuth hypothesis, *Cold Spring Harbour Symposia on Quantitative Biology*, **25**, 413–27, 1960

BRÜCKNER, G. H. Untersuchungen zur Tiersoziologie, insbesondere zur Auflösung der Familie, *Zeitschrift für Psychologie*, **128/1–3,** 1933.

BRUNSWIK, E. *Wahrnehmung und Gegenstandswelt, Psychologie vom Gegenstand her*, Leipzig and Vienna, 1934

BÜHLER, C. Das Problem des Instinktes, *Zeitschrift für Psychologie*, **103,** 1927

BÜHLER, K. *Zukunft der Psychologie*, Vienna, 1936.

CARMICHAEL, L. The development of behaviour in vertebrates experimentally removed from the influence of external stimulation, *Psychological Review*, **33** (1926); also in **34** and **35**

COBURN, C. A. The behavior of the Crow, *Journal of Animal Behavior*, **4,** 1914

CRAIG, W. Appetites and aversions as constituents of instincts, *Biological Bulletin*, **34/2,** 1918

CRAIG, W. A note on Darwin's work on the expression of emotions, etc., *Journal of Abnormal and Social Psychology*, December 1921, March 1922

CRAIG, W. The voices of Pigeons regarded as a means of social control, *The American Journal of Sociology*, **14,** 1908

CRAIG, W. The expression of emotion in the Pigeons: I. The Blond Ring-Dove (*Turtur risorius*), *The Journal of Comparative Neurology and Psychology*, **19/1,** 1909.

CRAIG, W. Observations on young Doves learning to drink, *The Journal of Animal Behavior*, **2/4,** 1912

CRAIG, W. Male Doves reared in isolation, *The Journal of Animal Behavior* **4/2,** 1914

CRAIG, W. Why do animals fight? *The International Journal of Ethics*, **31,** April 1921

CRANE, J. Comparative biology of salticid spiders at Rancho Grande, Venezuela. IV. An analysis of display, *Zoologica*, **34,** 159–214, 1949

CRANE, J. Combat, display and ritualisation in fiddler crabs, *Philosophical Transactions of the Royal Society*, London, B, **251,** 459–72, 1966

DOFLEIN, F. *Der Ameisenlöwe – Eine biologische, tierpsychologische und reflexbiologische Untersuchung*, Jena, 1916

EIBL–EIBESFELDT, I. Das Verhalten der Nagetiere, in Kükenthal: *Handbuch der Zoologie*, **8,** (10) 13, 1–88, 1958

EIBL-EIBESFELDT, I. *Grundriß der vergleichenden Verhaltensforschung*, München, 1967

ENGELMANN, W. Untersuchungen über die Schallokalisation bei Tieren, *Zeitschrift für Psychologie*, **105,** 1928

FABRICIUS, E. Zur Ethologie junger Anatiden, *Acta Zoologica Fenn.*, **68,** 1–178, 1951

FISCHER, H. Das Triumphgeschrei der Graugans (*Anser anser*), *Zeitschrift für Tierpsychologie*, **22,** 247–304, 1965

FLETCHER, R. *Instinct in man*, London, 1957

FRANZISKET, L. Die Bildung einer bedingten Hemmung bei Rückenmarksfröschen, *Zeitschrift für vergleichende Physiologie*, **37,** 161–8, 1955

FRIEDMANN, H. Social parasitism in birds, *Quarterly Review of Biology*, **3/4,** 1928

FRIEDMANN, H. The instinctive emotional life of birds, *The Psychoanalytical Review*, **21/3**, 4, 1934

FRISCH, O. V. Die Bedeutung der elterlichen Warnrufe für Brachvogel- und Limicolenkücken, *Zeitschrift für Tierpsychologie*, **15**, 381–2, 1958

GAMM, H. J. *Aggression und Friedensfähigkeit in Deutschland*, München, 1968

GOETHE, F. Beobachtungen und Untersuchungen zur Biologie der Silbermöwe (*Larus a. argentatus* Pontopp.) auf der Vogelinsel Memmertsand, *Journal für Ornithologie*, **85/1**, 1937

GRAHAM BROWN, T. The intrinsic factors in the act of progression in the mammal, *Proceedings of the Royal Society*, London, B, **84**, 1911

GRIFFIN, D. R. *Listening in the dark*, 2nd edition, New Haven, 1958

GROOS, K. *Die Spiele der Tiere*, 1907

GWINNER, E. Untersuchungen über das Ausdrucks- und Sozialverhalten des Kolkraben (*Corvus corax*), *Zeitschrift für Tierpsychologie*, **21**, 657–748, 1964

HAAS, A. Weitere Beobachtungen zum 'generischen Verhalten' bei Hummeln, *Zeitschrift für Tierpsychologie*, **22**, 305–20, 1956

HEDIGER, H. Zur Biologie und Psychologie der Flucht bei Tieren, *Biologisches Zentralblatt*, **54**, 21–40, 1934

HEDIGER, H. Zur Biologie und Psychologie der Zahmheit, *Archiv für Psychologie*, **93**, 1935

HEINROTH, O. Beiträge zur Biologie, namentlich Ethologie und Psychologie der Anatiden, *Verhandlungen des V. Internationalen Ornithologen-Kongresses*, Berlin, 1910

HEINROTH, O. Reflektorische Bewegungen bei Vögeln, *Journal für Ornithologie*, **66**, 1918

HEINROTH, O. Zahme und scheue Vögel, *Der Naturforscher*, **1**, 1924

HEINROTH, O. Über bestimmte Bewegungsweisen der Wirbeltiere, *Sitzungsbericht der Gesellschaft naturforschender Freunde*, Berlin, 1930

HEINROTH, O. und M. *Die Vögel Mitteleuropas*, Berlin-Lichtenfelde, 1924–8

HERRICK, F. H. *Instinct, Western Res. University Bulletin*, 22/6

HERRICK, F. H. *Wild birds at home*, New York and London, 1935

HESS, E. H. Imprinting, an effect of early experience, *Science*, **130**, 133–41, 1959

HINDE, R. A. The following response of Moorhens and Coots, *British Journal of Animal Behaviour*, **3**, 121–2, 1955

HINGSTON, R. W. G. *The meaning of animal colour and adornment*, London, 1932

HOLST, E. V. Alles oder Nichts, Block, Alternans, Bigemini und verwandte Phänomene als Eigenschaften des Rückenmarks, *Pflügers Archiv für die gesamte Physiologie*, **236/4**, 5, 6, 1935

HOLST, E. V. Versuche zur Theorie der relativen Koordination, *Pflügers Archiv für die gesamte Physiologie*, **237/1**, 1936

HOLST, E. V. Vom Dualismus der motorischen und der automatisch-rhythmischen Funktion im Rückenmark und vom Wesen des automatischen Rhythmus, *Pflügers Archiv für die gesamte Physiologie*, **237/3**, 1936

HOLST, E. V. and SAINT-PAUL, U. V. Vom Wirkungsgefüge der Triebe, *Die Naturwissenschaften*, **18**, 409–22, 1960

HOWARD, E. *An introduction to bird behaviour*, Cambridge, 1928

HOWARD, E. *The nature of a bird's world*, Cambridge, 1935

HULL, C. L. *Principles of Behavior*, New York, 1943

HUXLEY, J. S. The courtship of the Great Crested Grebe, *Proceedings of the Zoological Society*, London, 1914

HUXLEY, J. S. A natural experiment on the territorial instinct, *British Birds*, **27/10,** 1954

HUXLEY, J. S. and HOWARD, E. Field studies and psychology: a further correlation, *Nature*, **133,** 1934

IMMELMANN, K. Zur ontogenetischen Gesangsentwicklung bei Prachtfinken, *Zoologische Anzeiger Supplement*, **30,** 320–32, 1967

JENNINGS, H. S. *Behaviour of the lower organisms*, 2nd edition, New York, 1915

KATZ, D. *Hunger und Appetit*, Leipzig, 1931

KATZ, D. and REVESZ, G. Experimentelle psychologische Untersuchungen an Hühnern, *Zeitschrift für Psychologie*, **50,** 1909

KIRKMAN, F. B. *Bird behaviour*, London, 1937

KITZLER, G. Die Paarungsbiologie einiger Eidechsen, *Zeitschrift für Tierpsychologie*, **4,** 353, 1942

KLOPFER, P. H. An analysis of learning in young Anatidae, *Ecology*, **40,** 90–102, 1959

KOEHLER, O. Die Ganzheitsbetrachtung in der modernen Biologie, *Verhandlungen der Königsberger gelehrten Gesellschaft*, 1933

KOEHLER, O. and ZAGARUS, A. Beiträge zum Brutverhalten des Halsbrandregenpfeifers (*Charadrius hiaticula* L.), *Beiträge zur Fortpflanzung der Vögel*, **13,** 1937

KÖHLER, W. Intelligenzprüfungen an Anthropoiden, *Abhandlungen der Preußischen Akademie*, Phys.-mathem. Kl., 1915

KONISHI, M. Effects of deafening on song development in two species of Juncos, *Condor*, **66,** 85–102, 1964

KONISHI, M. The role of auditory feedback in the control of vocalisation in the White-Crowned Sparrow, *Zeitschrift für Tierpsychologie*, **22,** 77–783, 1965

KORTLANDT, A. Eine Übersicht über die angeborenen Verhaltensweisen des mitteleuropäischen Kormorans, *Archives neerlandaises de Zoologie*, **4,** 401–42, 1940

KORTLANDT, A. and KOOIJ, M. Protohominid behaviour in primates, *Symposia of the Zoological Society*, London, **10,** 61–88, 1963

KRAMER, G. Bewegungsstudien an Vögeln des Berliner Zoologischen Gartens, *Journal für Ornithologie*, **78/3,** 1930

KRAMER, G. Experiments on birds' orientation and their interpretation, *Ibis*, **96,** 173–85, 1957

LEYHAUSEN, P. Verhaltensstudien an Katzen. Beih., 2. *Zeitschrift für Tierpsychologie*, 1956

LEYHAUSEN, P. Über die Funktion der relativen Stimmungshierarchie (dargestellt am Beispiel der phylogenetischen und ontogenetischen Entwicklung des Beutefangs von Raubtieren), *Zeitschrift für Tierpsychologie*, **22,** 412–94, 1965

LISSMAN, H. Die Umwelt des Kampffisches *Betta splendens* Regan, *Zeitschrift für vergleichende Physiologie*, **18/65**

LORENZ, K. Beobachtungen an Dohlen, *Journal für Ornithologie*, **75/4**, 1927

LORENZ, K. Über den Begriff der Instinkthandlung, *Folia Biotheoretica*, **2**, 1937

LORENZ, K. A contribution to the comparative sociology of colony-nesting birds, *Proceedings of the VIIIth International Ornithological Congress*, London, 1934

LORENZ, K. *Evolution and modification of behaviour*, Chicago, 1965

LORENZ, K. Innate bases of learning, in: *On the biology of learning*, New York, 1969

LORENZ, K. and SAINT-PAUL, U. V. Die Entwicklung des Spießens und Klemmens bei den drei Würgerarten *Lanius collurio*, *L. senator* und *L. excubitor*, *Journal für Ornithologie*, **109**, 137–56, 1968

MAKKINK, G. F. *Einige Beobachtungen über die Säbelschnäbler* (*Recurvirostra avosetta L.*), **21/1**, 1932

MAKKINK, G. F. An attempt at an ethogram of the European Avocet (*Recurvirostra avosetta* L.) with ethological and psychological remarks, *Ardea*, **25**, 1–60, 1936

MCDOUGALL, W. *An outline of psychology*, London, 1923

MCDOUGALL, W. *An introduction to social psychology*, Boston, 1923

MCDOUGALL, W. The use and abuse of instinct in social psychology, *The Journal of Abnormal and Social Psychology*, **16/5**, 6 December 1921–March 1922

MEYER-HOLZAPFEL, M. Triebbedingte Ruhezustände als Ziel von Appetenzhandlungen, *Die Naturwissenschaften*, **28**, 273–80, 1940

MONTAGU, F. M. A. *Man and aggression*, New York, 1968

MORGAN, LL. *Instinkt und Erfahrung*, Berlin, 1913

MORRIS, D. *The naked ape*, London, 1967

NICE, M. M. Zur Naturgeschichte des Singammers, *Journal für Ornithologie*, **81/4**, 1933; **82/1**, 1934

NICOLAI, J. Der Brutparasitismus der Viduinae als ethologisches Problem. Prägungsphänomene als Faktoren der Rassen-und Artbildung, *Zeitschrift für Tierpsychologie*, **21**, 129–204, 1964

NOBLE, G. K. and BRADLEY, H. T. The mating behavior of the lizards: its bearing on the theory of sexual selection, *Annals of the New York Academy of Sciences*, **35/2**, 1933

NOBLE, G. K. and BRADLEY, H. T. Experimenting with the courtship of lizards, *Natural History*, **34/1**, 1933

OEHLERT, B. Kampf und Paarbildung einiger Cichliden, *Zeitschrift für Tierpsychologie*, **15**, 141–74, 1958

PECKHAM, G. W. and E. G. Observations on sexual selection in spiders of the family Attidae, *Occasional Papers of the National History Society of Wisconsin*, Milwaukee, 1889

PERACCA, M. C. Osservazioni sulla riproduzione della *Iguana tuberculata Boll. Mus. Zool. Anat. Comparat. Reg. Univ. Torino*, **6/110**

PORTIELJE, J. A. *Zur Ethologie, beziehungsweise Psychologie von Botaurus stellaris*, **15/1**, 2, 1926

PORTIELJE, J. A. Versuch einer verhaltenspsychologischen Deutung des Balzgebarens der Kampfschnepfe (*Philomachus pugnax* L.), *Proceedings of the VIIth International Ornithological Congress*, Amsterdam, 1930

PORTIELJE, J. A. *Zur Ethologie, beziehungsweise Psychologie von Phalacrocorax carbo subcormoranus*, **16/2**, 3, 1927

REVESZ, G., *see* KATZ, D.

RUSSEL, E. R. *The behaviour of animals*, London, 1934

SAUER, F. Zugorientierung einer Mönchsgrasmücke (*Sylvia atricapilla*, unter künstlichem Sternenhimmel, *Die Naturwissenschaften*, **43**, 231–2) 1956

SAUER, F. Further studies on the stellar orientation of nocturnally migrating birds, *Psychologische Forschung*, **26**, 224–44, 1961

SCHEIN, W. M. On the irreversibility of imprinting, *Zeitschrift für Tierpsychologie*, **20**, 462–7, 1963

SCHJELDERUP-EBBE, TH. Zur Sozialpsychologie des Haushuhnes, *Zeitschrift für Psychologie*, **87**, 1922/23

SCHJELDERUP-EBBE, TH. Zur Sozialpsychologie der Vögel, *Zeitschrift für Psychologie*, 1924: *Psychologische Forschung*, **88**, 1923

SCHLEIDT, M. Untersuchungen über die Auslösung des Kollerns beim Truthahn, *Zeitschrift für Tierpsychologie*, **11**, 417–35, 1954

SCHLEIDT, W. M. Die historische Entwicklung der Begriffe 'Angeborenes auslösendes Schema' und 'Angeborener Auslösemechanismus', *Zeitschrift für Tierpsychologie*, **19**, 697–722, 1962

SCHLEIDT, W. M. Wirkungen äußerer Faktoren auf das Verhalten, *Fortschritte de Zoologie*, **16**, 469–99, 1964

SCHLEIDT, W. M., SCHLEIDT, M. and MAGG, M. Störungen der Mutter-Kind-Beziehung bei Truthühnern durch Gehörverlust, *Behaviour*, **16**, 254–60, 1960

SCHUTZ, F. Sexuelle Prägung bei Anatiden, *Zeitschrift für Tierpsychologie*, **22**, 50–103, 1965

SEITZ, A. Die Paarbildung bei einigen Cichliden. I. Die Paarbildung bei *Astatotilapia strigigena* Pfeffer, *Zeitschrift für Tierpsychologie*, **4**, 1940

SEITZ, A. Die Paarbildung bei einigen Cichliden. II. Die Paarbildung bei *Hemichromis bimaculatus* Gill, *Zeitschrift für Tierpsychologie*, **5**, 1943

SEITZ, A. Untersuchungen über die Kumpanverhältnisse des jungen Brachvogels (*Numenius arquata*), *Zeitschrift für Tierpsychologie*, **7**, 402–17, 1950

SELOUS, E. Observations tending to throw light on the question of sexual selection in birds, including a day to day diary on the breeding habits of the Ruff, *Machetes pugnax*, *The Zoologist*, Fourth Series, **10/114**, 1905

SELOUS, E. Observational diary on the nuptual habits of the Blackcock, *Tetrao tetrix*, *The Zoologist*, **13**, 1909

SELOUS, E. An observational diary of the domestic life of the Little Grebe or Dabchick, *Wild Life*, **7**

SELOUS, E. Schaubalz und geschlechtliche Auslese beim Kampfläufer (*Philomachus pugnax*), *Journal für Ornithologie*, **77/2**, 1929

SIEWERT, H. Bilder aus dem Leben eines Sperberpaares zur Brutzeit, *Journal für Ornithologie*, **77/2**, 1930

SIEWERT, H. Der Schreiadler, *Journal für Ornithologie*, **80/1**, 1932

SIEWERT, H. Beobachtungen am Horst des Schwarzen Storches (*Ciconia nigra* L.), *Journal für Ornithologie*, **80/4**, 1932

SIEWERT, H. Die Brutbiologie des Hühnerhabichts, *Journal für Ornithologie*, **81/1**, 1933

SPITZ, R. A. *Hospitalism. The psychoanalytical study of the child*, **1**, 53–74, New York, 1945

SPITZ, R. A. *The first year of life*, New York, 1965

STORR, A. *Human aggression*, London, 1968

STRESEMANN, E. Aves, in KÜKENTHAL: *Handbuch der Zoologie*, **7**, Part 2, Berlin and Leipzig, 1927–34

SUNKEL, W. Bedeutung optischer Eindrücke der Vögel für die Wahl ihres Aufenthaltsortes, *Zeitschrift für wissenschaftliche Zoologie*, **132**, 1928

TINBERGEN, N. Waarnemingen en proeven over de sociologie van een zilvermeeuwenkolonie, *De levende Natuur*, 1935

TINBERGEN, N. Zur Soziologie der Silbermöwe (*Larus a. argentatus* Pontopp.), *Beiträge zur Fortflanzung der Vögel*, **12/3**, 1936

TINBERGEN, N. Die Übersprungbewegung, *Zeitschrift für Tierpsychologie*, **4**, 1–40, 1940

TINBERGEN, N. *The study of instinct*, London, 1951

TOLMAN, E. C. *Purposive behaviour in animals and men*, New York, 1932

UEXKÜLL, J. VON. *Umwelt und Innenwelt der Tiere*, Berlin, 1909

UEXKÜLL, J. VON. *Theoretische Biologie*, Berlin, 1928

UEXKÜLL, J. VON. *Streifzüge durch die Umwelten von Tieren und Menschen*, Berlin, 1934

VERWEY, J. Die Paarungsbiologie des Fischreihers, *Zoologische Jahrbücher* (Abteilung für allgemeine Zoologie), **48**, 1930

VOLKELT, H. *Die Vorstellungen der Tiere – Arbeiten zur Entwicklungspsychologie*, F. Krueger, II, 1914

VOLKELT, H. Tierpsychologie als genetische Ganzheitspsychologie, *Zeitschrift für Tierpsychologie*, **1/1**, 1937

WERNER, H. *Entwicklungspsychologie*, Leipzig, 1933

WHITMAN, C. O. *Animal behavior*, 16th lecture from *Biological Lectures from the Marine Biological Laboratory*, Woods Hole, Mass., 1898

WINTERBOTTOM, J. M. Studies in sexual phenomena: VI. Communal display in birds. VII. Transference of male secondary display characters to the female, *Proceedings of the Zoological Society of London*, 1929

WYLIE, PH. *The magic animal*, New York, 1968

Subject Index

Author Index